DIGESTIVE DISEASES - RESEARCH AND CLINICAL DEVELOPMENTS

FLAVONOIDS IN THE FIGHT AGAINST UPPER GASTROINTESTINAL TRACT CANCERS

Digestive Diseases - Research and Clinical Developments

Additional books in this series can be found on Nova's website under the Series tab.

Additional e-books in this series can be found on Nova's website under the eBooks tab.

Nutrition and Diet Research Progress

Additional books in this series can be found on Nova's website under the Series tab.

Additional e-books in this series can be found on Nova's website under the eBooks tab.

DIGESTIVE DISEASES - RESEARCH AND CLINICAL DEVELOPMENTS

FLAVONOIDS IN THE FIGHT AGAINST UPPER GASTROINTESTINAL TRACT CANCERS

KATRIN SAK

nova
biomedical
New York

Library of Congress Cataloging-in-Publication Data

ISBN: 978-1-53613-570-1

Published by Nova Science Publishers, Inc. † New York

With the deep respect and love to my parents

CONTENTS

PREFACE

Flavonoids are plant secondary metabolites with a basic polyphenolic skeleton and their anticancer properties have greatly attracted the interest of scientists already for several years. This structurally diverse group of phytochemicals can be found abundantly in various edible plant products, such as fruits and vegetables, tea and cocoa, nuts, spices, grains and medicinal herbs. In the past decades, numerous experimental studies have shown antiproliferative, proapoptotic, antiinflammatory, antiangiogenic, antiinvasive and antimetastatic activities of these compounds in different cancerous systems. Because of close contacts and interactions of ingested food components with the epithelial lining of organ sites of the upper gastrointestinal tract, flavonoids can directly affect the development and progression of malignant neoplasms in the oral cavity, esophagus and stomach. This book gives a comprehensive and contemporary survey about the different anticancer actions of various natural and semisynthetic flavonoids in experimental models of oral, pharyngeal, esophageal and gastric cancers, involving the data obtained from studies of both cell lines as well as laboratorial animals. Also, the effects of flavonoids on the therapeutic responses of conventional cancer treatment modalities, i.e., different chemotherapeutic drugs and irradiation, are presented and analyzed. Moreover, the current epidemiological knowledge about the role of dietary intake of flavonoids on the risk of carcinogenesis in the upper digestive tract in populations with different dietary and lifestyle habits is described. By virtue of the high topicality of the subject but also the thoroughness of its discussion, this book could be interesting and attractive to basic scientists working in the field of plant chemicals, molecular cancer research or design and development of novel anticancer drugs; synthetic chemists in different pharmaceutical companies; health workers and nutritional counselors; medical doctors working with cancer patients; people suffering from any type of upper digestive tract tumors and their family members; and anyone who is interested in the natural tools for the fight against malignant neoplasms and whose purpose is to consciously prevent carcinogenesis within the upper gastrointestinal tract.

INTRODUCTION

Cancer is an increasing heath problem and the second global leading cause of death. According to the data of GLOBOCAN 2012, there were 14.1 million new cancer cases and 8.2 million cancer deaths in 2012 worldwide, whereas despite the increasing efforts made in cancer research, there are still over 32.6 million people living with cancer around the world. Moreover, considering the steadily rising proportion of aged people, the estimated cancer incidence is projected to alarmingly increase to 23.6 million new cases by 2030, i.e., by 68%, if the present smoking habits and adoption of unhealthy lifestyles last.

Upper gastrointestinal tract cancers involve a group of etiologically diverse tumors evolved in the oral cavity, pharynx, esophagus, stomach, and duodenum. Occurrence of these neoplasms has remarkably increased in the past decades making these malignancies a serious health problem all over the world. On account of the direct contacts and interactions of the epithelium of digestive tract with the food components, the development and growth of these cancers may be more closely related to the daily diet. To date, it is generally accepted that high intake of fruits and vegetables can be related to the lower risk of carcinogenesis in the digestive tract revealing thus a certain chemopreventive potential. Therefore, in the recent years, studies of anticancer properties of polyphenolic plant secondary metabolites occurring abundantly in various edible plant products, flavonoids, have come into the focus of biomedical research. Because a long history of usage of these phytochemicals as constituents of food items or medicinal herbs, flavonoids are considered generally safe. Hence, the potential preventive and therapeutic activities of flavonoids against carcinogenesis in organ sites of upper gastrointestinal tract are highly attractive.

Cancer is a long-lasting complex multifactorial disease that develops via multistep process engaging numerous cellular signaling systems. The three major stages of carcinogenesis are initiation with formation of irreversible genetic change, promotion of altered cells to premalignant lesion, and further progression to malignant neoplasm characterized by uncontrolled cell division, invasion and metastasis to distant sites. Unraveling the molecular mechanisms by which flavonoids intervene in cellular networks of these carcinogenesis steps and thereby impede the malignant growth can make it possible to expand the current anticancer treatment strategies and develop novel targeted therapeutics in the future. Identification of highly efficient and tumor-selective anticancer flavonoid structures out of the diversity of these natural phytochemicals would be a challenge to the whole scientific society.

Despite the high topicality of the issue of anticancer effects of flavonoids, no books have still been published on the preventive and therapeutic activities of these plant secondary metabolites against the development and progression of upper gastrointestinal tract malignancies; likewise, the comprehensive reviews on this topic are largely lacking. Therefore, in the current book, the data published so far about the different *in vitro* and *in vivo* activities of various natural and synthetic flavonoids in the diverse preclinical models of oral, esophageal and gastric cancers are compiled and analyzed in the chapters 4, 5 and 6, respectively. Additionally, the known effects of flavonoids on the antitumor responses of conventional cancer treatment modalities (different chemotherapeutic drugs or irradiation) are presented. Moreover, also the present epidemiological knowledge is summarized in the chapter 7. To somewhat limit the immense amount of original source material, only the activities of single flavonoid molecules, but not their mixtures like flavanol-rich green tea extracts or different flavanolignans-containing silymarin, are considered. In this way, the protective role of dietary flavonoids in combating upper gastrointestinal tract cancers might be attractive to the rather wide readership: basic scientists working in the field of phytochemicals, molecular cancer research or design and development of novel drugs; chemists in the different pharmaceutical companies; health workers and nutritional counselors; medical doctors working with cancer patients; people suffering from any type of upper digestive tract malignancies and their family members; anyone interesting in the anticancer arsenal provided us by the nature through biosynthesis of phytochemicals.

Chapter 1

MAJOR FEATURES OF CANCER

Cancers are a set of diseases that can develop in any organ of the body and are characterized by disturbance of the balance between cell proliferation and cell death that in normal conditions maintains healthy tissue homeostasis [1, 2]. The complex nature of cancer derives from multistep functional changes in numerous cellular processes driving the progressive transformation of normal cells into highly malignant state. At that, cancer cells acquire a series of distinct physiological alterations or capabilities that make the malignant growth possible. Moreover, these hallmarks allow to explain what has gone wrong in molecular mechanisms of cancer cells [1, 3, 4]. The major features of malignant neoplasms are depicted in Figure 1.1 and described below in this chapter.

1.1. UNCONTROLLED CELL GROWTH AND ABERRANT PROLIFERATION

Probably the most important feature of malignant cells is the capability of sustaining proliferation leading to formation and growth of primary neoplasm [2, 5, 6]. In normal mammalian cells, signaling of a firmly regulated cell cycle controls the cell division cycles, cell growth and proliferation, maintaining thus healthy tissue homeostasis. Dysregulation of the cell cycle progression and disruption of its checkpoints (cell cycle aperiodicity) are critical features for malignant cells to grow; therefore, cancer is often considered as a cell cycle disorder [1, 7-10]. In eukaryotic cells, the cell cycle consists of four sequential phases: G0/G1, S (DNA synthesis), G2 and M (mitosis). Checkpoints to the cell cycle occur in the G1 and G2 phases. The cell cycle progression is tightly regulated by the sequential activation of a group of serine-threonine kinases, cyclin-dependent kinases (CDKs), which is dependent on their association with cyclins. In the mid-G1 phase, cell cycle progression is controlled by a cyclin D-CDK4/CDK6 complex. In the late-G1 to early S phase, progression is controlled by the cyclin E-CDK2 complex. In the G2 phase, cell cycle progression is controlled by the cyclin A/B-CDK1 complex leading the cells to transit from the G2 phase into mitosis [2, 8, 11-13]. Therefore, inhibition of the cell cycle progression (and thereby suppression of proliferation) is a rational therapeutic target for the control of cancer growth, and agents with the ability to induce cell cycle arrest provide promise not only in the management of malignant neoplasms but also in chemoprevention. For instance, it has become clear that the cell cycle progression is blocked in the G2/M phase in response to DNA damage and in this phase, the cells are also most sensitive to the toxic effects of irradiation [9, 11, 13].

To fuel rapid and uncontrolled growth and proliferation, cancer cells must reprogram their energy metabolism [3]. Differently from normal cells where oxidative phosphorylation is the main source of energy, cancer cells apply preferentially glycolytic pathway, developed initially as an adaptive mechanism to survive under hypoxic conditions, but remains the main power supplier even in the abundance of oxygen (aerobic glycolysis) [14, 15].

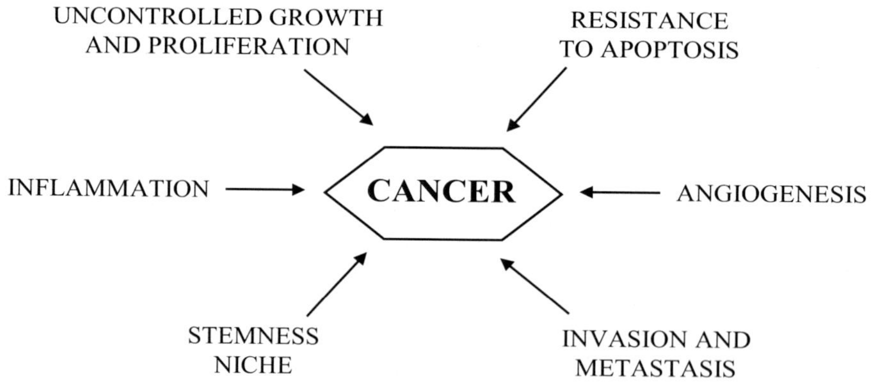

Figure 1.1. Major characteristics of cancer.

1.2. Impaired Apoptosis

The ability of cancer cell populations to increase in number results from not only by the rate of cell proliferation but also by the rate of cell death. Therefore, besides the deregulated cellular proliferation, evasion of cell death is another important characteristic of malignant neoplasms allowing aberrant cells to survive and eventually leading to tumor formation [8, 16]. Apoptosis as a programmed cell death is a normal physiological process by which damaged or abnormal cells are effectively removed protecting thereby the body against cancer development by eradicating the cells which have transformed precancerous or cancerous [11, 17, 18]. This is a tightly regulated complex suicidal process involving an orchestrated set of cellular biochemical events accompanied by the well-defined characteristic features, including typical morphological changes (cell shrinkage, membrane blebbing), chromatin condensation, and DNA fragmentation, resulting in formation of apoptotic bodies [17-19]. Defects in apoptotic death are crucial for initiation and progression of cancer.

Apoptosis predominantly occurs through two classical signaling pathways: the extrinsic mitochondria-independent membrane death receptor (DR) pathway and the intrinsic mitochondrial pathway [20, 21]. In the extrinsic pathway, cell surface death receptors (Fas, DR3, DR4, DR5) interact with their ligands leading to sequential activation of caspase-8 and caspase-3 and poly(ADP-ribose) polymerase (PARP), an enzyme that is involved in DNA repair and genome integrity [12, 20]. The mitochondrial pathway is triggered by the release of cytochrome c due to the loss of mitochondrial transmembrane potential after various intracellular stress signals, such as excessive accumulation of reactive oxygen species (ROS), and involves the sequential activation of caspase-9 and caspase-3 resulting in cleavage of PARP and ultimately the cell death [4, 11, 12, 20, 22-26]. The mitochondria-mediated apoptotic pathway is tightly regulated by B-cell lymphoma-2 (Bcl-2) family proteins which consist of proapoptotic and antiapoptotic members. Proapoptotic proteins (Bad, Bid, Bim, Bax, Bak) create pores in the mitochondrial outer membrane rendering the membrane permeable, through which cytochrome c is released from mitochondria into the cytosol; antiapoptotic proteins (Bcl-2, Bcl-xL) maintain the mitochondrial membrane structure; and therefore, the ratio of pro- and antiapoptotic proteins (Bax/Bcl-2) is a pivotal determinant of a cell for undergoing intrinsic cell death [11, 23, 27]. In cancer cells, Bcl-2 protein is commonly overexpressed and indicates poor prognosis [28, 29]. Both intrinsic and extrinsic apoptotic cascades are orchestrated by caspases, a family of cysteine proteases, which can cleave a number of structural and regulatory proteins and are divided into initiator caspases (-2, -8, -9, -10) and effector caspases (-3, -6, -7), whereas caspase-3 is a key mediator of apoptotic cell death [11, 18, 30].

As evasion of apoptosis may extensively contribute to promotion and progression of malignant neoplasms, inducing of apoptotic cascades in malignant cells has become a major focus of not only management of established cancer but also its chemoprevention. Therefore, modulation of apoptosis is an important target in development of anticancer agents [4, 17, 31, 32]. Moreover, it is expected that more detailed understanding of apoptotic mechanisms could lead to novel innovative molecular targeted therapies selective to tumor cells [4, 17, 21, 33].

Another cell death mechanism, autophagy, reveals context-dependent roles in cancer cells being either tumor-suppressive or tumor-promoting supporting the survival of tumor cells by providing an alternative resource of nutrients and energy and mitigating stress [34]. Such cytoprotective role of autophagy can contribute to evolving resistance and therefore, inhibitors of autophagy can even potentiate the proapoptotic effects of anticancer agents and suppress the tumor growth [34-36].

1.3. Angiogenesis

Solid tumors cannot grow beyond 2-3 mm^3 or progress and metastasize without new vasculature that supplies oxygen and nutrients to the neoplasm [1, 37]. Angiogenesis is the process of formation of new blood vessels from preexisting vessels, involving sprouting, proliferation, migration, invasion and assembly of endothelial cells [1, 38]. This highly complex and dynamic process is regulated by a number of pro- and antiangiogenic molecules, among which the vascular endothelial growth factor (VEGF) is the most potent proangiogenic factor [38, 39]. This growth factor is often overexpressed in various human malignancies being triggered mostly by hypoxia. Moreover, expression of VEGF is considered as a predictive and prognostic determinant for clinical outcome [38, 40, 41].

Numerous attempts have been made to block or delay cancer growth and metastasis by antiangiogenic therapy. At that, targeting VEGF was expected to be a promising novel approach for clinical management. However, the efficacy of antivascular treatment has still revealed only limited success that could be contributed to the tumor progression by triggering formation of vascular mimicry [37, 42]. This is a characteristic feature of aggressive cancer cells to form *de novo* vascular networks without the presence of endothelial cells to supply oxygen and nutrients to malignant tissue, showing a poor prognosis [1, 37].

1.4. Invasion and Metastatic Spread

The most fatal hallmark of malignant tumors comprises the ability of cancer cells to invade surrounding tissue and spread from the primary neoplasm to distant sites to form secondary growths or metastases [1, 5, 6, 16]. Metastasis is the leading cause of mortality among cancer patients being responsible for over 90% of malignancy-associated deaths [1, 43, 44]. Metastasis is a complex multistep process that involves a cascade of distinct steps: detachment of cancer cells from the primary tumor and local invasion through the extracellular matrix (ECM), including basement membrane (BM) and adjacent stromal components; intravasation into blood or lymphatic vessels; survival in the circulation; extravasation at distant organs; formation of micrometastases; and colonization in the distant sites [1, 5, 16, 45]. The malignant conversion of immotile and adhesive epithelial cells involves a switch to a migratory fibroblast-like phenotype, i.e., transformation into mesenchymal cells. During this epithelial to mesenchymal transition (EMT), the expression and function of proteins promoting cell-cell contact (such as E-cadherin) are lost in cancerous cells, whereas invasion-promoting mesenchymal biomarkers are concomitantly acquired (N-cadherin, vimentin, fibronectin). Extensive evidence supports a pivotal role of EMT in the development, invasion, migration and metastasis of epithelial tumors [1, 39, 46-50]. Basement membrane is the first physical barrier of the ECM against malignant invasion separating the epithelial and stromal cell compartments [12, 51]. Different proteolytic enzymes participate in the degradation of ECM, among which matrix metalloproteinases (MMPs) are the key players facilitating migration and invasion of cancer cells into tumor vasculature. In particular, the zinc-dependent endopeptidases MMP-2 and MMP-9 exert critical roles in decomposition of the type IV collagen that is the principal component of basement membrane [5, 44, 45, 47, 48, 52-55]. Indeed, upregulated expression and elevated enzyme activity of MMPs have been described in almost every type of human cancers being correlated with tumor aggressiveness, rapid disease progression and poor overall survival rates of patients [1, 16, 44, 51, 54, 56]. Furthermore, invasion and metastasis cascade is essentially affected by mutual interactions between malignant cells and neighboring microenvironmental components, including fibroblasts, blood cells and inflammatory cells, producing a variety of factors supporting cancer cell locomotion [51, 52]. Once in circulation, tumor cells must resist respective harsh conditions to survive and extravasate at remote target organs [1]. During all this process, migratory cancer cells need to evolve strategies to escape recognition and destruction by immune system [57].

As metastasis represents the final stage of carcinogenesis, it is crucial to develop further therapeutic interventions to prevent or retard the invasive and metastatic processes, including downregulation of expression or enzymatic activity of MMPs [44, 45, 47, 48, 55]. It is evident that a better comprehension of the complicated molecular mechanisms underlying the cancer spread could open new avenues for identification of novel targeted therapeutics considering also the interactions between malignant cells and microenvironmental components.

1.5. Inflammatory Microenvironment

Inflammation is a basic component of the tumor microenvironment supplying bioactive molecules, i.e., proinflammatory mediators such as cytokines and prostaglandins, and promoting all stages of carcinogenesis, including metastasis [1, 3,

25]. The roles of inflammation in cancer progression are very complex and involve multiple molecular mechanisms which are still far to be fully understood. Different proinflammatory cytokines are generated by both malignant cells as well as stromal cells neighboring cancerous tissues favoring the growth and survival of malignancy [58]. At that, cyclooxygenase (COX) enzymes play a key role in the biosynthesis of prostaglandins, whereas inducible COX-2 has been linked to promotion, progression and spread of the tumor cells. Indeed, increased levels of COX-2 and concomitant overproduction of prostaglandins are observed in various human malignant neoplasms [25, 30, 59-61]. Targeting complex molecular network linking inflammation and cancer offers a promising therapeutic approach to prevention and treatment of malignancies.

1.6. Cancer Stem Cells

According to cancer stem cells (CSCs) theory, CSCs constitute a unique subpopulation within malignant neoplasm which is able to initiate new tumors, possess self-renewal capacity, pluripotency and multidrug resistance, being thereby responsible for the disease relapse in cancer patients. Moreover, CSCs have higher migratory and invasive potential than other cancer cells contributing thus to malignant growth and metastasis [62, 63]. Therefore, identification and manipulation of CSCs provide a novel attractive strategy for clinical management of cancer with targeting the cancer roots.

All these processes characteristic and needful for development and progression of cancer, i.e., fast uncontrolled growth and proliferation, evasion of apoptosis, neovascularization, invasion and migration, inflammation, and stemness, could be considered as potential subjects for development of targeted anticancer therapeutics, either by using synthetic methods and strategies or identifying the novel lead compounds from diverse natural sources.

References

[1] Fouad YA, Aanei C. Revisiting the hallmarks of cancer. *Am J Cancer Res* 2017; 7: 1016-36.

[2] Zhang Y, Li Q, Ge Y, Chen Y, Chen J, Dong Y, Shi W. Silibinin triggers apoptosis and cell-cycle arrest of SGC7901 cells. *Phytother Res* 2013; 27: 397-403.

[3] Hanahan D, Weinberg RA. Hallmarks of cancer: the next generation. *Cell* 2011; 144: 646-74.

[4] Sun ZJ, Chen G, Hu X, Zhang W, Liu Y, Zhu LX, Zhou Q, Zhao YF. Activation of PI3K/Akt/IKK-alpha/NF-kappaB signaling pathway is required for the apoptosis-evasion in human salivary adenoid cystic carcinoma: its inhibition by quercetin. *Apoptosis* 2010; 15: 850-63.

[5] Lin SH, Shih YW. Antitumor effects of the flavone chalcone: inhibition of invasion and migration through the FAK/JNK signaling pathway in human gastric adenocarcinoma AGS cells. *Mol Cell Biochem* 2014; 391: 47-58.

[6] Ramachandran L, Manu KA, Shanmugam MK, Li F, Siveen KS, Vali S, Kapoor S, Abbasi T, Surana R, Smoot DT, Ashktorab H, Tan P, Ahn KS, Yap CW, Kumar AP, Sethi G. Isorhamnetin inhibits proliferation and invasion and induces apoptosis through the modulation of peroxisome proliferator-activated receptor γ activation pathway in gastric cancer. *J Biol Chem* 2012; 287: 38028-40.

[7] Zhang Q, Zhao XH, Wang ZJ. Cytotoxicity of flavones and flavonols to a human esophageal squamous cell carcinoma cell line (KYSE-510) by induction of G2/M arrest and apoptosis. *Toxicol In Vitro* 2009; 23: 797-807.

[8] Wang YX, Cai H, Jiang G, Zhou TB, Wu H. Silibinin inhibits proliferation, induces apoptosis and causes cell cycle arrest in human gastric cancer MGC803 cells via STAT3 pathway inhibition. *Asian Pac J Cancer Prev* 2014; 15: 6791-8.

[9] Cui HB, Na XL, Song DF, Liu Y. Blocking effects of genistein on cell proliferation and possible mechanism in human gastric carcinoma. *World J Gastroenterol* 2005; 11: 69-72.

[10] Lin X, Tian L, Wang L, Li W, Xu Q, Xiao X. Antitumor effects and the underlying mechanism of licochalcone A combined with 5-fluorouracil in gastric cancer cells. *Oncol Lett* 2017; 13: 1695-701.

[11] Lee DH, Park KI, Park HS, Kang SR, Nagappan A, Kim JA, Kim EH, Lee WS, Hah YS, Chung HJ, An SJ, Kim GS. Flavonoids isolated from Korea Citrus aurantium L. induce G2/M phase arrest and apoptosis in human gastric cancer AGS cells. *Evid Based Complement Alternat Med* 2012; 2012: 515901.

[12] Lin E, Lin WH, Wang SY, Chen CS, Liao JW, Chang HW, Chen SC, Lin KY, Wang L, Yang HL, Hseu YC. Flavokawain B inhibits growth of human squamous carcinoma cells: Involvement of apoptosis and cell cycle dysregulation in vitro and in vivo. *J Nutr Biochem* 2012; 23: 368-78.

[13] Tan C, Qian X, Ge Y, Yang B, Wang F, Guan Z, Cai J. Oroxylin A could be a promising radiosensitizer for esophageal squamous cell carcinoma by inducing G2/M arrest and activating apoptosis. *Pathol Oncol Res* 2017; 23: 323-8.

[14] Yao S, Wang X, Li C, Zhao T, Jin H, Fang W. Kaempferol inhibits cell proliferation and glycolysis in esophagus squamous cell carcinoma via targeting EGFR signaling pathway. *Tumour Biol* 2016; 37: 10247-56.

[15] Chen F, Zhuang M, Zhong C, Peng J, Wang X, Li J, Chen Z, Huang Y. Baicalein reverses hypoxia-induced 5-FU resistance in gastric cancer AGS cells through suppression of glycolysis and the PTEN/Akt/HIF-1α signaling pathway. *Oncol Rep* 2015; 33: 457-63.

[16] Bao L, Liu F, Guo HB, Li Y, Tan BB, Zhang WX, Peng YH. Naringenin inhibits proliferation, migration, and invasion as well as induces apoptosis of gastric cancer SGC7901 cell line by downregulation of AKT pathway. *Tumour Biol* 2016; 37: 11365-74.

[17] Hong L, Ying SH. Ethanol extract and isolated constituents from Artemisia dracunculus inhibit esophageal squamous cell carcinoma and induce apoptotic cell death. *Drug Res* (Stuttg) 2015; 65: 101-6.

[18] Moon JY, Cho M, Ahn KS, Cho SK. Nobiletin induces apoptosis and potentiates the effects of the anticancer drug 5-fluorouracil in p53-mutated SNU-16 human gastric cancer cells. *Nutr Cancer* 2013; 65: 286-95.

[19] Hseu YC, Lee MS, Wu CR, Cho HJ, Lin KY, Lai GH, Wang SY, Kuo YH, Kumar KJ, Yang HL. The chalcone flavokawain B induces G2/M cell-cycle arrest and apoptosis in human oral carcinoma HSC-3 cells through the intracellular ROS generation and downregulation of the Akt/p38 MAPK signaling pathway. *J Agric Food Chem* 2012; 60: 2385-97.

[20] Kim JS, Park MR, Lee SY, Kim DK, Moon SM, Kim CS, Cho SS, Yoon G, Im HJ, You JS, Oh JS, Kim SG. Licochalcone A induces apoptosis in KB human oral cancer cells via a caspase-dependent FasL signaling pathway. *Oncol Rep* 2014; 31: 755-62.

[21] Yang Z, Zhao Y, Yao Y, Li J, Wang W, Wu X. Equol induces mitochondria-dependent apoptosis in human gastric cancer cells via the sustained activation of ERK1/2 pathway. *Mol Cells* 2016; 39: 742-9.

[22] Wu D, Zhang J, Wang J, Li J, Liao F, Dong W. Hesperetin induces apoptosis of esophageal cancer cells via mitochondrial pathway mediated by the increased intracellular reactive oxygen species. *Tumour Biol* 2016; 37: 3451-9.

[23] Wang X, Xia M. 5-Hydroxy-3,6,7,8,3`,4`-hexamethoxyflavone, a polymethoxyflavone, exerts antitumor effect on PI3K/Akt signaling pathway in human gastric cancer cell BGC-7901. *J Recept Signal Transduct Res* 2016; 36: 471-7.

[24] Zhang S, Li T, Zhang Y, Xu H, Li Y, Zi X, Yu H, Li J, Jin CY, Liu HM. A new brominated chalcone derivative suppresses the growth of gastric cancer cells in vitro and in vivo involving ROS mediated up-regulation of DR5 and 4 expression and apoptosis. *Toxicol Appl Pharmacol* 2016; 309: 77-86.

[25] Chen F, Qin X, Xu G, Gou S, Jin X. Reversal of cisplatin resistance in human gastric cancer cells by a wogonin-conjugated Pt(IV) prodrug via attenuating Casein

Kinase 2-mediated Nuclear Factor-κB pathways. *Biochem Pharmacol* 2017; 135: 50-68.

[26] Zhu Y, Jiang Y, Shi L, Du L, Xu X, Wang E, Sun Y, Guo X, Zou B, Wang H, Wang C, Sun L, Zhen Y. 7-O-Geranylquercetin induces apoptosis in gastric cancer cells via ROS-MAPK mediated mitochondrial signaling pathway activation. *Biomed Pharmacother* 2017; 87: 527-38.

[27] Song H, Bao J, Wei Y, Chen Y, Mao X, Li J, Yang Z, Xue Y. Kaempferol inhibits gastric cancer tumor growth: An in vitro and in vivo study. *Oncol Rep* 2015; 33: 868-74.

[28] Zhang H, Zhong X, Zhang X, Shang D, Zhou YI, Zhang C. Enhanced anticancer effect of ABT-737 in combination with naringenin on gastric cancer cells. *Exp Ther Med* 2016; 11: 669-73.

[29] Xu GY, Tang XJ. Troxerutin (TXN) potentiated 5-fluorouracil (5-Fu) treatment of human gastric cancer through suppressing STAT3/NF-κB and Bcl-2 signaling pathways. *Biomed Pharmacother* 2017; 92: 95-107.

[30] Zikri NN, Riedl KM, Wang LS, Lechner J, Schwartz SJ, Stoner GD. Black raspberry components inhibit proliferation, induce apoptosis, and modulate gene expression in rat esophageal epithelial cells. *Nutr Cancer* 2009; 61: 816-26.

[31] Lee YM, Kim YC, Choi BJ, Lee DW, Yoon JH, Kim EC. Mechanism of sappanchalcone-induced growth inhibition and apoptosis in human oral cancer cells. *Toxicol In Vitro* 2011; 25: 1782-8.

[32] Luo Y, Yu H, Yang Y, Tian W, Dong K, Shan J, Ma X. A flavonoid compound from Chrysosplenium nudicaule inhibits growth and induces apoptosis of the human stomach cancer cell line SGC-7901. *Pharm Biol* 2016; 54: 1133-9.

[33] Wang P, Zhang K, Zhang Q, Mei J, Chen CJ, Feng ZZ, Yu DH. Effects of quercetin on the apoptosis of the human gastric carcinoma cells. *Toxicol In Vitro* 2012; 26: 221-8.

[34] Singh SS, Vats S, Chia AY, Tan TZ, Deng S, Ong MS, Arfuso F, Yap CT, Goh BC, Sethi G, Huang RY, Shen HM, Manjithaya R, Kumar AP. Dual role of autophagy in hallmarks of cancer. *Oncogene* 2018; 37: 1142-58.

[35] Du F, Feng Y, Fang J, Yang M. MicroRNA-143 enhances chemosensitivity of quercetin through autophagy inhibition via target GABARAPL1 in gastric cancer cells. *Biomed Pharmacother* 2015; 74: 169-77.

[36] Moon JY, Cho SK. Nobiletin induces protective autophagy accompanied by ER-stress mediated apoptosis in human gastric cancer SNU-16 cells. *Molecules* 2016; 21: E914.

[37] Zang M, Hu L, Zhang B, Zhu Z, Li J, Zhu Z, Yan M, Liu B. Luteolin suppresses angiogenesis and vasculogenic mimicry formation through inhibiting Notch1-

VEGF signaling in gastric cancer. *Biochem Biophys Res Commun* 2017; 490: 913-9.

[38] Lu N, Gao Y, Ling Y, Chen Y, Yang Y, Gu HY, Qi Q, Liu W, Wang XT, You QD, Guo QL. Wogonin suppresses tumor growth in vivo and VEGF-induced angiogenesis through inhibiting tyrosine phosphorylation of VEGFR2. *Life Sci* 2008; 82: 956-63.

[39] Lei CS, Hou YC, Pai MH, Lin MT, Yeh SL. Effects of quercetin combined with anticancer drugs on metastasis-associated factors of gastric cancer cells: in vitro and in vivo studies. *J Nutr Biochem* 2018; 51: 105-13.

[40] Lee H, Lee JH, Jung KH, Hong SS. Deguelin promotes apoptosis and inhibits angiogenesis of gastric cancer. *Oncol Rep* 2010; 24: 957-63.

[41] Xiao W, Wu K, Yin M, Han S, Ding Y, Qiao A, Lu G, Deng B, Bo P, Gong W. Wogonin inhibits tumor-derived regulatory molecules by suppressing STAT3 signaling to promote tumor immunity. *J Immunother* 2015; 38: 167-84.

[42] Lin YH, Chen ZR, Lai CH, Hsieh CH, Feng CL. Active targeted nanoparticles for oral administration of gastric cancer therapy. *Biomacromolecules* 2015; 16: 3021-32.

[43] Yan X, Rui X, Zhang K. Baicalein inhibits the invasion of gastric cancer cells by suppressing the activity of the p38 signaling pathway. *Oncol Rep* 2015; 33: 737-43.

[44] Xia Y, Lian S, Khoi PN, Yoon HJ, Joo YE, Chay KO, Kim KK, Do Jung Y. Chrysin inhibits tumor promoter-induced MMP-9 expression by blocking AP-1 via suppression of ERK and JNK pathways in gastric cancer cells. *PLoS One* 2015; 10: e0124007.

[45] Hou C, Li W, Li Z, Gao J, Chen Z, Zhao X, Yang Y, Zhang X, Song Y. Synthetic isoliquiritigenin inhibits human tongue squamous carcinoma cells through its antioxidant mechanism. *Oxid Med Cell Longev* 2017; 2017: 1379430.

[46] Choi EJ, Oh HM, Wee H, Choi CS, Choi SC, Kim KH, Han WC, Oh TY, Kim SH, Jun CD. Eupatilin exhibits a novel anti-tumor activity through the induction of cell cycle arrest and differentiation of gastric carcinoma AGS cells. *Differentiation* 2009; 77: 412-23.

[47] Chen PN, Chu SC, Kuo WH, Chou MY, Lin JK, Hsieh YS. Epigallocatechin-3 gallate inhibits invasion, epithelial-mesenchymal transition, and tumor growth in oral cancer cells. *J Agric Food Chem* 2011; 59: 3836-44.

[48] Shen H, Zeng G, Tang G, Cai X, Bi L, Huang C, Yang Y. Antimetastatic effects of licochalcone A on oral cancer via regulating metastasis-associated proteases. *Tumour Biol* 2014; 35: 7467-74.

[49] Lu L, Chen J, Tang H, Bai L, Lu C, Wang K, Li M, Yan Y, Tang L, Wu R, Ye Y, Jin L, Liang Z. EGCG suppresses ERK5 activation to reverse tobacco smoke-

triggered gastric epithelial-mesenchymal transition in BALB/c mice. *Nutrients* 2016; 8: E380.

[50] Zhang X, Zheng L, Sun Y, Wang T, Wang B. Tangeretin enhances radiosensitivity and inhibits the radiation-induced epithelial-mesenchymal transition of gastric cancer cells. *Oncol Rep* 2015; 34: 302-10.

[51] Lim YC, Park HY, Hwang HS, Kang SU, Pyun JH, Lee MH, Choi EC, Kim CH. (-)-Epigallocatechin-3-gallate (EGCG) inhibits HGF-induced invasion and metastasis in hypopharyngeal carcinoma cells. *Cancer Lett* 2008; 271: 140-52.

[52] Koh YW, Choi EC, Kang SU, Hwang HS, Lee MH, Pyun J, Park R, Lee Y, Kim CH. Green tea (-)-epigallocatechin-3-gallate inhibits HGF-induced progression in oral cavity cancer through suppression of HGF/c-Met. *J Nutr Biochem* 2011; 22: 1074-83.

[53] Li H, Huang D, Gao Z, Chen Y, Zhang L, Zheng J. Scutellarin inhibits the growth and invasion of human tongue squamous carcinoma through the inhibition of matrix metalloproteinase-2 and -9 and αvß6 integrin. *Int J Oncol* 2013; 42: 1674-81.

[54] Kim S, Choi MG, Lee HS, Lee SK, Kim SH, Kim WW, Hur SM, Kim JH, Choe JH, Nam SJ, Yang JH, Kim S, Lee JE, Kim JS. Silibinin suppresses TNF-alpha-induced MMP-9 expression in gastric cancer cells through inhibition of the MAPK pathway. *Molecules* 2009; 14: 4300-11.

[55] Kim HS, Kim MH, Jeong M, Hwang YS, Lim SH, Shin BA, Ahn BW, Jung YD. EGCG blocks tumor promoter-induced MMP-9 expression via suppression of MAPK and AP-1 activation in human gastric AGS cells. *Anticancer Res* 2004; 24: 747-53.

[56] Lin CW, Chen PN, Chen MK, Yang WE, Tang CH, Yang SF, Hsieh YS. Kaempferol reduces matrix metalloproteinase-2 expression by down-regulating ERK1/2 and the activator protein-1 signaling pathways in oral cancer cells. *PLoS One* 2013; 8: e80883.

[57] Cheng CW, Shieh PC, Lin YC, Chen YJ, Lin YH, Kuo DH, Liu JY, Kao JY, Kao MC, Way TD. Indoleamine 2,3-dioxygenase, an immunomodulatory protein, is suppressed by (-)-epigallocatechin-3-gallate via blocking of gamma-interferon-induced JAK-PKC-delta-STAT1 signaling in human oral cancer cells. *J Agric Food Chem* 2010; 58: 887-94.

[58] Park BB, Yoon Js, Kim Es, Choi J, Won Yw, Choi Jh, Lee YY. Inhibitory effects of eupatilin on tumor invasion of human gastric cancer MKN-1 cells. *Tumour Biol* 2013; 34: 875-85.

[59] Shi C, Fan LY, Cai Z, Liu YY, Yang CL. Cellular stress response in Eca-109 cells inhibits apoptosis during early exposure to isorhamnetin. *Neoplasma* 2012; 59: 361-9.

[60] Li YS, Wu LP. Li KH, Liu YP, Xiang R, Zhang SB, Zhu LY, Zhang LY. Involvement of nuclear factor κB (NF-κB) in the downregulation of cyclooxygenase-2 (COX-2) by genistein in gastric cancer cells. *J Int Med Res* 2011; 39: 2141-50.
[61] Cheong E, Ivory K, Doleman J, Parker ML, Rhodes M, Johnson IT. Synthetic and naturally occurring COX-2 inhibitors suppress proliferation in a human oesophageal adenocarcinoma cell line (OE33) by inducing apoptosis and cell cycle arrest. *Carcinogenesis* 2004; 25: 1945-52.
[62] Huang W, Wan C, Luo Q, Huang Z, Luo Q. Genistein-inhibited cancer stem cell-like properties and reduced chemoresistance of gastric cancer. *Int J Mol Sci* 2014; 15: 3432-43.
[63] Yu D, Shin HS, Lee YS, Lee D, Kim S, Lee YC. Genistein attenuates cancer stem cell characteristics in gastric cancer through the downregulation of Gli1. *Oncol Rep* 2014; 31: 673-8.

Chapter 2

FLAVONOIDS AS NATURAL POLYPHENOLS WITH ANTICANCER PROPERTIES

2.1. FLAVONOIDS AND THEIR SUBCLASSES

Natural plant-derived products play an important role in reduction of many health risks by inhibiting or delaying different human chronic pathologies, such as cardiovascular diseases, neurodegenerative disorders, diabetes, but also cancer across the world [1-3]. It has been estimated that incidence of more than two-thirds of all human malignancies could be prevented through modifications in lifestyle, including dietary habits [4-7]. Indeed, epidemiological studies have suggested that high daily dietary intake of fruits and vegetables is associated with a lower risk of various forms of cancer, including malignant neoplasms in the upper gastrointestinal tract, i.e., oral cavity, esophagus and stomach [4, 8-14]. At that, low consumption of fruits and vegetables can account for even 15% to 40% of cancers in the digestive tract [11, 15]. These health-promoting effects of plant foods have been attributed to their non-nutritional components, so-called bioactive phytochemicals or nutraceuticals, which are consumed in hundreds of agents every day [4, 16]. Among these compounds, flavonoids may explain, at least in part, the decreased prevalence of cancer [17-21]. These abundant polyphenolic agents comprise a large group of low-molecular weight plant secondary metabolites which cannot be synthesized by humans, and were originally known only for their role in plant pigmentation [11, 22, 23]. However, numerous preclinical experimental studies have demonstrated that in human systems, flavonoids exert diverse biological activities, such as antioxidant, antiinflammatory, antibacterial, antiviral, immunomodulatory, antimutagenic, antiproliferative, and anticarcinogenic effects [18,

20, 24-26]. Therefore, the research on plant flavonoids as potential anticancer substances has considerably increased in the recent decades.

To date, more than 9000 structurally different flavonoids have been derived from plants which exist in the nature either as free aglycones or sugar conjugates [3, 19]. Flavonoids as benzo-γ-pyrone derivatives contain a diphenylpropane (C6-C3-C6) backbone in which two phenolic rings (A and B) are linked by a heterocyclic, oxygen-containing ring C [27-30] (Figure 2.1). Such basic skeleton gives rise to a wide variety of structurally different compounds, whereas hydroxyl groups at aromatic rings contribute to the antioxidant activity of flavonoids [12]. Based on their structural complexity (hydroxylation pattern and functional substitutions), flavonoids are further divided into six subclasses: flavanols or catechins, flavanones, flavones, flavonols, isoflavones, and anthocyanidins [19, 27, 29, 31]. The major representatives of flavanols are catechin, epicatechin, epicatechin gallate, epigallocatechin, and epigallocatechin gallate; flavanones comprise naringenin, hesperetin, eriodictyol, and liquiritigenin; the most common flavones are apigenin, chrysin, luteolin, baicalein, wogonin, tangeretin, and nobiletin; flavonols include quercetin, myricetin, kaempferol, and fisetin; isoflavones subclass contains genistein, daidzein, biochanin A, and formononetin; and six well-known anthocyanidins represent cyanidin, pelargonidin, petunidin, peonidin, delphinidin, and malvidin. Structures of these compounds are depicted in Figure 2.1. In contrast to other flavonoids, anthocyanidins occur in the nature as cations carrying a positive charge in the C-ring of the molecule [32, 33]. At that, chalcones are natural precursors for synthesis of flavonoids in plants [34, 35].

General skeleton of flavonoids

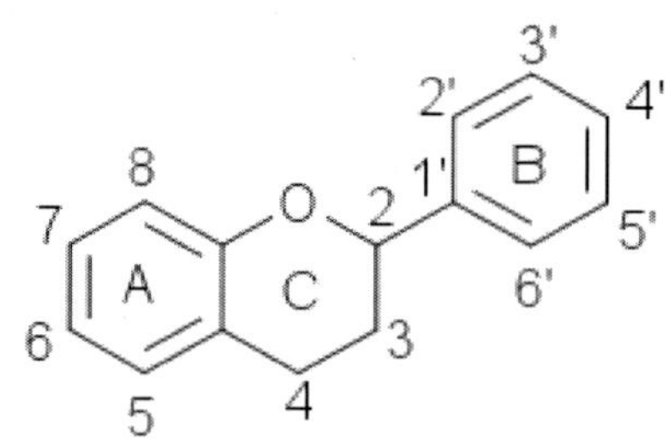

FLAVANOLS

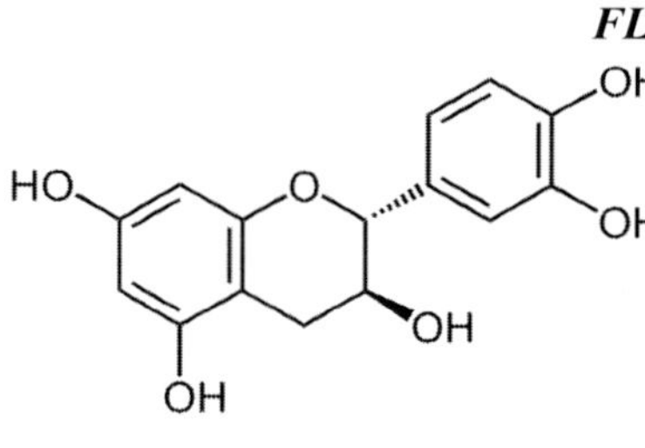

(+)-Catechin (C)

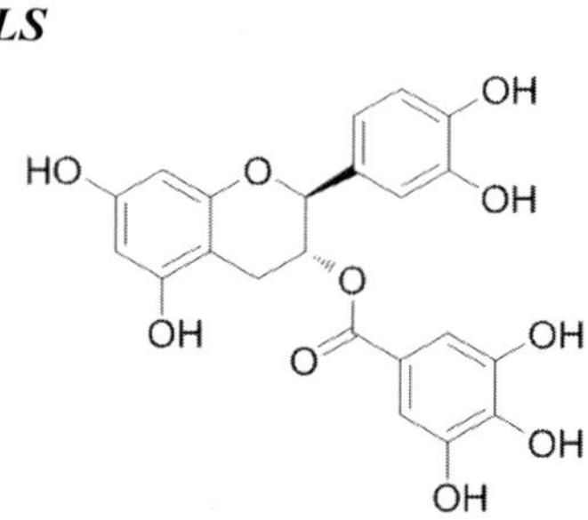

(-)-Catechin gallate (CG)

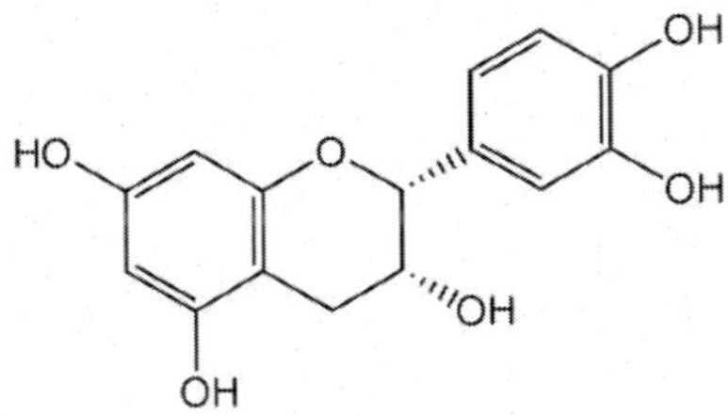

(-)-Epicatechin (EC)

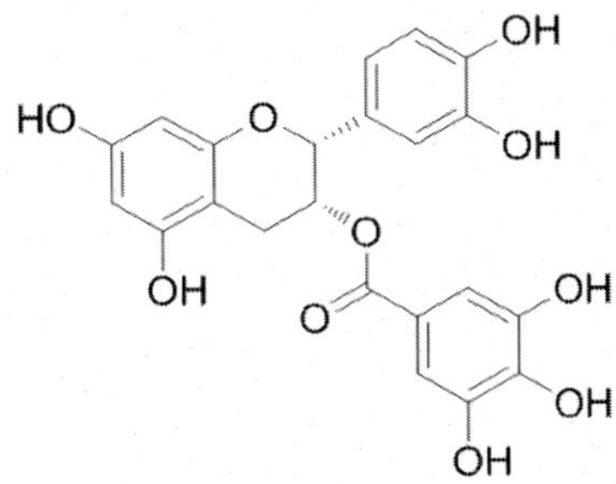

(-)-Epicatechin gallate (ECG)

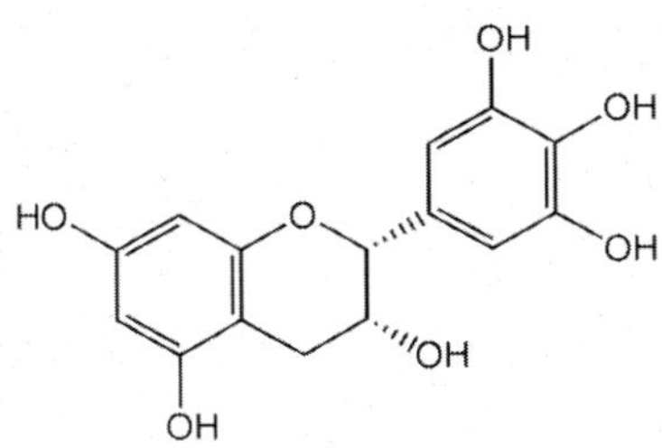

(-)-Epigallocatechin (EGC)

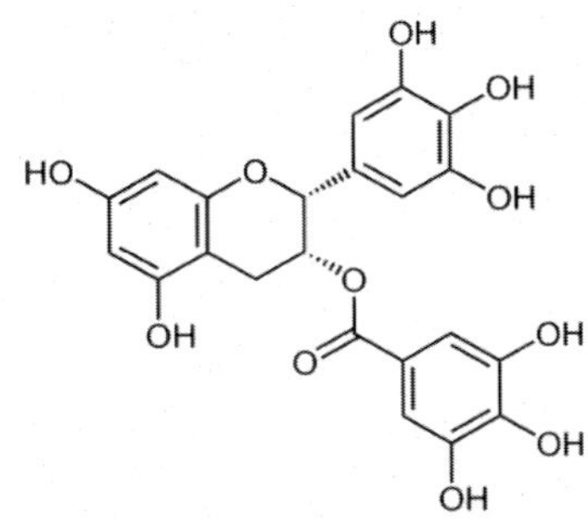

(-)-Epigallocatechin gallate (EGCG)

FLAVANONES

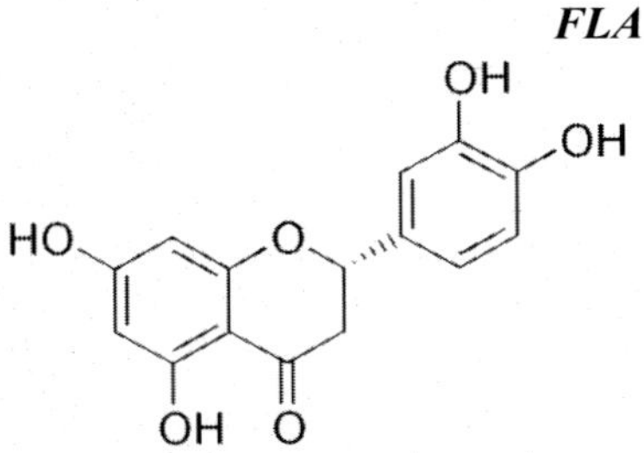

Eriodictyol

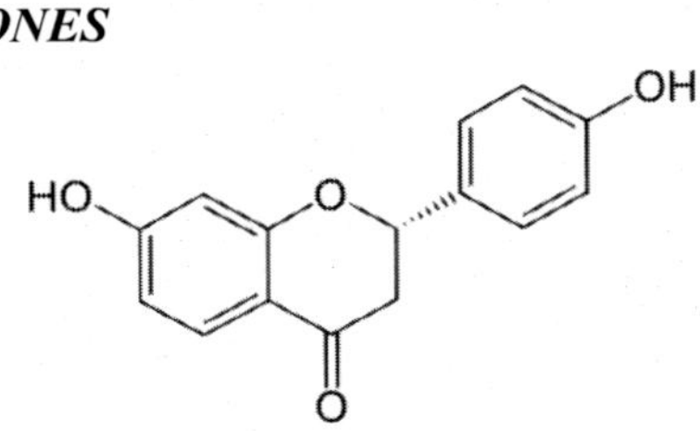

Liquiritigenin

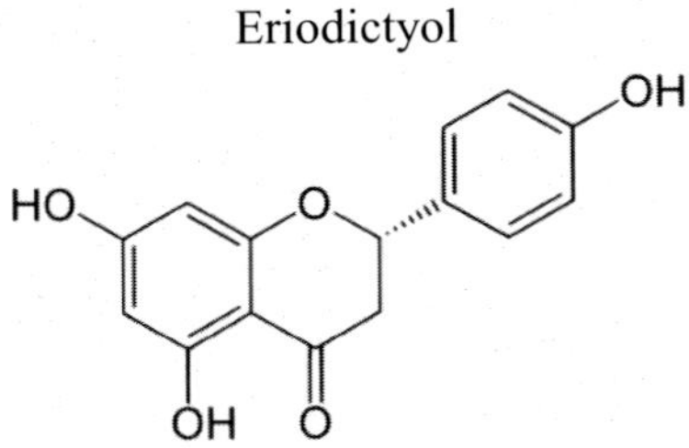

Naringenin

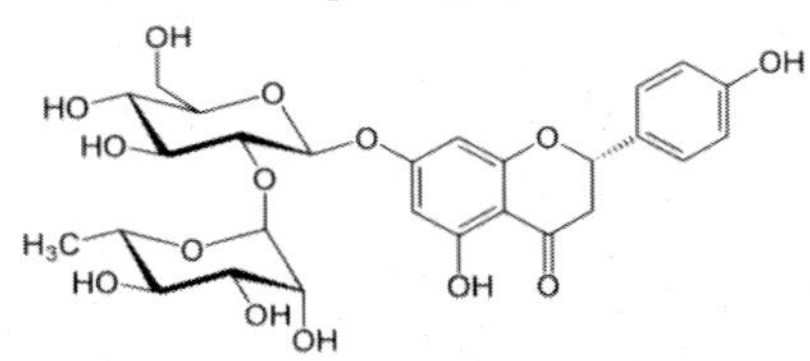

Naringin

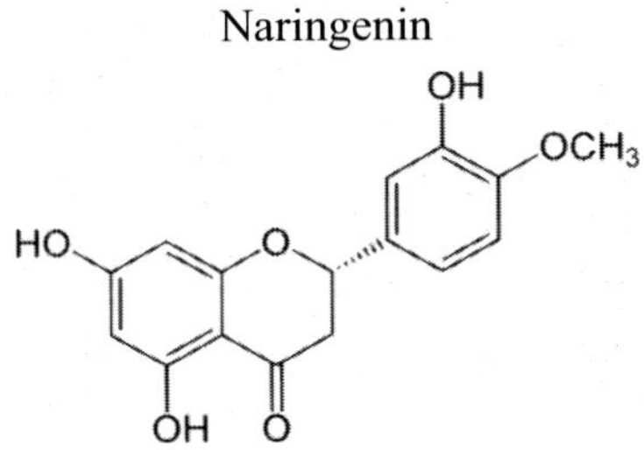

Hesperetin

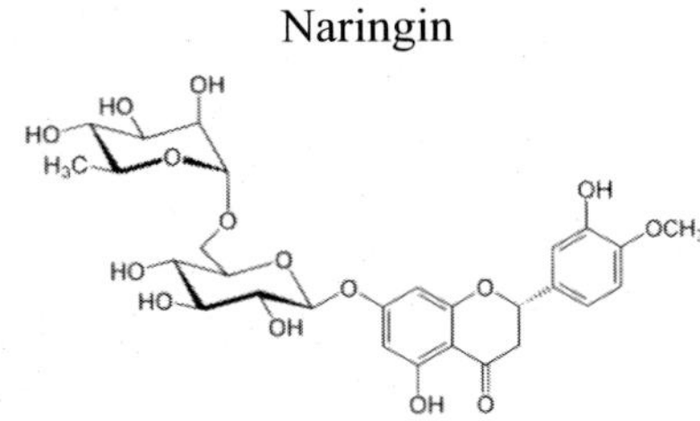

Hesperidin

Figure 2.1. (Continued).

FLAVONES

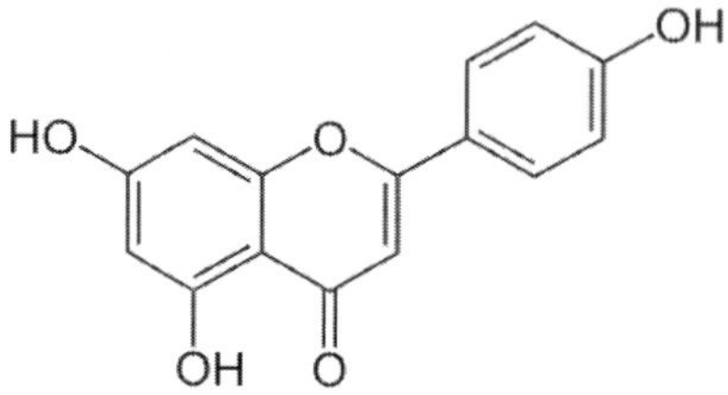

Apigenin

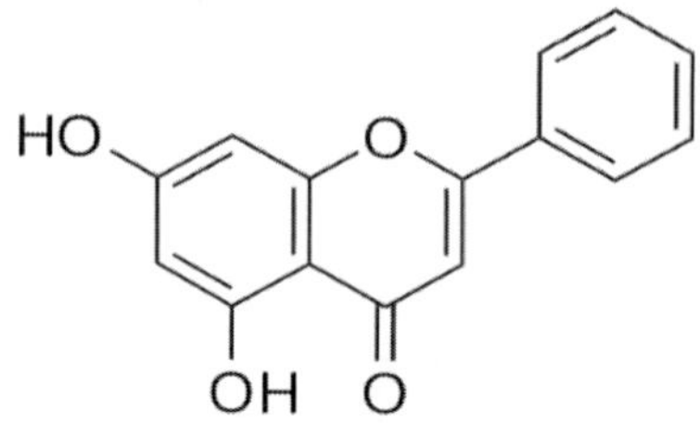

Chrysin

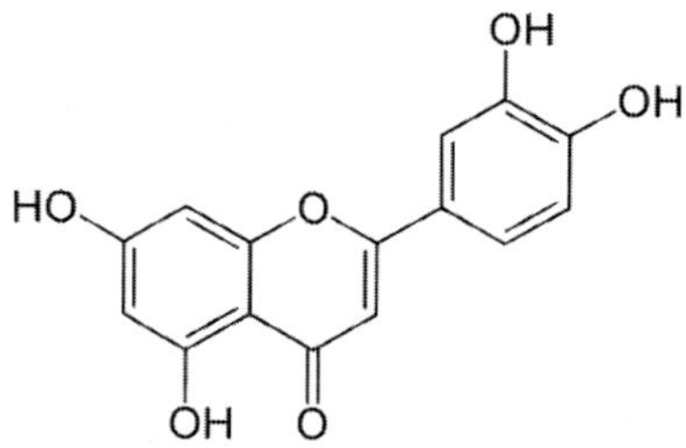

Luteolin

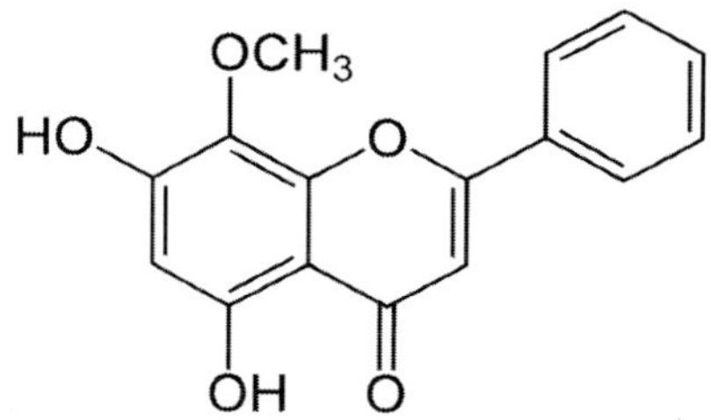

Wogonin

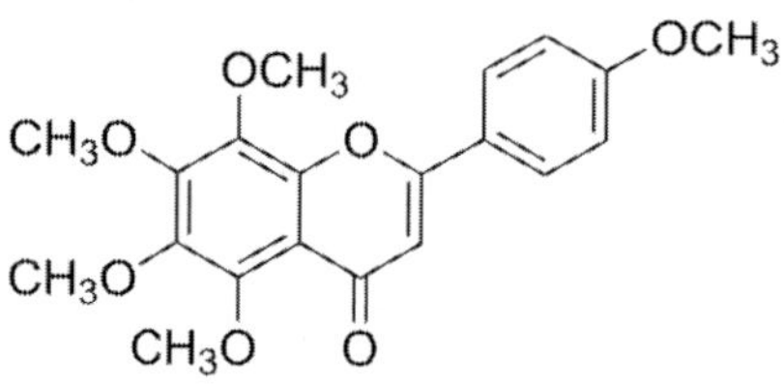

Tangeretin

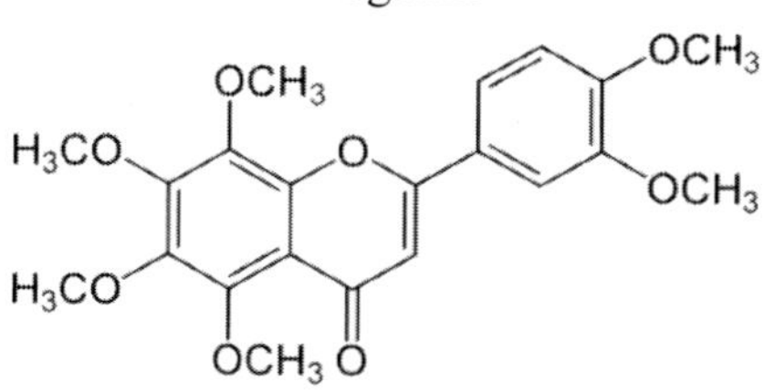

Nobiletin

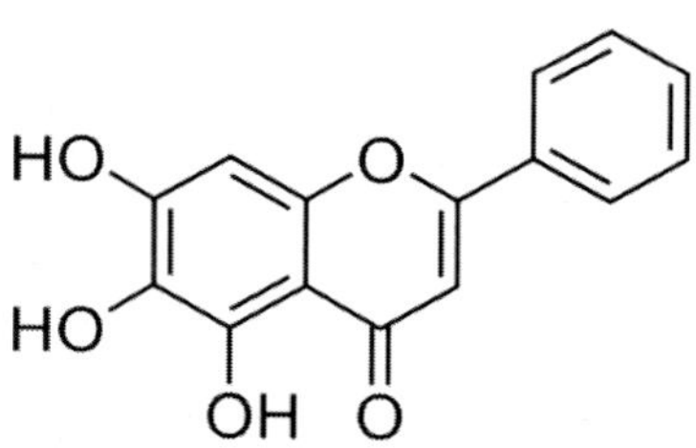

Baicalein

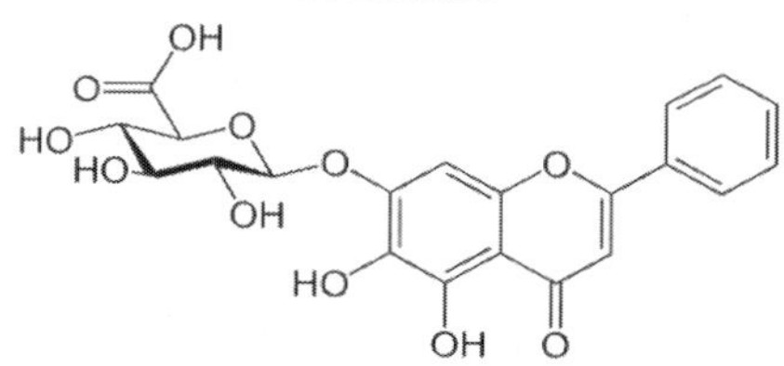

Baicalin

FLAVONOLS

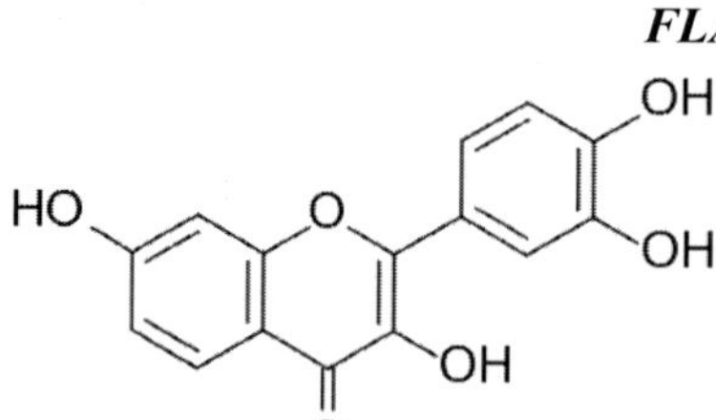

Fisetin

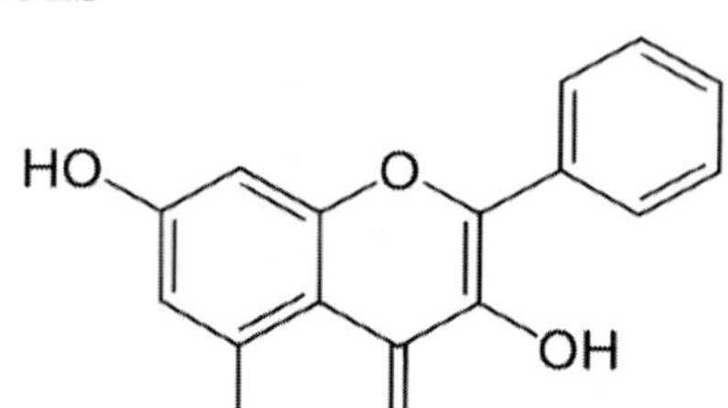

Galangin

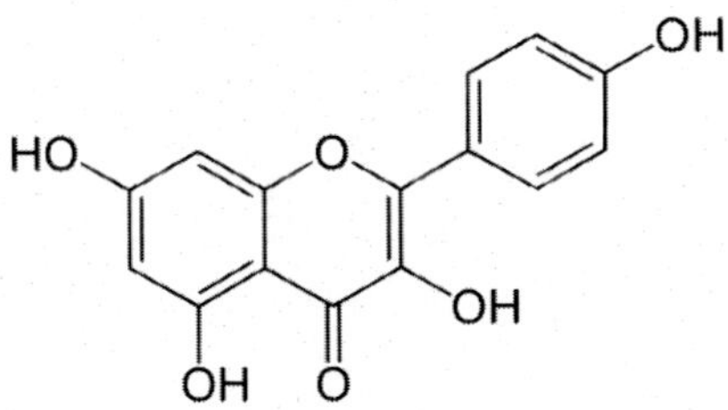

Kaempferol

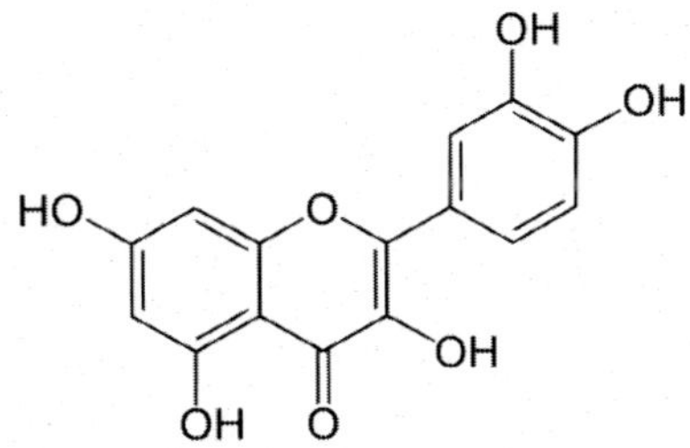

Quercetin

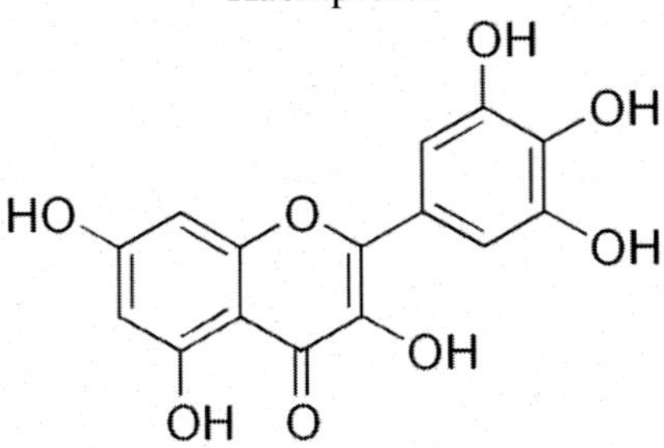

Myricetin

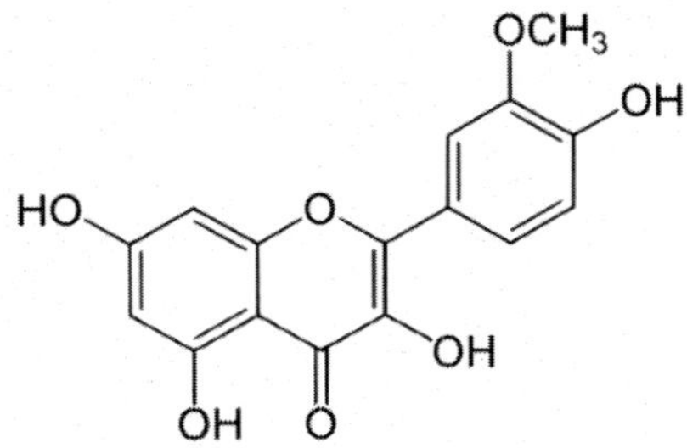

Isorhamnetin

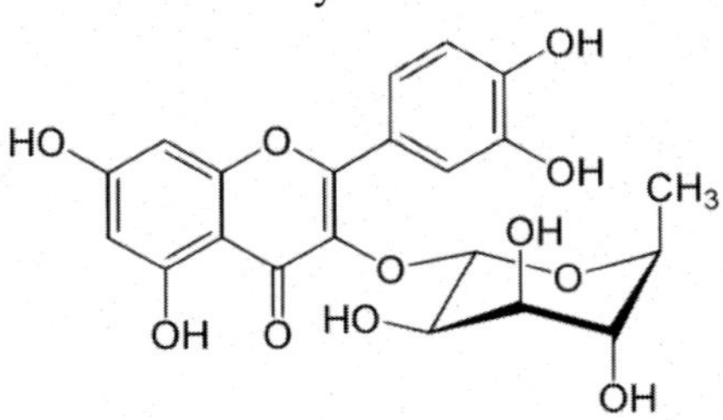

Quercitrin

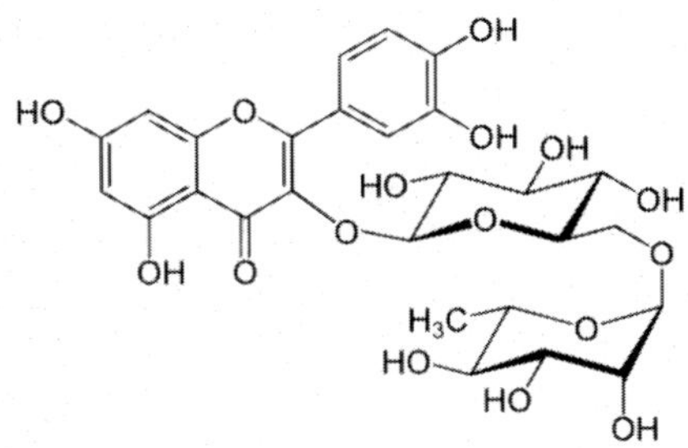

Rutin

ISOFLAVONES

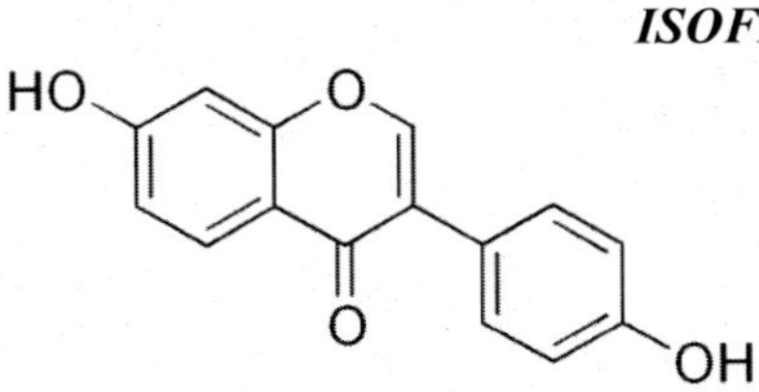

Daidzein

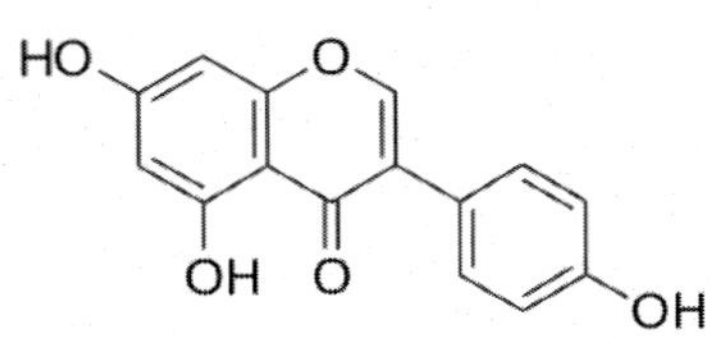

Genistein

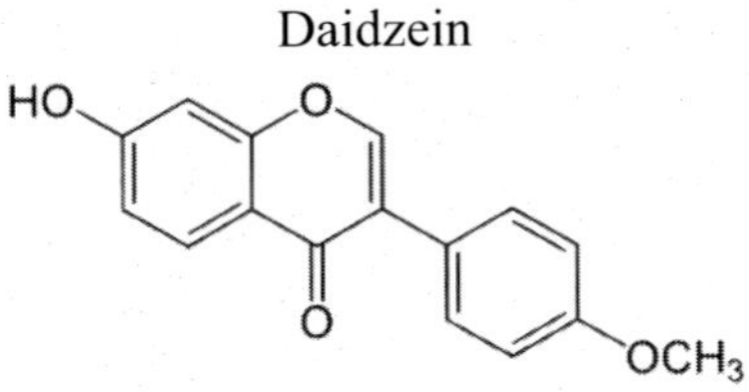

Formononetin

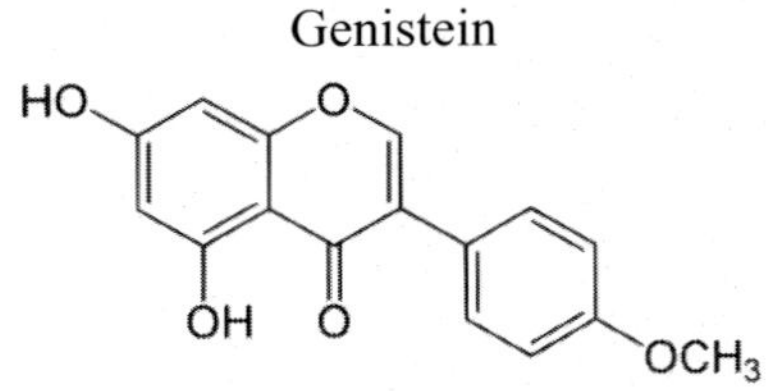

Biochanin A

Figure 2.1. (Continued).

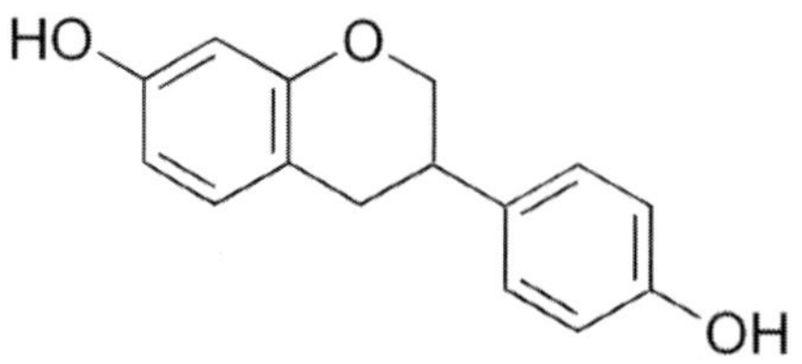

Equol

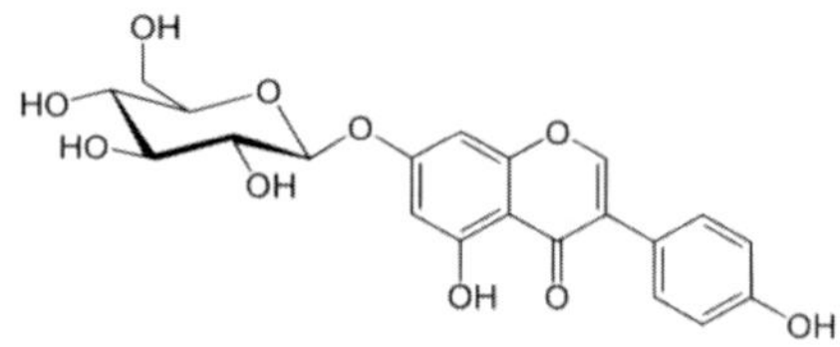

Genistin

ANTHOCYANIDINS

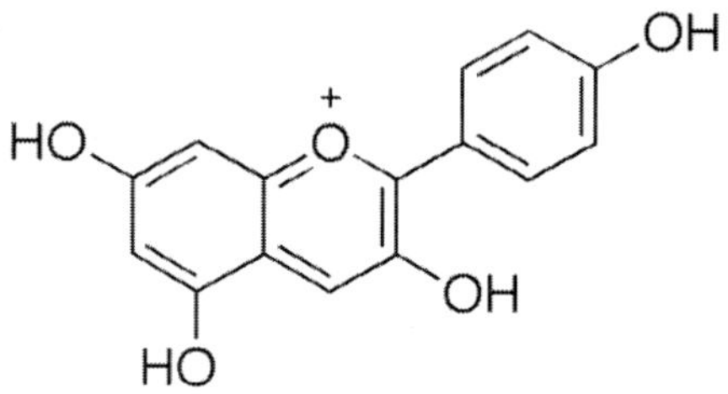

Pelargonidin

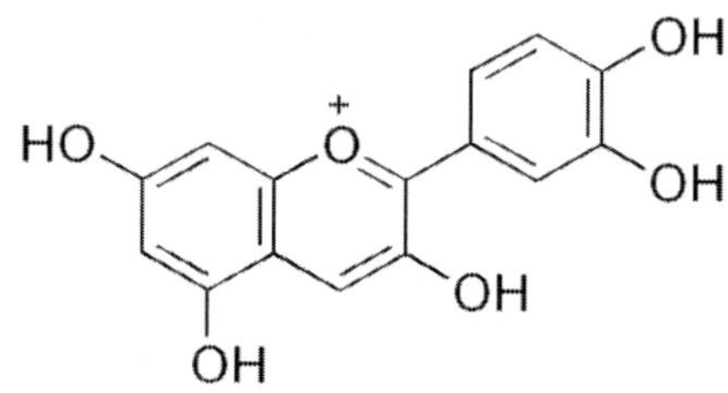

Cyanidin

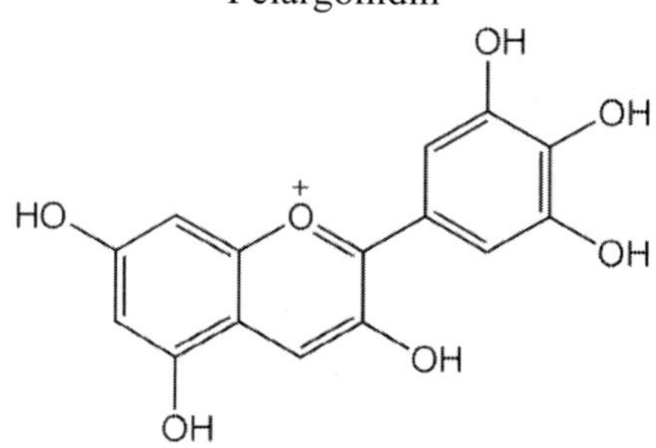

Delphinidin

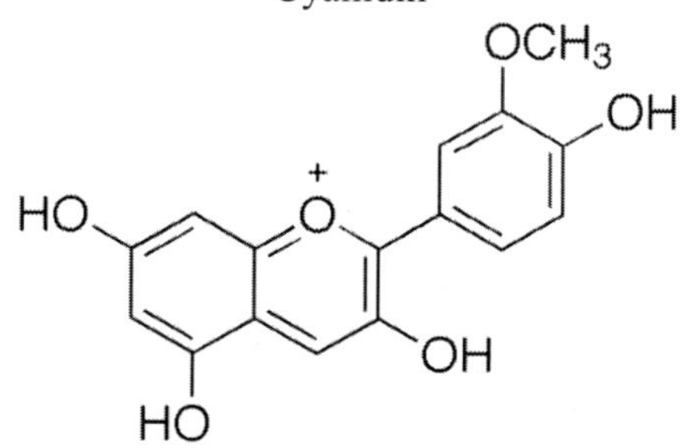

Peonidin

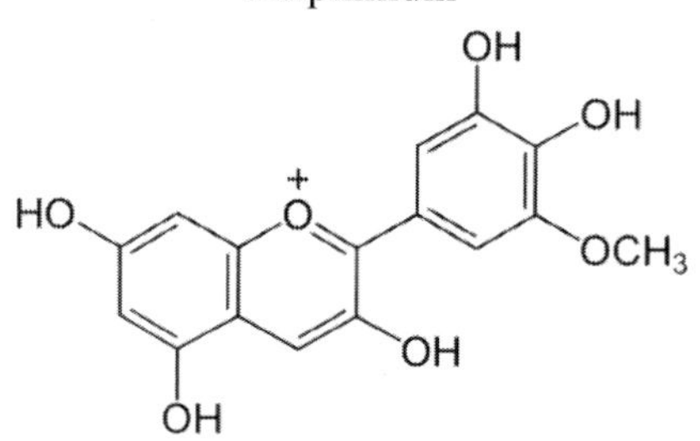

Petunidin

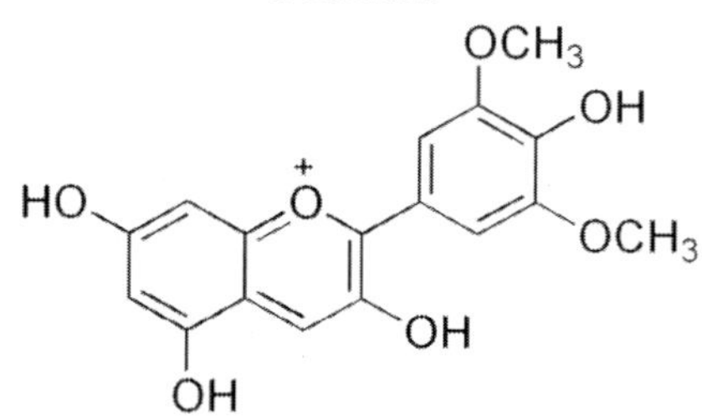

Malvidin

Figure 2.1. Structures of the most common natural flavonoids.

The intake of flavonoids varies significantly in amounts and subclasses among individuals due to differences in their dietary habits [27]. Although the average daily intake of flavonoids can reach even 1 g, it can be also only about twenty mg dependent on geographical location, culture, and religious creeds of people [26, 36, 37]. The main dietary sources of flavonoids are fruits and vegetables, whereas consumption of these compounds from fruits is estimated to be about three times higher than from vegetables [27]. Other common sources of flavonoids include plant-derived beverages, such as tea, juices, wine, beer, and cocoa products, spices, whole grains, seeds, nuts, and medicinal herbs [19, 38-41]. At that, each flavonoid subclass has somewhat different plant sources

[42]. Flavanols are mostly found in green and white tea, cocoa, dark chocolate, and berries [43, 44]. Flavanones are present at high doses in different citrus fruits and their food products [44]. Flavones can be found in different spices and herbs, such as oregano, peppermint, and thyme [44]. Dietary flavonols exist in abundance in various vegetables, including kale, onions, cabbage, radish, peppers, and fennel, but also in several berries [20, 44]. Soy and soy foods are practically the only dietary source of isoflavones [44]. Anthocyanidins occur predominantly in various berries, such as black raspberries, blueberries, bilberries, black currants, and cranberries, contributing to their reddish and purple colors [32, 45, 46]. The most important natural sources of major dietary flavonoids are listed in Table 2.1.

Table 2.1. Major dietary sources of the most common natural flavonoids (composed based on USDA Database for the Flavonoid Content of Selected Foods, Release 3.1 and USDA Database for the Isoflavone Content of Selected Foods, Release 2.1)

Flavonoid	Major dietary sources
FLAVANOLS	
Catechin (C)	Cacao beans, cocoa powder, dark chocolate; green tea; carob flour; blueberries, blackberries, cranberries, black grapes; plums, white peaches; broad beans
Epicatechin (EC)	Cacao beans, cocoa powder, dark chocolate; soybeans, broad beans; grape seeds, blueberries; apples, sour cherries; green tea
Epicatechin gallate (ECG)	Green tea, white tea, oolong tea, black tea; carob flour
Epigallocatechin (EGC)	Cacao beans; green tea, white tea, black tea, oolong tea; broad beans; plums
Epigallocatechin gallate (EGCG)	Green tea, white tea, oolong tea, black tea; carob flour
FLAVANONES	
Eriodictyol	Oregano, peppermint; lemons
Naringenin	Oregano, rosemary; kumquats, grapefruit, pummelo, oranges, tangerines; artichokes
Hesperetin	Limes, lemons, oranges, pummelo, tangerines; peppermint
FLAVONES	
Apigenin	Parsley, vinespinach, celery, artichokes; kumquats, juniper berries; oregano, peppermint
Luteolin	Oregano, thyme, sage, peppermint; juniper berries; radicchio, celery, peppers, saw-thistle
FLAVONOLS	
Kaempferol	Capers, chives, dill weed, tarragon; kale, mustard greens, arugulo, ginger, onions, watercress, Chinese cabbage, radish, cress, turnip, dock, endive, chives; beans; Chia seeds
Quercetin	Capers, dill weed, oregano; lovage, dock, radish, coriander, peppers, fennel, onions, radicchio, watercress, hartwort, corn poppy, hawthorn, kale, drumstick, saw-thistle, asparagus; juniper berries, elderberries, chokeberries, bog whortleberries; carob flour, cowpeas; bee pollen; Chia seeds; buckwheat
Myricetin	Fennel, parsley, dock; goji berries, bog whortleberries, cranberries, black currants; carob flour
Isorhamnetin	Dill weed, tarragon; sea buckthorn berries, sour cherries, elderberries; kale, mustard greens, fennel, chives, asparagus, hartwort, onions

Table 2.1. (Continued)

Flavonoid	Major dietary sources
ISOFLAVONES	
Daidzein	Soy and soy-based foods (soy flour, soy milk curd, natto, miso, miso soup, tofu, tempeh, soy yogurt); red clover
Genistein	Soy and soy-based foods (soy flour, tofu, soy milk curd, miso, miso soup, natto, tempeh, soy yogurt); red clover
Formononetin	Red clover, soybeans, clover sprouts, alfalfa seeds
Biochanin A	Chickpeas, kala chana
ANTHOCYANIDINS	
Pelargonidin	Strawberries, black raspberries, acerola; radishes
Cyanidin	Black raspberries, plums, elderberries, chokeberries, acai berries, golden currants, blackberries, arctic bramble berries, bilberries, red currants, black currants, cranberries, raspberries, lingonberries, sour cherries, sweet cherries, cedar bay cherries, grapes, jostaberries, blueberries, acerola; peppers, red cabbage, radicchio
Delphinidin	Bilberries, black currants, grapes, blueberries, jambuls, plums, red currants, cranberries, bananas, jostaberries; eggplant, radicchio; cowpeas, black beans
Peonidin	Cranberries, bilberries, blueberries, plums, jambuls; cowpeas
Petunidin	Bilberries, blueberries, jambuls; cowpeas, black beans
Malvidin	Blueberries, bilberries, grapes, jambuls; cowpeas, black beans

2.2. Flavonoids as Potential Chemopreventive and Chemotherapeutic Agents

Chemoprevention is a promising strategy for prevention and controlling of cancer by using specific natural or synthetic agents. The term chemoprevention was introduced by Michael B. Sporn in 1976, whereas bioactive phytochemicals are considered as attractive chemopreventive agents [16, 47]. In chemoprevention, bioactive substances can suppress, retard or reverse the promotion and progression stages of carcinogenesis and prevent transformation of premalignant lesions into invasive phenotype [16, 17, 48-53]. At that, chemoprevention might decrease the cancer incidence in general population, but can also inhibit the risk of recurrence of malignancy among cancer patients, meaning thus a better quality of life [4, 53]. An ideal chemopreventive agent must reveal only minimal or no adverse side effects, besides being efficient in multiple disease stages with known action mechanisms, providing possibilities for peroral administration, possessing low cost and easy availability [48, 49, 53, 54]. As the current cancer therapies (chemotherapy and radiotherapy) are mostly associated with a high level of toxicity, the general safety of dietary polyphenolic flavonoids makes them highly interesting candidates for modulation of cancer process [1, 53-58]. Indeed, accumulating evidence has suggested that many flavonoids possess strong chemopreventive potential, exerting numerous biological activities, such as controlling the growth and proliferation of cancerous cells, regulation of the cell cycle progression, induction of apoptotic death, modulation of inflammatory pathways, inhibition of angiogenesis, and suppression of invasion, without affecting normal tissues [4, 8, 10, 38, 51, 56, 59]. Moreover, a number of studies have shown that flavonoids reveal pleiotropic action mechanisms by simultaneously affecting multiple cellular targets, and inhibit different stages of carcinogenesis, i.e., initiation, promotion and progression [4, 60, 61]. At that, the effects of flavonoids depend on both their structural features, but also on the cell line characteristics and type of cancer, making necessary to systematically study each individual flavonoid in different malignant cells [38, 62, 63]. The chemopreventive activities of flavonoids have been investigated in numerous cellular and animal models, whereas in humans, these effects seem to be further influenced by several cancer risk factors, including smoking status and alcohol consumption [20]. Chemopreventive approach through ingestion of certain plant-based food items rich in flavonoids might be especially important in the world areas with higher cancer incidence due to more intensive exposure of inhabitants to certain carcinogens [64, 65]. Hence, chemoprevention strategy has received progressively more attention in the recent years and identification of novel compounds which will meet the requirements for chemopreventive agents is considered to be a crucial task in combating cancer.

Natural flavonoids interfering with carcinogenesis by virtue of their important biological activities are of potential clinical value and have a wide range of therapeutic applications [26, 39, 66]. Indeed, the majority (more than 60%) of the approved and currently clinically used anticancer drugs is of natural origin [3, 62, 67-70]. Therefore, the search and investigation of anticancer agents from dietary and medicinal plants as leads for development of novel chemotherapeutic drugs must certainly continue. At that, the remarkable global biodiversity along with modern analytical and chemical methods gives promise of more successful fighting against cancer in the future.

References

[1] Ren K, Zhang W, Wu G, Ren J, Lu H, Li Z, Han X. Synergistic anti-cancer effects of galangin and berberine through apoptosis induction and proliferation inhibition in oesophageal carcinoma cells. *Biomed Pharmacother* 2016; 84: 1748-59.

[2] Krishnakumar N, Sulfikkarali NK, Manoharan S, Nirmal RM. Screening of chemopreventive effect of naringenin-loaded nanoparticles in DMBA-induced hamster buccal pouch carcinogenesis by FT-IR spectroscopy. *Mol Cell Biochem* 2013; 382: 27-36.

[3] Gunda SK, Kongaleti SF, Shaik M. Natural flavonoid derivatives as oral human epidermoid carcinoma cell inhibitors. *Int J Comput Biol Drug Des* 2015; 8: 19-39.

[4] Ullah MF, Bhat SH, Husain E, Abu-Duhier F, Hadi SM, Sarkar FH, Ahmad A. Pharmacological intervention through dietary nutraceuticals in gastrointestinal neoplasia. *Crit Rev Food Sci Nutr* 2016; 56: 1501-18.

[5] Weisburg JH, Weissman DB, Sedaghat T, Babich H. In vitro cytotoxicity of epigallocatechin gallate and tea extracts to cancerous and normal cells from the human oral cavity. *Basic Clin Pharmacol Toxicol* 2004; 95: 191-200.

[6] Chen SF, Nien S, Wu CH, Liu CL, Chang YC, Lin YC. Reappraisal of the anticancer efficacy of quercetin in oral cancer cells. *J Chin Med Assoc* 2013; 76: 146-52.

[7] Gonzalez CA, Sala N, Rokkas T. Gastric cancer: epidemiologic aspects. *Helicobacter* 2013; 18 Suppl 1: 34-8.

[8] Sun L, Subar AF, Bosire C, Dawsey SM, Kahle LL, Zimmerman TP, Abnet CC, Heller R, Graubard BI, Cook MB, Petrick JL. Dietary flavonoid intake reduces the risk of head and neck but not esophageal or gastric cancer in US men and women. *J Nutr* 2017; 147: 1729-38.

[9] Petrick JL, Steck SE, Bradshaw PT, Trivers KF, Abrahamson PE, Engel LS, He K, Chow WH, Mayne ST, Risch HA, Vaughan TL, Gammon MD. Dietary intake of flavonoids and oesophageal and gastric cancer: incidence and survival in the United States of America (USA). *Br J Cancer* 2015; 112: 1291-300.

[10] Bo Y, Sun J, Wang M, Ding J, Lu Q, Yuan L. Dietary flavonoid intake and the risk of digestive tract cancers: a systematic review and meta-analysis. *Sci Rep* 2016; 6: 24836.

[11] Rossi M, Bosetti C, Negri E, Lagiou P, La Vecchia C. Flavonoids, proanthocyanidins, and cancer risk: a network of case-control studies from Italy. *Nutr Cancer* 2010; 62: 871-7.

[12] Ekambaram G, Rajendran P, Magesh V, Sakthisekaran D. Naringenin reduces tumor size and weight lost in N-methyl-N`-nitro-N-nitrosoguanidine-induced gastric carcinogenesis in rats. *Nutr Res* 2008; 28: 106-12.

[13] Tang S, Hu J, Meng Q, Dong X, Wang K, Qi Y, Chu C, Zhang X, Hou L. Daidzein induced apoptosis via down-regulation of Bcl-2/Bax and triggering of the mitochondrial pathway in BGC-823 cells. *Cell Biochem Biophys* 2013; 65: 197-202.

[14] Tanaka T, Makita H, Ohnishi M, Hirose Y, Wang A, Mori H, Satoh K, Hara A, Ogawa H. Chemoprevention of 4-nitroquinoline 1-oxide-induced oral carcinogenesis by dietary curcumin and hesperidin: comparison with the protective effect of beta-carotene. *Cancer Res* 1994: 54: 4653-9.

[15] Lim YC, Lee SH, Song MH, Yamaguchi K, Yoon JH, Choi EC, Baek SJ. Growth inhibition and apoptosis by (-)-epicatechin gallate are mediated by cyclin D1 suppression in head and neck squamous carcinoma cells. *Eur J Cancer* 2006; 42: 3260-6.

[16] Bhavana SM, Lakshmi CR. Oral oncoprevention by phytochemicals – a systematic review disclosing the therapeutic dilemma. *Adv Pharm Bull* 2014; 4 Suppl 1: 413-20.

[17] Iriti M, Varoni EM. Chemopreventive potential of flavonoids in oral squamous cell carcinoma in human studies. *Nutrients* 2013; 5: 2564-76.

[18] Lim JC, Park SY, Nam Y, Nguyen TT, Sohn UD. The protective effect of eupatilin against hydrogen peroxide-induced injury involving 5-lipoxygenase in feline esophageal epithelial cells. *Korean J Physiol Pharmacol* 2012; 16: 313-20.

[19] Cui L, Liu X, Tian Y, Xie C, Li Q, Cui H, Sun C. Flavonoids, flavonoid subclasses, and esophageal cancer risk: a meta-analysis of epidemiologic studies. *Nutrients* 2016: 8: E350.

[20] Xie Y, Huang S, Su Y. Dietary flavonols intake and risk of esophageal and gastric cancer: a meta-analysis of epidemiological studies. *Nutrients* 2016; 8: 91.

[21] Park CH, Hahm ER, Lee JH, Jung KC, Yang CH. Inhibition of beta-catenin-mediated transactivation by flavanone in AGS gastric cancer cells. *Biochem Biophys Res Commun* 2005; 331: 1222-8.

[22] Rossi M, Garavello W, Talamini R, La Vecchia C, Franceschi S, Lagiou P, Zambon P, Dal Maso L, Bosetti C, Negri E. Flavonoids and risk of squamous cell esophageal cancer. *Int J Cancer* 2007; 120: 1560-4.

[23] Sabarwal A, Agarwal R, Singh RP. Fisetin inhibits cellular proliferation and induces mitochondria-dependent apoptosis in human gastric cancer cells. *Mol Carcinog* 2017; 56: 499-514.

[24] Fan C, Yang Y, Liu Y, Jiang S, Di S, Hu W, Ma Z, Li T, Zhu Y, Xin Z, Wu G, Han J, Li X, Yan X. Icariin displays anticancer activity against human esophageal cancer cells via regulating endoplasmic reticulum stress-mediated apoptotic signaling. *Sci Rep* 2016; 6: 21145.

[25] Woo HD, Kim J. Dietary flavonoid intake and risk of stomach and colorectal cancer. *World J Gastroenterol* 2013; 19: 1011-9.

[26] Tanaka T, Makita H, Kawabata K, Mori H, Kakumoto M, Satoh K, Hara A, Sumida T, Fukutani K, Tanaka T, Ogawa H. Modulation of N-methyl-N-amylnitrosamine-induced rat oesophageal tumourigenesis by dietary feeding of diosmin and hesperidin, both alone and in combination. *Carcinogenesis* 1997; 18: 761-9.

[27] Ding Y, Yao H, Yao Y, Fai LY, Zhang Z. Protection of dietary polyphenols against oral cancer. *Nutrients* 2013; 5: 2173-91.

[28] Zang W, Wang T, Wang Y, Li M, Xuan X, Ma Y, Du Y, Liu K, Dong Z, Zhao G. Myricetin exerts anti-proliferative, anti-invasive, and pro-apoptotic effects on esophageal carcinoma EC9706 and KYSE30 cells via RSK2. *Tumour Biol* 2014; 35: 12583-92.

[29] Zhang Q, Zhao XH, Wang ZJ. Cytotoxicity of flavones and flavonols to a human esophageal squamous cell carcinoma cell line (KYSE-510) by induction of G2/M arrest and apoptosis. *Toxicol In Vitro* 2009; 23: 797-807.

[30] Raha S, Yumnam S, Hong GE, Lee HJ, Saralamma VV, Park HS, Heo JD, Lee SJ, Kim EH, Kim JA, Kim GS. Naringin induces autophagy-mediated growth inhibition by downregulating the PI3K/Akt/mTOR cascade via activation of MAPK pathways in AGS cancer cells. *Int J Oncol* 2015; 47: 1061-9.

[31] Borska S, Chmielewska M, Wysocka T, Drag-Zalesinska M, Zabel M, Dziegiel P. In vitro effect of quercetin on human gastric carcinoma: targeting cancer cells death and MDR. *Food Chem Toxicol* 2012; 50: 3375-83.

[32] Zikri NN, Riedl KM, Wang LS, Lechner J, Schwartz SJ, Stoner GD. Black raspberry components inhibit proliferation, induce apoptosis, and modulate gene expression in rat esophageal epithelial cells. *Nutr Cancer* 2009; 61: 816-26.

[33] Atnip AA, Sigurdson GT, Bomser J, Giusti MM. Time, concentration, and pH-dependent transport and uptake of anthocyanins in a human gastric epithelial (NCI-N87) cell line. *Int J Mol Sci* 2017; 18: E446.

[34] Lou C, Yang G, Cai H, Zou M, Xu Z, Li Y, Zhao F, Li W, Tong L, Wang M, Cai B. 2`,4`-Dihydroxychalcone-induced apoptosis of human gastric cancer MGC-803 cells via down-regulation of survivin mRNA. *Toxicol In Vitro* 2010; 24: 1333-7.

[35] Lin SH, Shih YW. Antitumor effects of the flavone chalcone: inhibition of invasion and migration through the FAK/JNK signaling pathway in human gastric adenocarcinoma AGS cells. *Mol Cell Biochem* 2014; 391: 47-58.

[36] Yoshimizu N, Otani Y, Saikawa Y, Kubota T, Yoshida M, Furukawa T, Kumai K, Kameyama K, Fujii M, Yano M, Sato T, Ito A, Kitajima M. Anti-tumour effects of nobiletin, a citrus flavonoid, on gastric cancer include: antiproliferative effects, induction of apoptosis and cell cycle deregulation. *Aliment Pharmacol Ther* 2004; 20 Suppl 1: 95-101.

[37] Sak K. Cytotoxicity of dietary flavonoids on different human cancer types. *Pharmacogn Rev* 2014; 8: 122-46.

[38] Zhang Q, Zhao XH, Wang ZJ. Flavones and flavonols exert cytotoxic effects on a human oesophageal adenocarcinoma cell line (OE33) by causing G2/M arrest and inducing apoptosis. *Food Chem Toxicol* 2008; 46: 2042-53.

[39] Lou C, Wang M, Yang G, Cai H, Li Y, Zhao F, Yang H, Tong L, Cai B. Preliminary studies on anti-tumor activity of 2`,4`-dihydroxychalcone isolated

from Herba Oxytropis in human gastric cancer MGC-803 cells. *Toxicol In Vitro* 2009; 23: 906-10.

[40] Han Y, Yang X, Zhao N, Peng J, Gao H, Qiu X. Alpinumisoflavone induces apoptosis in esophageal squamous cell carcinoma by modulating miR-370/PIM1 signaling. *Am J Cancer Res* 2016; 6: 2755-71.

[41] Ramachandran L, Manu KA, Shanmugam MK, Li F, Siveen KS, Vali S, Kapoor S, Abbasi T, Surana R, Smoot DT, Ashktorab H, Tan P, Ahn KS, Yap CW, Kumar AP, Sethi G. Isorhamnetin inhibits proliferation and invasion and induces apoptosis through the modulation of peroxisome proliferator-activated receptor γ activation pathway in gastric cancer. *J Biol Chem* 2012; 287: 38028-40.

[42] Bobe G, Peterson JJ, Gridley G, Hyer M, Dwyer JT, Brown LM. Flavonoid consumption and esophageal cancer among black and white men in the United States. *Int J Cancer* 2009; 125: 1147-54.

[43] Pierini R, Kroon PA, Guyot S, Ivory K, Johnson IT, Belshaw NJ. Procyanidin effects on oesophageal adenocarcinoma cells strongly depend on flavan-3-ol degree of polymerization. *Mol Nutr Food Res* 2008; 52: 1399-407.

[44] Lagiou P, Samoli E, Lagiou A, Peterson J, Tzonou A, Dwyer J, Trichopoulos D. Flavonoids, vitamin C and adenocarcinoma of the stomach. *Cancer Causes Control* 2004; 15: 67-72.

[45] Zhang Y, Vareed SK, Nair MG. Human tumor cell growth inhibition by nontoxic anthocyanidins, the pigments in fruits and vegetables. *Life Sci* 2005; 76: 1465-72.

[46] Shih PH, Yeh CT, Yen GC. Effects of anthocyanidin on the inhibition of proliferation and induction of apoptosis in human gastric adenocarcinoma cells. *Food Chem Toxicol* 2005; 43: 1557-66.

[47] Ye F, Zhang GH, Guan BX, Xu XC. Suppression of esophageal cancer cell growth using curcumin, (-)-epigallocatechin-3-gallate and lovastatin. *World J Gastroenterol* 2012; 18: 126-35.

[48] Messadi DV, Sato K. Oral cancer chemoprevention: current status and future direction. *J Calif Dent Assoc* 2016; 44: 101-11.

[49] Lee UL, Choi SW. The chemopreventive properties and therapeutic modulation of green tea polyphenols in oral squamous cell carcinoma. *ISRN Oncol* 2011; 2011: 403707.

[50] de Moura CF, Noguti J, de Jesus GP, Ribeiro FA, Garcia FA, Gollucke AP, Aguiar O Jr, Ribeiro DA. Polyphenols as a chemopreventive agent in oral carcinogenesis: putative mechanisms of action using in-vitro and in-vivo test systems. *Eur J Cancer Prev* 2013; 22: 467-72.

[51] Hseu YC, Lee MS, Wu CR, Cho HJ, Lin KY, Lai GH, Wang SY, Kuo YH, Kumar KJ, Yang HL. The chalcone flavokawain B induces G2/M cell-cycle arrest and apoptosis in human oral carcinoma HSC-3 cells through the intracellular ROS

generation and downregulation of the Akt/p38 MAPK signaling pathway. *J Agric Food Chem* 2012; 60: 2385-97.

[52] Khafif A, Schantz SP, al-Rawi M, Edelstein D, Sacks PG. Green tea regulates cell cycle progression in oral leukoplakia. *Head Neck* 1998; 20: 528-34.

[53] Majumdar D, Jung KH, Zhang H, Nannapaneni S, Wang X, Amin AR, Chen Z, Chen ZG, Shin DM. Luteolin nanoparticle in chemoprevention: in vitro and in vivo anticancer activity. *Cancer Prev Res* (Phila) 2014; 7: 65-73.

[54] Amin AR, Wang D, Zhang H, Peng S, Shin HJ, Brandes JC, Tighiouart M, Khuri FR, Chen ZG, Shin DM. Enhanced anti-tumor activity by the combination of the natural compounds (-)-epigallocatechin-3-gallate and luteolin: potential role of p53. *J Biol Chem* 2010; 285: 34557-65.

[55] Maggioni D, Biffi L, Nicolini G, Garavello W. Flavonoids in oral cancer prevention and therapy. *Eur J Cancer Prev* 2015; 24: 517-28.

[56] Gohulkumar M, Gurushankar K, Rajendra Prasad N, Krishnakumar N. Enhanced cytotoxicity and apoptosis-induced anticancer effect of silibinin-loaded nanoparticles in oral carcinoma (KB) cells. *Mater Sci Eng C Mater Biol Appl* 2014; 41: 274-82.

[57] Chen F, Qin X, Xu G, Gou S, Jin X. Reversal of cisplatin resistance in human gastric cancer cells by a wogonin-conjugated Pt(IV) prodrug via attenuating Casein Kinase 2-mediated Nuclear Factor-κB pathways. *Biochem Pharmacol* 2017; 135: 50-68.

[58] Babu TM, Rammohan A, Baki VB, Devi S, Gunasekar D, Rajendra W. Development of novel HER2 inhibitors against gastric cancer derived from flavonoid source of Syzygium alternifolium through molecular dynamics and pharmacophore-based screening. *Drug Des Devel Ther* 2016; 10: 3611-32.

[59] Song H, Bao J, Wei Y, Chen Y, Mao X, Li J, Yang Z, Xue Y. Kaempferol inhibits gastric cancer tumor growth: An in vitro and in vivo study. *Oncol Rep* 2015; 33: 868-74.

[60] Koh YW, Choi EC, Kang SU, Hwang HS, Lee MH, Pyun J, Park R, Lee Y, Kim CH. Green tea (-)-epigallocatechin-3-gallate inhibits HGF-induced progression in oral cavity cancer through suppression of HGF/c-Met. *J Nutr Biochem* 2011; 22: 1074-83.

[61] Lee JH, Park CH, Jung KC, Rhee HS, Yang CH. Negative regulation of beta-catenin/Tcf signaling by naringenin in AGS gastric cancer cell. *Biochem Biophys Res Commun* 2005; 335: 771-6.

[62] Moghaddam G, Ebrahimi SA, Rahbar-Roshandel N, Foroumadi A, Antiproliferative activity of flavonoids: influence of the sequential methoxylation state of the flavonoid structure. *Phytother Res* 2012; 26: 1023-8.

[63] Xu YX, Wang B, Zhao XH. In vitro effects and the related molecular mechanism of galangin and quercetin on human gastric cancer cell line (SGC-7901). *Pak J Pharm Sci* 2017; 30: 1279-87.

[64] Stoner GD, Wang LS, Chen T. Chemoprevention of esophageal squamous cell carcinoma. *Toxicol Appl Pharmacol* 2007; 224: 337-49.

[65] Lu L, Chen J, Tang H, Bai L, Lu C, Wang K, Li M, Yan Y, Tang L, Wu R, Ye Y, Jin L, Liang Z. EGCG suppresses ERK5 activation to reverse tobacco smoke-triggered gastric epithelial-mesenchymal transition in BALB/c mice. *Nutrients* 2016; 8: E380.

[66] Wang L, Feng J, Chen X, Guo W, Du Y, Wang Y, Zang W, Zhang S, Zhao G. Myricetin enhance chemosensitivity of 5-fluorouracil on esophageal carcinoma in vitro and in vivo. *Cancer Cell Int* 2014; 14: 71.

[67] Zeng G, Shen H, Yang Y, Cai X, Xun W. Licochalcone A as a potent antitumor agent suppresses growth of human oral cancer SCC-25 cells in vitro via caspase-3 dependent pathways. *Tumour Biol* 2014; 35: 6549-55.

[68] Johnson TL, Lai MB, Lai JC, Bhushan A. Inhibition of cell proliferation and MAP kinase and Akt pathways in oral squamous cell carcinoma by genistein and biochanin A. *Evid Based Complement Alternat Med* 2010; 7: 351-8.

[69] Hong L, Ying SH. Ethanol extract and isolated constituents from Artemisia dracunculus inhibit esophageal squamous cell carcinoma and induce apoptotic cell death. *Drug Res* (Stuttg) 2015; 65: 101-6.

[70] Duarte N, Lage H, Abrantes M, Ferreira MJ. Phenolic compounds as selective antineoplastic agents against multidrug-resistant human cancer cells. *Planta Med* 2010; 76: 975-80.

Chapter 3

METABOLIC BIOCONVERSION AND BIOAVAILABILITY OF FLAVONOIDS IN THE HUMAN BODY

Although development of novel anticancer agents from natural products has emerged as a highly attractive strategy in the fight against malignant neoplasms in the past decades, flavonoids are recognized as xenobiotics by the human body being rapidly eliminated from the systemic circulation [1-3]. The diet-derived fraction of phytochemicals that reaches bloodstream for delivery to target cells and tissues where the biological activities are exerted is defined as oral bioavailability, whereas a successful application of flavonoids in the management of cancer is mainly limited by their poor overall oral bioavailability [2, 4-10]. Therefore, further studies on the pharmacokinetics and bioavailability of these plant secondary metabolites are crucial.

To reach effective doses and display biological activities at their target sites, ingested flavonoids have to overcome a number of biological barriers presented by the complex structures of gastrointestinal tract [2]. Most flavonoids, with the exception of flavanols, exist in the dietary food sources as different glycosides [10-15]. Such glycosylation process is generally the final step in biosynthesis of polyphenolic substances and is necessary for storage and accumulation of these phytochemicals in plants [15]. Also, flavonoid sugar conjugates are typically rather stable under cooking conditions and after ingestion, these glycosides pass through the stomach and reach intact the small intestine [13, 15]. Because of their high hydrophilicity, flavonoid glycosides are unable to absorb per se and must first undergo hydrolysis to their respective aglycones. This deglycosylation process occurs in the small intestine, being conducted mostly by the enzymatic action of lactase phlorizin hydrolase (LPH) [3, 10-14, 16]. Due to the increased hydrophobicity, the released aglycones can then absorb through the intestinal epithelial cell membrane by passive diffusion [3, 10, 13, 15]. At that, some compounds are effluxed back to the intestinal lumen by the action of the members of the ATP-binding cassette (ABC) family of membrane transporters, including multidrug resistance protein (MRP) and P-glycoprotein (P-gp) [3, 10, 13, 15]. Once flavonoids are absorbed through the intestinal wall, phase I and II metabolic enzymes catalyze their conversion and conjugation, after which these metabolites are transported to the liver via the hepatic portal vein [10, 13, 14]. In the liver, flavonoids undergo further extensive bioconversion generating glucuronides and sulfated and methylated metabolites through the respective action of UDP-glucuronosyltransferases, sulfotransferases and catechol-O-methyltransferases, as a result of which these dietary polyphenols enter the bloodstream mainly as respective derivatives [3, 10, 13, 15]. At that, some metabolites are recycled back into the small intestine via the bile excretion [3, 15]. Despite an important absorption of flavonoids in the small intestine, a substantial amount of unabsorbed flavonoids passes from the small intestine to the large intestine where the colonic microorganisms degrade these molecules to smaller products, including phenolic acids, before removal of them by fecal excretion [3, 10, 13]. The simplified scheme of this complex metabolic conversion of dietary flavonoids is depicted in Figure 3.1.

The absorption of dietary flavonoids is primarily determined by their structural characteristics and physicochemical properties, such as low solubility [17, 18]. In addition, several additional factors play important roles in bioavailability of these plant secondary metabolites, including their dosage and food matrix in which flavonoids are consumed, but also age, gender, genetic peculiarities and intestinal microbial population of the host organism, revealing thus a remarkable interindividual variability [10, 14, 15, 19-22]. However, the current knowledge about these complex mechanisms is still rather limited. Furthermore, there is no doubt that to resolve the bioactivity issue of flavonoids, the processes of their absorption and metabolism need to be unraveled in detail.

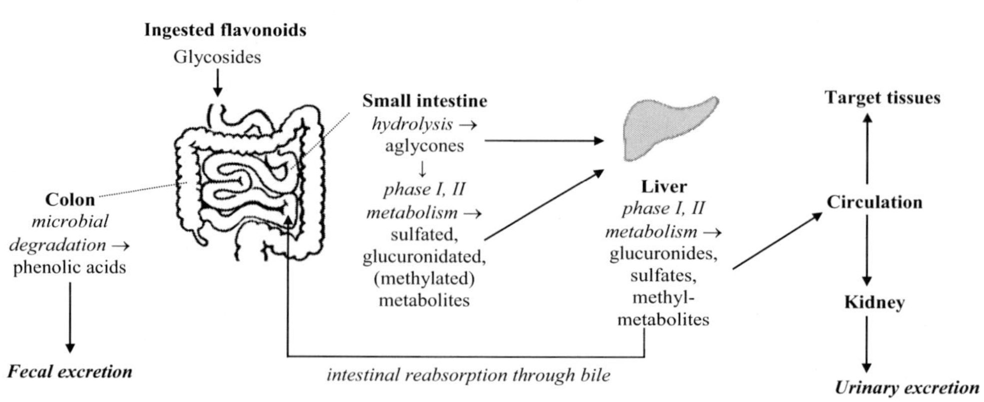

Figure 3.1. A simplified scheme of metabolic bioconversion of dietary flavonoids.

It is evident that the actual biological effects exerted by flavonoids in the human body are determined by their various metabolites and can be substantially different from those of the activities of parent aglycones in the experimental model systems. To date, only a few of this metabolic diversity have been studied concerning their potential anticancer properties revealing both increased as well as reduced cytotoxic potential [23, 24].

As a result of this complicated biotransformation, the plasma concentrations of flavonoids rarely exceed 1-5 μM [13, 14, 16]. At that, different dietary flavonoids reveal also rather different bioavailability, ranging from 0.2-0.9% for tea flavanols to 20% for quercetin and isoflavones [10, 16, 25-27]. Indeed, the plasma concentrations of epigallocatechin gallate (EGCG) after drinking several cups of green tea or receiving single oral dose of EGCG supplement remain only in the range of 0.1-1.0 μM [7, 26, 28, 29]. The blood levels of genistein can reach 4 μM following consumption of soy-rich diets [30] and the plasma concentrations of quercetin may exceed 8 μM after intake of the diets abundant in this flavonol [31]. Moreover, the bioavailability seems considerably depend also on the certain food sources of flavonoids, being in the case of quercetin superior from onions as compared to apples [14]. In addition, also the absorption profiles of flavonoids are variable, revealing maximum plasma levels between 0.5 and 9 hours after ingestion [14, 20, 21]. While flavanols exhibit peak plasma concentrations at one to two hours after tea drinking [32, 33], half-lives of isoflavones in the blood remain 7.7 to 9.5 hours [22]. For apigenin, the elimination half-time in plasma is higher than other flavonoids promoting thus the potential accumulation of this flavone in the target tissues [34]. Altogether, these data show that extrapolation of results obtained with relatively high micromolar doses of flavonoids in various cell line studies to the human body, where considerably lower plasma concentrations are achievable after oral intake of flavonoid-rich diets, could not be relevant and certainly needs caution [2, 16, 35-37]. However, it remains possible that the concentrations of flavonoids in target tissues can be somewhat different from those circulating in the blood [38].

Because the sites of upper gastrointestinal tract (oral cavity, esophagus, stomach) directly contact and interact with dietary phytochemicals, carcinogenesis in these organs may be more closely related to the ingestion of flavonoids [6, 7, 19, 27, 35]. Indeed, mucosal epithelia of upper digestive tract are accessible and exposed to high doses of flavonoids either through the diet or consumption of food supplements, by virtue of a ready local bioavailability. Accordingly, the concentrations of EGCG in saliva can be greater than 25 μM after drinking green tea and reach even 180 μM by chewing or holding green tea leaves in the mouth for some minutes [26, 36, 39-42]. The tea flavanols would be in close proximity also to the esophageal and gastric epithelia, whereas intragastric concentrations of EGCG of about 40 μM have been measured after drinking green tea beverage [27, 42]. In practice, these are the doses at which flavonoids efficiently exert their effects against carcinogenesis. Furthermore, certain flavonoid glycosides (spiraeoside, genistin, anthocyanins) can be hydrolyzed in the human oral cavity to free aglycones with a remarkable interindividual variability, displaying then *in loco* diverse biological activities [2, 12]. Such enzymatic deglycosylation probably continues to some extent through the passage of the plant-derived food items throughout the upper digestive tract. Therefore, depending on the ingested amounts of plant-based food products, it can be expected that over the course of the day the mucosal epithelia of upper gastrointestinal tract are repeatedly exposed to biologically relevant doses of different flavonoids (i.e., deglycosylated but unmetabolised aglycones) to exert their diverse anticancer effects.

REFERENCES

[1] Sulfikkarali N, Krishnakumar N, Manoharan S, Nirmal RM. Chemopreventive efficacy of naringenin-loaded nanoparticles in 7,12-dimethylbenz(a)anthracene induced experimental oral carcinogenesis. *Pathol Oncol Res* 2013; 19: 287-96.

[2] Iriti M, Varoni EM. Chemopreventive potential of flavonoids in oral squamous cell carcinoma in human studies. *Nutrients* 2013; 5: 2564-76.

[3] Crozier A, Del Rio D, Clifford MN. Bioavailability of dietary flavonoids and phenolic compounds. *Mol Aspects Med* 2010; 31: 446-67.

[4] Walle T, Ta N, Kawamori T, Wen X, Tsuji PA, Walle UK. Cancer chemopreventive properties of orally bioavailable flavonoids – methylated versus unmethylated flavones. *Biochem Pharmacol* 2007; 73: 1288-96.

[5] Yang SF, Yang WE, Chang HR, Chu SC, Hsieh YS. Luteolin induces apoptosis in oral squamous cancer cells. *J Dent Res* 2008; 87: 401-6.

[6] Haghiac M, Walle T. Quercetin induces necrosis and apoptosis in SCC-9 oral cancer cells. *Nutr Cancer* 2005; 53: 220-31.

[7] Ullah MF, Bhat SH, Husain E, Abu-Duhier F, Hadi SM, Sarkar FH, Ahmad A. Pharmacological intervention through dietary nutraceuticals in gastrointestinal neoplasia. *Crit Rev Food Sci Nutr* 2016; 56: 1501-18.

[8] Lin YH, Chen ZR, Lai CH, Hsieh CH, Feng CL. Active targeted nanoparticles for oral administration of gastric cancer therapy. *Biomacromolecules* 2015; 16: 3021-32.

[9] Zhu Y, Jiang Y, Shi L, Du L, Xu X, Wang E, Sun Y, Guo X, Zou B, Wang H, Wang C, Sun L, Zhen Y. 7-O-Geranylquercetin induces apoptosis in gastric cancer cells via ROS-MAPK mediated mitochondrial signaling pathway activation. *Biomed Pharmacother* 2017; 87: 527-38.

[10] Viskupicova J, Ondrejovic M, Sturdik E. Bioavailability and metabolism of flavonoids. *J Food Nutr Res* 2008; 47: 151-62.

[11] Browning AM, Walle UK, Walle T. Flavonoid glycosides inhibit oral cancer cell proliferation – role of cellular uptake and hydrolysis to the aglycones. *J Pharm Pharmacol* 2005; 57: 1037-42.

[12] Walle T, Browning AM, Steed LL, Reed SG, Walle UK. Flavonoid glucosides are hydrolyzed and thus activated in the oral cavity in humans. *J Nutr* 2005; 135: 48-52.

[13] Gonzales GB. In vitro bioavailability and cellular bioactivity studies of flavonoids and flavonoid-rich plant extracts: questions, considerations and future perspectives. *Proc Nutr Soc* 2017; 76: 175-81.

[14] Hollman PC. Absorption, bioavailability, and metabolism of flavonoids. *Pharm Biol* 2004; 42 Suppl: 74-83.

[15] Sak K, Everaus H. Nanotechnological approach to improve the bioavailability of dietary flavonoids with chemopreventive and anticancer properties. In: Grumezescu AM, ed. Nutraceuticals. *Nanotechnology in the Agri-Food Industry*, Vol. 4. Elsevier 2016, pp. 427-79.

[16] Bo Y, Sun J, Wang M, Ding J, Lu Q, Yuan L. Dietary flavonoid intake and the risk of digestive tract cancers: a systematic review and meta-analysis. *Sci Rep* 2016; 6: 24836.

[17] Ding Y, Yao H, Yao Y, Fai LY, Zhang Z. Protection of dietary polyphenols against oral cancer. *Nutrients* 2013; 5: 2173-91.

[18] Xiang HL, Liu F, Quan MF, Cao JG, Lv Y. 7-Difluoromethoxyl-5,4`-di-n-octylgenistein inhibits growth of gastric cancer cells through downregulating forkhead box M1. *World J Gastroenterol* 2012; 18: 4618-26.

[19] Xie Y, Huang S, Su Y. Dietary flavonols intake and risk of esophageal and gastric cancer: a meta-analysis of epidemiological studies. *Nutrients* 2016; 8: 91.

[20] Sun L, Subar AF, Bosire C, Dawsey SM, Kahle LL, Zimmerman TP, Abnet CC, Heller R, Graubard BI, Cook MB, Petrick JL. Dietary flavonoid intake reduces the risk of head and neck but not esophageal or gastric cancer in US men and women. *J Nutr* 2017; 147: 1729-38.

[21] Petrick JL, Steck SE, Bradshaw PT, Trivers KF, Abrahamson PE, Engel LS, He K, Chow WH, Mayne ST, Risch HA, Vaughan TL, Gammon MD. Dietary intake of flavonoids and oesophageal and gastric cancer: incidence and survival in the United States of America (USA). *Br J Cancer* 2015; 112: 1291-300.

[22] Hara A, Sasazuki S, Inoue M, Miura T, Iwasaki M, Sawada N, Shimazu T, Yamaji T, Tsugane S, Japan Public Health Center-Based Prospective Study Group. Plasma isoflavone concentrations are not associated with gastric cancer risk among Japanese men and women. *J Nutr* 2013; 143: 1293-8.

[23] Sak K. Anticancer action of sulfated flavonoids as phase II metabolites. In: Grumezescu AM, Holban AM, eds. Food Bioconversion. *Handbook of Food Bioengineering*, Vol. 2. Elsevier 2017, pp. 207-36.

[24] Sak K. The Val158Met polymorphism in COMT gene and cancer risk: role of endogenous and exogenous catechols. *Drug Metab Rev* 2017; 49: 56-83.

[25] Woo HD, Kim J. Dietary flavonoid intake and risk of stomach and colorectal cancer. *World J Gastroenterol* 2013; 19: 1011-9.

[26] Babich H, Zuckerbraun HL, Weinerman SM. In vitro cytotoxicity of (-)-catechin gallate, a minor polyphenol in green tea. *Toxicol Lett* 2007; 171: 171-80.

[27] Horie N, Hirabayashi N, Takahashi Y, Miyauchi Y, Taguchi H, Takeishi K. Synergistic effect of green tea catechins on cell growth and apoptosis induction in gastric carcinoma cells. *Biol Pharm Bull* 2005; 28: 574-9.

[28] Masuda M, Suzui M, Weinstein IB. Effects of epigallocatechin-3-gallate on growth, epidermal growth factor receptor signaling pathways, gene expression, and chemosensitivity in human head and neck squamous cell carcinoma cell lines. *Clin Cancer Res* 2001; 7: 4220-9.

[29] Lim YC, Park HY, Hwang HS, Kang SU, Pyun JH, Lee MH, Choi EC, Kim CH. (-)-Epigallocatechin-3-gallate (EGCG) inhibits HGF-induced invasion and metastasis in hypopharyngeal carcinoma cells. *Cancer Lett* 2008; 271: 140-52.

[30] Alhasan SA, Pietrasczkiwicz H, Alonso MD, Ensley J, Sarkar FH. Genistein-induced cell cycle arrest and apoptosis in a head and neck squamous cell carcinoma cell line. *Nutr Cancer* 1999; 34: 12-9.

[31] Wang P, Zhang K, Zhang Q, Mei J, Chen CJ, Feng ZZ, Yu DH. Effects of quercetin on the apoptosis of the human gastric carcinoma cells. *Toxicol In Vitro* 2012; 26: 221-8.

[32] Sasazuki S, Inoue M, Miura T, Iwasaki M, Tsugane S, Japan Public Health Center-based Prospective Study Group. Plasma tea polyphenols and gastric cancer risk: a case-control study nested in a large population-based prospective study in Japan. *Cancer Epidemiol Biomarkers Prev* 2008; 17: 343-51.

[33] Sun CL, Yuan JM, Lee MJ, Yang CS, Gao YT, Ross RK, Yu MC. Urinary tea polyphenols in relation to gastric and esophageal cancers: a prospective study of men in Shanghai, China. *Carcinogenesis* 2002; 23: 1497-503.

[34] Masuelli L, Marzocchella L, Quaranta A, Palumbo C, Pompa G, Izzi V, Canini A, Modesti A, Galvano F, Bei R. Apigenin induces apoptosis and impairs head and neck carcinomas EGFR/ErbB2 signaling. *Front Biosci* (Landmark Ed) 2011; 16: 1060-8.

[35] Kato K, Long NK, Makita H, Toida M, Yamashita T, Hatakeyama D, Hara A, Mori H, Shibata T. Effects of green tea polyphenol on methylation status of RECK

gene and cancer cell invasion in oral squamous cell carcinoma cells. *Br J Cancer* 2008; 99: 647-54.

[36] Weisburg JH, Weissman DB, Sedaghat T, Babich H. In vitro cytotoxicity of epigallocatechin gallate and tea extracts to cancerous and normal cells from the human oral cavity. *Basic Clin Pharmacol Toxicol* 2004; 95: 191-200.

[37] Lim YC, Lee SH, Song MH, Yamaguchi K, Yoon JH, Choi EC, Baek SJ. Growth inhibition and apoptosis by (-)-epicatechin gallate are mediated by cyclin D1 suppression in head and neck squamous carcinoma cells. *Eur J Cancer* 2006; 42: 3260-6.

[38] Ye F, Wu J, Dunn T, Yi J, Tong X, Zhang D. Inhibition of cyclooxygenase-2 activity in head and neck cancer cells by genistein. *Cancer Lett* 2004; 211: 39-46.

[39] Tao L, Forester SC, Lambert JD. The role of the mitochondrial oxidative stress in the cytotoxic effects of the green tea catechin, (-)-epigallocatechin-3-gallate, in oral cells. *Mol Nutr Food Res* 2014; 58: 665-76.

[40] Yamamoto T, Staples J, Wataha K, Lewis J, Lockwood P, Schoenlein P, Rao S, Osaki T, Dickinson D, Kamatani T, Schuster G, Hsu S. Protective effects of EGCG on salivary gland cells treated with gamma-radiation or cis-platinum(II)diammine dichloride. *Anticancer Res* 2004; 24: 3065-73.

[41] Babich H, Krupka ME, Nissim HA, Zuckerbraun HL. Differential in vitro cytotoxicity of (-)-epicatechin gallate (ECG) to cancer and normal cells from the human oral cavity. *Toxicol In Vitro* 2005; 19: 231-42.

[42] Gutierrez-Orozco F, Stephens BR, Neilson AP, Green R, Ferruzzi MG, Bomser JA. Green and black tea inhibit cytokine-induced IL-8 production and secretion in AGS gastric cancer cells via inhibition of NF-κB activity. *Planta Med* 2010; 76: 1659-65.

Chapter 4

ANTICANCER ACTION OF FLAVONOIDS IN ORAL CANCER MODELS

4.1. INTRODUCTION TO ORAL CANCER

Oral cancer is one of the most common and lethal subtype of head and neck malignancies representing any cancerous tissue growth located in the oral cavity (the mouth) and pharynx (the throat) [1-7]. The malignant neoplasms can develop in any part of the oral cavity, including tongue, lips, cheeks, teeth, gums, salivary glands, roof and floor of the mouth, oropharynx, or hypopharynx [6, 8-11]. The most common histological type of oral cancer is oral squamous cell carcinoma (OSCC) constituting more than 90% of malignant neoplasms of the mouth and being a major problem of public health [1, 8, 12-18].

Oral cancer is a significant global health burden with about 443000 new cases diagnosed each year worldwide, accounting for 3.1% of all cancer cases and resulting in more than 241000 deaths based on the data of GLOBOCAN 2012 [19]. Both the incidence and mortality rate of oral cancer are higher in men than in women with approximately 314000 new cases diagnosed worldwide in men and 129000 cases in women [6, 20-22]. This type of cancer occurs mostly in people older than 40 years with an average diagnosis age of 60 years [6, 23]. Coming from the continuously raising proportion of elderly people around the world, a 50% increase in the incidence rate has been estimated over the next 20 years [23]. The incidence and mortality rate of oral cancer vary extensively (more than 20-fold) according to geographical areas being less frequent in western countries (3-4% of all malignancies) but accounting even for 40-50% of all new cancer cases in South and Southeast Asia [3, 17, 21, 23-26]. In Europe and America, incidence and mortality rate increased until the late 1980s and have thereafter leveled off or slightly decreased during the last decades in the most industrialized countries [3, 21]. On the contrary, about two-thirds of oral cancer cases occur in developing countries with the highest rates reported in South Asia and Asia Pacific regions, especially India, Taiwan, Pakistan, Bangladesh, Sri Lanka, South Vietnam, the Philippines, Papua New Guinea, and Australia, where the incidence level is steadily growing up [3, 24, 27-30]. In fact, oral cancer is one of the leading causes of cancer-related deaths in India, Taiwan and South Asian countries [8, 24, 26, 27, 31, 32]. Besides the different exposure levels to known etiologic agents, also the availability and accessibility of health services can contribute to this wide geographical variety in oral cancer incidence and mortality rate [20, 22, 29, 33]. Moreover, while oropharyngeal and hypopharyngeal cancers represent as many or even more cases than malignancies of the oral cavity in European countries, cancerous neoplasms of the tongue, floor of the mouth and other parts of the oral cavity account for the majority of cases in India [21].

Oral carcinogenesis is believed to be a multifactorial process. The two most important risk factors in etiology of oral cavity and pharyngeal cancer include tobacco smoking and alcohol consumption [1, 6, 21, 32, 34-37]. The prevalence of oral cancer among smokers is about 3-10-fold higher than in non-smokers and all forms of smoking and smokeless tobacco, including the use of tobacco snuff, are related to the development of oral cancer [6, 21, 35]. At that, tobacco smoke from cigarettes causes cancer in the mouth and throat, pipe smoking brings about lesions on the lips where the pipe comes into contact with the soft tissue and causes oral cancer, chewing tobacco can produce cancerous growth in lips, cheeks and gums [6, 21]. The risk is higher for heavy and long-term smokers, whereas smoking cessation is followed by a decline in oral cancer risk [21, 34]. In India and Taiwan, the rapid increase in oral cancer incidence is mostly attributed to chewing of tobacco and betel quid, smoking of bidi and cigarettes [8, 21, 25, 27, 28, 33]. The other major risk factor for oral carcinogenesis, i.e., consumption of alcoholic beverages, increases the susceptibility to oral cancer development for 6-10-fold when heavy drinkers are compared to abstainers and very light drinkers [6, 21]. Although ethanol is the main factor responsible for carcinogenicity of alcoholic beverages, it is unclear whether different types of alcoholic beverages may have different influences on oral carcinogenesis and whether spirits are related to higher risks than wine or beer [21]. However, frequent consumption of liquor has been particularly related to the development of metastatic oral cancer [6]. Moreover, it has been suggested that individuals carrying the fast-metabolizing alcohol dehydrogenase type 3 (ADH3) allele may be more vulnerable to the effect of chronic alcohol consumption and are therefore also at increased risk to oral carcinogenesis [1]. The exposure to both risk factors together, i.e., tobacco smoking and alcohol drinking, has synergistic action on oral carcinogenesis with a combined effect accounting for 75-90% of all oral cancer cases [1, 20, 21, 35, 36, 38]. In fact, when smoking of 20 cigarettes daily for more than 20 years is combined with alcohol abuse, the elevation of oral cancer risk is about 40-fold [39]. As men tend to be more likely to indulge in these deleterious habits (the use of tobacco and alcohol) they are also more likely to develop oral cancer as compared to women [6, 13]. In addition, exposure to ultraviolet light, i.e., radiation from the sun, can increase the risk of lip carcinogenesis and when combined with tobacco use, the risk of lip cancer may be increased by 20% [6, 40]. Furthermore, recent studies have clearly shown that infection with sexually transmitted human papillomavirus (HPV), especially the high-risk subtype HPV16, also contributes to oral cancer development by initiating malignancies among young non-smoking people and promoting the proliferative potential and progression of cancerous cells among smokers [1, 6, 35, 40-42]. Indeed, HPV infection is the most common cause of oral cancer in those patients who have never used tobacco products and plays a role in cancer development with higher incidence in oropharyngeal tissues rather than oral mucosa membrane [6, 34, 42]. Furthermore, nutrition and certain diet have also been related to the susceptibility for development of oral cancer [43].

Despite several advancements achieved in cancer diagnostics and therapy, the overall five-year survival rate of oral cancer has not significantly improved over the past few decades and has remained around 50%, being lower than for most other malignancies, including cervical, testicular, laryngeal, skin, and thyroid cancer [1, 6, 12, 14, 15, 18, 35, 36, 44-46]. The poor prognosis and low survival rate of oral cancer patients are due to extensive local tumor invasion, frequent regional lymph node metastasis in the neck, a high risk for loco-regional relapse, relative resistance to conventional chemotherapeutic drugs, and development of distant metastases [1, 3, 4, 16, 20, 46-51]. This aggressive cancer type has a high propensity to early metastasize and spread to other organs suggesting that metastasis is a major clinical challenge for a successful cancer treatment to improve patients' outcome [6, 14, 50, 52]. Indeed, the incidence of occult neck metastases was determined to be as high as 42% for stage I/II oral cancer [2]. Moreover, oral cancer is often associated with a high recurrence rate and about 20-30% of patients apparently cured by primary treatment develop a second primary tumor within a few years, generally in the head and neck region or esophagus or lung [7, 35, 38, 53, 54]. As the oral cavity and pharynx have many important functions in the human body assisting phonation, breathing, mastication, digestion and swallowing, cancerous growth in this region has devastating consequences affecting the ability of patients to talk, breathe, swallow and eat, representing an important health and social problem and resulting in a serious decline in quality of life [3, 6, 55]. Therefore, the demand for new therapeutic strategies, i.e., searching novel molecular targets and identifying preventive agents with antiinvasive and antimigratory (antimetastatic) properties, is crucial to improve the treatment outcome of oral cancer in the future [2, 14, 36, 43, 44, 50, 56, 57].

Oral carcinogenesis is a stepwise process arising through initiation phase (accumulation of irreversible genetic alterations), promotion (development of a visible but often unnoticed premalignant lesion inside the mouth) and progression (development of some of these lesions into a highly malignant tumor) [47, 58-60]. Premalignant lesions are well established clinical entities, such as oral leukoplakia, erythroplakia, erosive lichen planus and submucous fibrosis, carrying a wide potential to undergo malignant transformation: 0.13-34% for leukoplakia and 50-90% for erythoplakia [1, 17, 18, 39, 58, 59]. This variability arises partly as a result of differences in tobacco exposure and history as well as dietary habits [34, 47]. Intervention in the promotion step and targeting preneoplastic lesions by chemopreventive agents may be a good option to prevent malignant transformation and tumor progression with the ultimate goal to decrease the oral cancer incidence, morbidity and loss of function [18, 61]. Although the visual examination of oral cavity is an easy way to discover malignant lesions, early signs of oral cancer are often not visible and remain unnoticed until the advanced stages of cancer [6, 14]. Moreover, there is no comprehensive screening program still evolved for early detection of oral cancer [6, 61].

Although early diagnosis and removal of premalignant lesions can lead to the best treatment outcomes, this is rarely achieved because oral cancers are usually detected only at advanced stage when selection of the appropriate therapeutic modalities is rather complicated [6, 23, 62]. The most extensively used treatment modalities include surgery, radiation and chemotherapy, either alone or in combination, with the main goals to eliminate the malignant tissue and relieve the cancer symptoms [6]. In early stage disease (stages I and II), surgery is applied to resect the tumor along with any affected surrounding tissues, being often combined with radiation therapy [6, 7, 63-67]. However, a tumor depth of more than 5 mm implicates a significant increase in the risk of local recurrence and mostly requires more aggressive clinical approaches [7, 68]. Treatment of advanced oral cancers (stages III and IV) is a challenging task due to both local recurrence as well as regional and/or distant metastases. At this stage, chemotherapy is often used in combination with radiation and/or surgery [7, 64, 66, 69]. However, to date there are no effective chemotherapeutics available for treatment of unresectable oral cancer and drugs are usually indicated only for palliative purposes [46, 70, 71]. In general, 5-fluorouracil (5-FU) and/or cisplatin (CDDP) are most commonly administrated, but their use is associated with severe side effects and emergence of drug resistance resulting in an important therapeutic failure [15, 65, 72-77]. Adverse effects, including hair loss, nausea, oral mucositis, xerostomia, fever, leukopenia and anorexia, remain serious problems accompanying the use of anticancer drugs [37, 46, 65, 73, 77, 78]. Complications of radiation and chemotherapy can even produce another primary tumor in the irradiated field [71]. Moreover, surgical resection can cause disfigurement and functional loss of oral functions, affecting breathing, speech and swallowing, leading to an important decline in quality of life and social isolation of patients [5, 11, 34, 45, 71, 78]. Altogether, there are limited possibilities for oral cancer treatment and current approaches have not succeeded to achieve satisfactory clinical outcome revealing only low survival rate and poor quality of life [31, 33, 37, 48, 53, 66, 76]. Therefore, the demand for preventive medications, new effective treatment strategies with fewer side effects and complementary methods is urgent to combat oral cancer with more success [6, 15, 31, 45, 46, 71]. For this, intensive research is carried out to identify novel targets and develop new chemopreventive or chemotherapeutic agents from either natural or synthetic products [24, 37, 45, 59].

The epidemiological case-control studies have suggested that diet and nutrition may play an important role in the development of oral cancer [3, 21, 40]. In western populations, selected aspects of diet have been related to 20-25% of malignancies in oral cavity and pharynx and this proportion might be even greater in some developing countries [21]. Based on a meta-analysis, the high intake of vegetables is associated with a 48% decrease in the risk of oral and pharyngeal cancers and the corresponding parameter for high fruits consumption is 45% [79]. Such beneficial effects were attributed to different micronutrients, including polyphenols and fibers, and may probably result from a combined action of various phytochemicals [79, 80]. Therefore, a poor diet with low amount of fruits and vegetables can be related to an elevation of the susceptibility to oral carcinogenesis and plant polyphenols, especially flavonoids, might express protective activities against oral neoplasms [3, 36, 39, 76, 81, 82]. The potential of flavonoids as chemopreventive and chemotherapeutic agents against oral carcinogenesis has been demonstrated in numerous preclinical studies performed both *in vitro* cell culture models as well as with experimental animals *in vivo* [1]. These dietary polyphenols can suppress the oral cancer incidence by inhibition of tumor growth, invasion and migration and induction of cell death, probably by affecting multiple cellular targets and acting through different pathways [20, 36]. Moreover, flavonoids might also overcome adverse effects characteristic to conventional chemotherapeutics and due to their natural origin, these plant secondary metabolites possess a safety profile [24, 45]. As oral cancers develop from clinically clearly defined premalignant lesions to invasive malignancies, flavonoids can potentially modulate different steps in this multistage process and therefore, their precise action mechanisms are certainly worth of further molecular studies. In sum, although the current knowledge is still too scant to make certain recommendations for prevention of oral cancer by effective nutritional intervention, consumption of a diet rich in flavonoids-containing fruits and vegetables, immunization against HPV at an adolescent age, limitation of alcohol drinking and cessation of tobacco use can all decrease the risk of development of oral malignancies [6, 40, 83]. There is no doubt that all these preventive measures require awareness and education of the general public.

4.2. Model Systems Used to Study the Effect of Flavonoids on Oral Carcinogenesis

An overview of the *in vitro* and *in vivo* model systems used for studies of the effects of flavonoids on oral carcinogenesis is presented in the Tables 4.1 and 4.2. A large variety of human oral cancer cell lines established from different regions of oral cavity (floor of the mouth, buccal mucosa, tonsil, gingiva, retromolar trigone, tongue, salivary gland) or pharynx (hypopharynx) has been investigated, derived from either primary tumors or their loco-regional metastases, and revealing different potential for invasiveness. Both HPV positive and negative cellular models have been studied, and besides chemosensitive lines also their chemoresistant counterparts have been used (towards cisplatin, vincristine or doxorubicin) (Table 4.1).

To better understand the role of flavonoids in oral carcinogenesis, it is important and imperative to investigate the action of these natural polyphenols in experimental animal models which reflect some typical features of human oral cancer development, before starting the clinical trials [27, 61]. Several *in vivo* studies have indeed demonstrated a suppressive effect of flavonoids on oral tumors either induced by carcinogenic chemicals in rodents (hamsters, mice, rats) or in xenograft mice subcutaneously inoculated with different human oral cancer cell lines (Table 4.2). The hamster buccal pouch carcinogenesis model is a well-known and most widely employed animal system to investigate the efficacy of flavonoids in inhibition of initiation and progression of oral tumors [25, 27, 29, 41, 61, 84]. Carcinomas are induced by the topical application of 7,12-dimethylbenz[a]anthracene (DMBA) to the right or left buccal pouch of male or female Syrian hamsters, whereas the tumors closely resemble human oral squamous cell carcinomas both histologically, morphologically and biochemically as well as at molecular level [17, 20, 27, 28, 32, 35, 38, 41]. Similarly to human oral carcinogenesis, DMBA-induced buccal pouch tumorigenesis also proceeds stepwise evolving from premalignant lesions into cancerous neoplasms, being therefore an appropriate model for studies of the chemopreventive properties of natural agents, such as flavonoids [25, 28, 60, 85]. Another experimental *in vivo* oral tumorigenesis model was developed by continuous application of a synthetic water-soluble carcinogen 4-nitroquinoline-1-oxide (4NQO) to male rats in the drinking water [29]. 4NQO causes many endophytic and ulcerated lesions in the rat oral cavity, particularly tongue, being thus very comparable to the oral lesions in humans [29, 38, 86] (Table 4.2).

Table 4.1. Human oral cancer cell lines used for *in vitro* studies with flavonoids

Cell line name	Tumor location	Characterization of cell line[1]	Ref.
38	Tonsil fossa	Chemoresistant	[137]
1483	Retromolar trigone (untreated)	Moderately differentiated SCC	[58, 94]
1986LN	Tongue, lymph node metastasis	SCC	[59]
686LN	Tongue, lymph node metastasis		[104]
A253	Submaxillary salivary gland	SCC	[77]
ACC-2	Salivary gland	Adenoid cystic carcinoma, low metastatic	[9]
ACC-M	Salivary gland	Adenoid cystic carcinoma, high metastatic	[9]
Ca9-22	Gingiva	SCC	[2, 13, 93, 103]
CAL27	Tongue	SCC, HPV16-negative	[2, 18, 24, 40, 50, 88, 89, 93, 98, 103, 105, 108, 125, 136]
CAR (CDDP-resistant CAL27)	Tongue	SCC, CDDP-resistant	[15]
FaDu	Hypopharynx	SCC	[4, 56, 107-109, 130, 138]
HN4	Tongue (right base)	SCC, invasive cells, moderate to poorly differentiated, keratinizing	[54, 78, 124, 127]
HN12	Oral cavity	SCC, metastatic	[78, 124, 127]
HN22	Oral cavity	SCC	[12, 14, 68]
HSC2	Floor of the oral cavity	SCC	[98, 106, 116, 139, 140]
HSC3	Tongue, cervical lymph node metastasis	SCC	[2, 13, 24, 46, 63, 100, 102, 103, 112, 116, 117, 120, 139]
HSC4	Tongue	SCC	[12, 14, 52, 68, 100, 139]
HSG	Submandibular salivary gland		[69, 105, 116, 140]
HSG1	Salivary gland		[98]

Table 4.1. (Continued)

Cell line name	Tumor location	Characterization of cell line[1]	Ref.
KB	Oral cavity	SCC, epidermoid	[4, 45, 88, 120, 122, 123, 128, 141]
KB/VCR (VCR-resistant KB)	Oral cavity	VCR-resistant, over-expressing P-gp	[118]
KB-A-1 (DOX-resistant KB)	Oral cavity	DOX-resistant	[135]
OC2	Buccal mucosa		[66, 101]
OEC-M1	Oral cavity	SCC	[13, 90, 96]
OSC2	Gingiva, cervical submandibular lymph node metastasis	SCC, p57 deficient, no expression of p16, mutated p53 (missence mutation in exon 8, codon 280)	[73, 87, 95, 97]
OSC4	Oral cavity, cervical metastatic lymph nodes	SCC, mutated p53 (silent mutation in exon 5, codon 174)	[73, 97]
PCI-13	Oral cavity		[109]
PCI-15B	Oral cavity, metastatic lymph node		[109]
SAS	Tongue	SCC, poorly differentiated	[2, 13, 33, 43, 52, 57, 71, 93, 103, 125]
OCSCs isolated from SAS	Oral cavity	SCC, stem cells	[64]
SAS7	Oral cavity	SCC	[46]
SCC4	Tongue	SCC, highly invasive	[2, 8, 10, 13, 16, 46, 66, 67, 103, 110, 111]
SCC9	Tongue	SCC	[18, 49, 76, 77, 81-83, 92, 100, 107, 114, 123]
PCI-15B	Oral cavity, metastatic lymph node		[109]
SCC15	Tongue	SCC	[108, 121]
SCC25	Tongue	SCC, well differentiated	[3, 22, 37, 43, 48, 51, 55, 65, 74, 83, 88, 92, 100, 115, 121]
DRSPs generated from SCC25	Tongue		[75]

Cell line name	Tumor location	Characterization of cell line[1]	Ref.
SCC84	Tongue (male, lateral, stage III)	HPV-negative	[42]
SCC1483	Retromolar trigone	SCC	[115]
SCC-QLL1	Oral cavity, metastatic lymph nodes	SCC	[115]
SGT	Salivary gland	Adenocarcinoma	[70]
SQCCY1	Oral cavity		[59, 137]
Tca8113	Tongue	SCC	[11, 126]
TSCCa	Tongue	SCC	[11]
Tu212	Hypopharynx		[104, 113, 137]
Tu686	Tongue		[59, 104]
TW1.5	Oral cavity		[2]
TW2.6	Oral cavity		[2, 125]
TW206	Tongue	SCC	[117]
UMSCC47	Tongue (male, lateral, stage III)	HPV16-positive	[42]
UMSCC29	Oral cavity (male, stage IV)	HPV-negative	[42]
UPCI-SCC90	Base of tongue (male)	HPV16-positive	[42]
YCU-H891	Hypopharynx	SCC	[72, 91]
YD-10B	Oral cavity	SCC	[102]

[1]CDDP, cisplatin; DOX, doxorubicin; DRSP, drug-resistant spheres; OCSC, oral cancer stem cells; P-gp, P-glycoprotein; SCC, squamous cell carcinoma; VCR, vincristine.

Table 4.2. Animal models used for *in vivo* studies of flavonoids on oral cancer

Species	Strain	Gender	Age, weeks	Weight, g	Induction of tumorigenesis[1]		Application route and site	Ref.
					Chemical	Implanted cell line[2]		
Hamsters	Syrian, Syrian golden	Male, female	4…20	60…135	DMBA, DMN-OAC		Topical treatment of right or left buccal pouch	[17, 25-28, 32, 41, 60, 61, 84, 85, 129, 131-133]
Mice	BALB/c, C3H/HeJ, C57BL/6, Harlan Sprague Dawley, NOD-SCID	Male, female	4…8	18…22	4NQO	ACC-M, FaDu, HSC3, KB, SAS, SCC4, SCC9, SCC25, SCC VII/SF, Tu212, YD-10B	Subcutaneous inoculation of cells into the right or left flank, right front axilla, buccal mucosa or tongue	[2, 4, 9, 13, 18, 49, 57, 66, 71, 75, 102, 104, 109, 113, 117, 120, 128, 130, 137]
Rats	F344	Male	4…6		4NQO		In drinking water	[38, 86, 134]

[1]4NQO, 4-nitroquinoline-1-oxide; DMBA, 7,12-dimethylbenz[a]anthracene; DMN-OAC, methyl-(acetoxymethyl)-nitrosamine.
[2]see Table 4.1; SCC VII/SF, murine OSCC cell line.

4.3. Effect of Different Flavonoids in Oral Cancer Development

4.3.1. *In Vitro* Results

The results describing growth inhibitory activities and molecular action mechanisms of various flavonoids in different human oral cancer cell lines are summarized in Table 4.3. When available, the comparable effects of these polyphenols also in normal oral cells or non-tumorigenic premalignant lines (MSK Leuk1, derived from leukoplakia specimen) are presented. At that, only the pure flavonoids (and not their mixtures) are considered.

4.3.1.1. Flavanols

Polyphenolic compounds constitute about 36% of the fresh green tea leaf dry weight representing the most investigated group of flavonoids in chemoprevention of oral cancer [1, 23, 34]. The four major polyphenols in green tea are epicatechin, epigallocatechin, epicatechin gallate and epigallocatechin gallate with epigallocatechin gallate (EGCG) as the predominant constituent (about 50% of green tea extract) [87, 88]. Multiple *in vitro* studies have demonstrated the antiproliferative and proapoptotic potential of green tea polyphenols in human oral cancer cell lines, whereas most of these works have focused on EGCG [35, 53]. Micromolar doses of EGCG have been shown to dose-dependently inhibit the growth of various oral cancer cell lines, including SSC4, SCC9, SCC25, Ca9-22, KB, OEC-M1, SAS, YCU-H891, CAL27, and cisplatin-resistant CAL27 (CAR) [15, 16, 43, 88-93] (Table 4.3). Such suppressive effects on cell viability can be mediated through different intracellular signaling mechanisms since EGCG targets multiple pathways resulting in the concerted inhibition of malignant cell growth.

In CAL27 cells, EGCG affected the expression and methylation status of different genes involved in the multiple cellular processes, including oxidative processes, transport, cell adhesion, apoptosis, and regulation of cell cycle [89]. Studies on growth inhibition of human oral cancer cell lines displayed certain differences in sensitivities to EGCG treatment as CAL27 cells were more sensitive than KB and SCC25 cells, showing thus heterogeneity in signaling networks in oral squamous cell carcinoma. At that, exposure to EGCG led to a significant cell cycle arrest at S and G2/M phases with the major affected signaling pathways of EGFR and Notch [88]. However, in 1483 cells, EGCG caused a cell cycle block in G1 phase, whereas the growth inhibitory efficacy decreased as the normal oral epithelial cells progressed towards malignancy [58, 94]. A G1 phase cell cycle arrest and inhibition of proliferation were observed also in SAS and SCC25 oral cancer cells treated with EGCG, caused by upregulation of B-cell translocation gene 2 (BTG2) expression mainly via the p38 MAPK signaling pathway [43]. In CAL27 and SAS cells, EGCG caused a significant growth inhibition by arresting the cell cycle at G1 phase and inducing apoptotic cell death via death receptor-regulated pathway. This process was mediated by suppression of constitutive and interleukin-6 (IL-6)-triggered JAK/STAT3 signaling, whereas elevated levels of IL-6 have been detected in the serum of oral cancer patients as compared to disease-free subjects [93]. In YCU-H891 cells, EGCG resulted in an increase of cells in G1 phase of the cell cycle and induced apoptosis via a mitochondrial pathway. EGCG caused an elevation of the proapoptotic protein Bax, decreased antiapoptotic proteins Bcl-2 and Bcl-XL, activated caspase-9, inhibited EGRF-related downstream signaling pathways and suppressed levels of phosphorylated HER-2 [72, 91]. The HER-2 protein is often overexpressed in head and neck squamous cell carcinomas being implicated with poor prognosis [91]. Treatment of OSC2 cells with EGCG led to stimulation of c-Jun N-terminal kinase (JNK) signaling pathway resulting in apoptotic cell death [95]. In CAR cells, EGCG stimulated the activity of caspases-3 and -9, increased the proapoptotic Bax protein levels and decreased the expression of antiapoptotic Bcl-2. Besides mitochondria-dependent apoptosis, EGCG also induced autophagic cell death through elevating Atgs, Beclin-1 and LC3B levels and these processes were mediated by blocking the AKT/STAT3 signaling as well as suppressing the expression of multidrug resistance 1 (MDR1) protein [15]. Both apoptotic and autophagic cell death pathways were activated also in SCC4 line in response to exposure to EGCG [16].

EGCG can exert its antitumor activity also by regulating glycolysis as the elevated glycolysis is one of the hallmarks of human cancerous growth. In both Tca8113 and TSCCa cells, this flavanol decreased the mitochondrial expression of glycolytic enzyme hexokinase 2 (HK2) and inhibited glucose consumption and lactate production. This process was mediated by Akt signaling pathway [11]. In OECM-1 cells, EGCG downregulated the expression and secretion of amyloid precursor protein (APP) that might also play some role in oral carcinogenesis [96].

Furthermore, in SCC9 and SCC25 cells, EGCG suppressed the cellular viability and induced apoptosis with a significant increase in intracellular reactive oxygen species (ROS) in SCC25 cells. At the same time, human nonmalignant gingival fibroblasts were almost resistant to the cytotoxic action of this flavanol revealing a robust induction of antioxidant defense genes in the response to EGCG [92]. This differential prooxidative action of EGCG was found to be associated with an important mitochondrial redox modulator, sirtuin 3 (SIRT3). In fact, EGCG decreased the expression and activity of this

Table 4.3. ***In vitro*** **antiproliferative effects and respective molecular mechanisms of flavonoids in human oral cancer cell lines**

Compound	Cell line	Cell growth inhibition IC_{50}, μM[1]	Assay method[2]	Assay time, h	Action mechanism[3]	Effect in human normal oral cells[4]	Ref.
CHALCONES							
3,3`,4`,5`-Tetramethoxy-chalcone (TMC)	HSC3	1	MTT	24	Apoptosis	No antiproliferative effect up to 6 μM in OF	[46]
	SAS7	4.5	MTT	24	Apoptosis		
	SCC4	3	MTT	24	Apoptosis through induction of DNA DSBs and p53-mitochondria-dependent pathway, migration inhibition by attenuating CCL5-CCR5 axis		
Butein	HSC2	26	MTT	24		IC_{50} 274 μM in HGF, 180 μM in HPC, 234 μM in HPLF	[116]
	HSC3	69	MTT	24			
	HSG	188	MTT	24			
Flavokawain B	HSC3	17.23	MTT	24	G2/M phase cell cycle arrest, apoptosis through mitochondrial, death receptor and ER stress-induced pathways, induction of oxidative stress, inhibition of PI3K/Akt activity, inactivation of p38 MAPK		[24]
	KB	105.52	TB	24	Apoptosis through death receptor and mitochondrial pathways, G2/M phase cell cycle arrest, increase in ROS generation, reduction of MMP-9 and u-PA expression and MMP-9 activity		[128]
		20.05	TB	48			
		15.12	TB	72			

Table 4.3. (Continued)

Compound	Cell line	Cell growth inhibition IC_{50}, μM[1]	Assay method[2]	Assay time, h	Action mechanism[3]	Effect in human normal oral cells[4]	Ref.
Isoliquiritigenin	HSC2	16	MTT	48		IC_{50} 164 μM in HGF, 98 μM in HPC, 144 μM in HPLF	[139]
	HSC3	37	MTT	48			
	HSC4	39	MTT	48			
	HSC2	79	MTT	24		IC_{50} 311 μM in HGF, 319 μM in HPC, 336 μM in HPLF	[116]
	HSC3	97	MTT	24			
	HSG	125	MTT	24			
	Ca9-22	135	MTT	24	G2/M phase cell cycle arrest, apoptosis, inhibition of DNA repair	>80% viability in OF and Smulow-Glickman cells at 200 μM	[13]
	HSC3	~50	MTT	24	G2/M phase cell cycle arrest, apoptosis, inhibition of DNA repair, decrease in migration, inhibition of MMP-2 and -9, reduction of colony formation		
	OEC-M1	~100	MTT	24	G2/M phase cell cycle arrest, apoptosis, inhibition of DNA repair, decrease in migration, reduction of colony formation		
	SAS	~100	MTT	24	G2/M phase cell cycle arrest, apoptosis, inhibition of DNA repair, decrease in migration, inhibition of MMP-2 and -9, reduction of colony formation		
	SCC4	~50	MTT	24	G2/M phase cell cycle arrest, apoptosis, inhibition of DNA repair		
	Tca8113	69.07	SRB	24	Suppression of intracellular ROS, arrest in S phase cell cycle at lower doses, G1 phase at higher doses, apoptosis, inhibition of adhesion, migration, invasion, decrease in MMP-2 and -9 activities		[126]
		39.18	SRB	48			
		37.74	SRB	72			

Compound	Cell line	Cell growth inhibition IC_{50}, μM[1]	Assay method[2]	Assay time, h	Action mechanism[3]	Effect in human normal oral cells[4]	Ref.
	Tca8113	69.07	SRB	24	Suppression of intracellular ROS, arrest in S phase cell cycle at lower doses, G1 phase at higher doses, apoptosis, inhibition of adhesion, migration, invasion, decrease in MMP-2 and -9 activities		[126]
		39.18	SRB	48			
		37.74	SRB	72			
Isoliquiritigenin 2`-methyl ether (ILME)	HN4	10…100	MTT	72 h	Activation of HO-1-associated pathway (involving MAPK, NF-κB, Nrf2) to induce apoptosis		[127]
	HN12	10…100	MTT	72 h	Activation of HO-1-associated pathway (involving MAPK, NF-κB, Nrf2) to induce apoptosis		[127]
Licochalcone A	HN22	17.87	MTS	48	Apoptosis through downregulation of Sp1 and ensuing target proteins (cyclin D1, survivin, Mcl-1)		[14]
	HSC4	20.42	MTS	48	Apoptosis through downregulation of Sp1 and ensuing target proteins (cyclin D1, survivin, Mcl-1)		[14]
	KB	~50	MTT	24	Apoptosis through extrinsic pathway (upregulation of FasL via ERK and p38 activation)	No effect up to 100 μM in NHOK	[45]
	SCC25	105.17	MTT	36	Apoptosis through death receptor and mitochondrial pathway, S and G2/M phase cell cycle arrest		[37]
	SCC25				Inhibition of adhesion, migration and invasion, decrease in MMP-2 activity, increase in TIMP-2 activity, increase in E-cadherin, decrease in N-cadherin		[48]

Table 4.3. (Continued)

Compound	Cell line	Cell growth inhibition IC_{50}, μM[1]	Assay method[2]	Assay time, h	Action mechanism[3]	Effect in human normal oral cells[4]	Ref.
Licochalcone B	HN22	15	Counting	48	Apoptosis through intrinsic and death receptor pathways, G1 phase cell cycle arrest		[12]
	HSC4	13	Counting	48	Apoptosis through intrinsic and death receptor pathways, G1 phase cell cycle arrest		[12]
Sappanchalcone	HN4	5.2	MTT	72	Apoptosis through intrinsic pathway, activation of p53-dependent, ERK, p38, JNK and NF-κB signaling		[78]
	HN12	25.3	MTT	72	Apoptosis through intrinsic pathway, activation of p53-dependent, ERK, p38, JNK and NF-κB signaling		[78]
FLAVANOLS							
Catechin (C)	HSC2	497±35.0	NR	72		IC_{50} >500 μM in HGF2	[106]
Catechin gallate (CG)	CAL27	62	NR	72		IC_{50} 132 μM in HGF1, 58 μM in S-G	[105]
	HSG	90	NR	72			
	HSC2	46±7.6	NR	72		IC_{50} 131±14.7 μM in HGF2	[106]
Epicatechin (EC)	HSC2	>500	NR	72		IC_{50} >>500 μM in HGF2	[106]
Epicatechin gallate (ECG)	HSC2	121±6.5	NR	24	Apoptosis	IC_{50} 474±45.1 μM (24h), 139±4.1 μM (48h), 100±9.5 μM (72h) in HGF2	[106]
		82±8.6	NR	48			
		67±5.4	NR	72			

Compound	Cell line	Cell growth inhibition IC_{50}, μM[1]	Assay method[2]	Assay time, h	Action mechanism[3]	Effect in human normal oral cells[4]	Ref.
Epigallocatechin (EGC)	HSC-2	277±13.2	NR	72		IC_{50} >500 μM in HGF2	[106]
Epigallocatechin gallate (EGCG)	686LN				Apoptosis		[104]
	1483	17.8±7.0	DNA fluorescence	120	G1 phase cell cycle arrest. Synergistic growth inhibition in combination with curcumin	IC_{50} 5.99±2.5 μM in NOE, 7.15±1.62 μM in MSK Leuk1, 9.01±1.09 μM in MSK Leuk1a (G1 arrest)	[58]
	1483	18	DNA fluorescence	120	G1 phase cell cycle arrest	IC_{50} 5.5 μM in NOE, 8 μM in MSK Leuk1, 11.5 μM in MSK Leuk1a (G1 arrest)	[94]
	Ca9-22	18.1	MTT	24			[93]
	Ca9-22				Suppression of IDO		[103]
	CAL27	16.15	MTT	24	Apoptosis through death receptor pathway, G1 phase cell cycle arrest, inhibition of constitutive and IL-6-induced JAK/STAT3 signaling		[93]
	CAL27	>100	MTT	24	Change of methylation status and expression of different genes		[89]
	CAL27	128.72	MTT	72	S and G2/M phase cell cycle arrest, action through EGFR and Notch signaling		[88]
	CAL27				Inhibition of migration and invasion; decrease in MMP-2 expression and activity		[136]
	CAL27				Suppression of IDO		[103]

Table 4.3. (Continued)

Compound	Cell line	Cell growth inhibition IC_{50}, μM[1]	Assay method[2]	Assay time, h	Action mechanism[3]	Effect in human normal oral cells[4]	Ref.
	CAL27	180±20	NR	24	Prooxidant: formation of H_2O_2, potentiation of Fe2+-induced lipid peroxidation, reduction of GSH, induction of cytoplasmic vacuolization	IC_{50} 440±50 μM (24h) and 171±15.6 μM (72h) in GN56 (prooxidant), 246±17.9 μM (72h) in HGF1, 80±4.9 μM (72h) in S-G	[98]
		45±0.8	NR	72			
	HSC2	78±5.6	NR	72			
	HSG1	420±30	NR	24	Prooxidant: formation of H_2O_2, potentiation of Fe2+-induced lipid peroxidation, reduction of GSH, induction of cytoplasmic vacuolization		
		109±5.7	NR	72			
	CAR	73.25±2.85	SRB	24	Apoptosis through mitochondrial pathway, autophagy, inhibition of AKT/STAT3 signaling, inhibition of MDR1 expression		[15]
		50.26±3.25	SRB	48			
		32.98±3.67	SRB	72			
	FaDu	>200	MTT	16			[4]
	FaDu		Counting	24, 72, 120	Inhibition of HGF-induced migration and invasion, decrease in MMP-9 and u-PA expression and activities, inhibition of Akt and Erk pathways		[56]
	HSC2	73±2.7	NR	72		IC_{50} 256±28.2 μM in HGF2	[106]
	HSC3	NA up to 100 μM	MTT	24, 48	Inhibition of migration and invasion		[102]

Compound	Cell line	Cell growth inhibition IC_{50}, μM[1]	Assay method[2]	Assay time, h	Action mechanism[3]	Effect in human normal oral cells[4]	Ref.
	HSC3				Suppression of IDO via blocking of γ-IFN-induced JAK-PKC-δ-STAT1 signaling		[103]
	HSC3				Inhibition of migration and invasion, decrease in MMP-2, -9 expression and activity, increase in RECK expression		[100]
	HSC4				Inhibition of migration and invasion, decrease in MMP-2 activity, increase in RECK expression		[100]
	KB		MTT	16	Apoptosis, inhibition of HGF-induced invasiveness, MMP-2, -9 expression, activation of JNK pathway		[4]
	KB	223.62	MTT	72	Action through EGFR and Notch signaling		[88]
	OC2	>60	MTT	24	Inhibition of migration and invasion, decrease in MMP-2, -9 and u-PA activities		[101]
	OEC-M1	>40	MTT	24	Decrease in MMP-13 expression and activity, decrease in MMP-2 and -9 activities		[90]
	OECM-1				Inhibition of APP expression and secretion		[96]
	OSC2	~200	MTT	24			[87]
	OSC2				Apoptosis mediated by activation of JNK signaling		[95]

Table 4.3. (Continued)

Compound	Cell line	Cell growth inhibition IC_{50}, μM[1]	Assay method[2]	Assay time, h	Action mechanism[3]	Effect in human normal oral cells[4]	Ref.
	OSC2	>200	MTT	72		Small increase in viability of NS-SV-AC, IC_{50} >>200 μM in NS-SV-DC	[73]
	OSC4	~200	MTT	72			
	OSC2	~200	MTT	24	Production of ROS	Inhibition of ROS production in NS-SV-AC	[97]
	OSC4	~200	MTT	24	Production of ROS		
	SAS	6.26	MTT	24	Apoptosis through death receptor pathway, G1 phase cell cycle arrest, inhibition of constitutive and IL-6-induced JAK/STAT3 signaling		[93]
	SAS	>25	Thymidine incorporation	24	G1 phase cell cycle arrest, upregulation of BTG2 expression via p38 and ERK pathways		[43]
	SAS				Suppression of IDO		[103]
	SCC4	52.3	RTCA	0…49	Apoptosis, autophagy		[16]
	SCC4				Suppression of IDO		[103]
	SCC9	NA up to 20 μM	MTT	24	Inhibition of migration and invasion, reduction of cell-matrix interactions, decrease in MMP-2 and u-PA activities, decrease in PMA-induced MMP-9 and u-PA activities	NA up to 20 μM in HGF	[49]
	SCC9	83	MTT	72	Apoptosis	IC_{50} >200 μM in HGF1	[92]

Compound	Cell line	Cell growth inhibition IC_{50}, μM[1]	Assay method[2]	Assay time, h	Action mechanism[3]	Effect in human normal oral cells[4]	Ref.
	SCC25	95	MTT	72	Apoptosis, increase in intracellular ROS, downregulation of antioxidant response-related genes	IC_{50} >200 μM in HGF1	[92]
	SCC9				Decrease in MMP-2, -9 expression and activity, increase in RECK expression		[100]
	SCC9				Increase in mitochondrial ROS, decrease in SIRT3 mRNA expression, changes in downstream gene expression	No effect on ROS and SIRT3 level in HGF1, increase in ROS and decrease in SIRT3 level in MSK-Leuk1	[83]
	SCC25				Increase in mitochondrial ROS, decrease in SIRT3 expression and activity, changes in downstream gene expression	No effect on ROS and SIRT3 level in HGF1, increase in ROS and decrease in SIRT3 level in MSK-Leuk1	[83]
	SCC25	>25	Thymidine incorporation	24	G1 phase cell cycle arrest, upregulation of BTG2 expression via p38 and ERK pathways		[43]
	SCC25	223.62	MTT	72	Action through EGFR and Notch signaling		[88]
	SCC25				Decrease in MMP-2 activity		[100]
	SGT	>200	MTT	24	Inhibition of invasion and migration via ß1 integrin-mediated signaling, decrease in MMP-2, -9 expression and activity		[70]

Table 4.3. (Continued)

Compound	Cell line	Cell growth inhibition IC_{50}, µM[1]	Assay method[2]	Assay time, h	Action mechanism[3]	Effect in human normal oral cells[4]	Ref.
	Tca8113				Apoptosis, inhibition of glycolysis by decrease in HK2 through Akt signaling		[11]
	TSCCa				Apoptosis, inhibition of glycolysis by decrease in HK2 through Akt signaling		[11]
	Tu212				Apoptosis. Synergistic effect with luteolin for apoptosis through death receptor, mitochondria-independent pathway		[104]
	Tu686				Apoptosis		[104]
	YCU-H891		Colony formation	48	Apoptosis through mitochondrial pathway, G1 phase cell cycle arrest, suppression of EGFR-related signaling, inhibition of ERK and Stat3 pathways		[72]
	YCU-H891	21.82	WST-1	72	Decrease in phosphorylated HER-2		[91]
	YD-10B	NA up to 100 µM	MTT	24, 48	Inhibition of migration and invasion by disrupting functional invadopodia formation and ECM degradation through decrease in Src, FAK, CTTN, MT1-MMP, MMP-2, -9 and RhoA activities		[102]

Compound	Cell line	Cell growth inhibition IC_{50}, μM[1]	Assay method[2]	Assay time, h	Action mechanism[3]	Effect in human normal oral cells[4]	Ref.
FLAVANOLIGNANS							
Silibinin	FaDu	~200	MTT	24			[138]
	KB	78.66	MTT	48	Increase in intracellular ROS, apoptosis		[122]
	SCC4	>100	MTT	24	Inhibition of invasion and motility, decrease in MMP-2 and u-PA activities, inhibition of ERK1/2 signaling	IC_{50} >100 μM in HGF	[10]
FLAVANONES							
Eriodictyol	KB	173.78					[141]
Liquiritigenin	HSC2	148	MTT	48		IC_{50} 566 μM in HGF, 273 μM in HPC, 538 μM in HPLF	[139]
	HSC3	191	MTT	48			
	HSC4	336	MTT	48			
Naringenin	SCC25	~200	MTT	48	G0/G1 phase cell cycle arrest, no apoptosis, inhibition of migration and invasiveness, inhibition of ERK1/2 and AKT signaling		[3]
FLAVONES							
5-Hydroxyflavone (5-HF)	SCC4	>100	MTT	24			[110]
7-Hydroxyflavone (7-HF)	SCC4	>100	MTT	24	Inhibition of invasion and migration, decrease in MMP-2 and u-PA activity, increase in TIMP-2 and PAI-1 levels		[110]
	SCC4	>100	MTT	24			[66]
7-Methoxyflavone (7-MF)	SCC9	36.5	BrdU incorporation	24			[107]
7,8-Dihydroxyflavone (7,8-DHF)	HN22		MTS	24, 48	Apoptosis through suppression of Sp1		[68]
	HSC4		MTS	24, 48	Apoptosis through suppression of Sp1		[68]

Table 4.3. (Continued)

Compound	Cell line	Cell growth inhibition IC_{50}, μM[1]	Assay method[2]	Assay time, h	Action mechanism[3]	Effect in human normal oral cells[4]	Ref.
5,7-Dimethoxyflavone (5,7-DMF)	FaDu	7.7	BrdU incorporation	24			[107]
	SCC9	8	BrdU incorporation	24	G1 phase cell cycle arrest		[107]
		~50…100	CV				
		~50…100	MTT				
	SCC9				Inhibition of BaP-DNA adduct formation, decrease in BaP-induced CYP1A1 expression, inhibition of CYP1B1 activity		[114]
7,4`-Dimethoxyflavone (7,4`-DMF)	SCC9	24.2	BrdU incorporation	24			[107]
3`,4`-Dimethoxyflavone (3`,4`-DMF)	SCC9				Inhibition of BaP-induced CYP1B1 expression		[114]
5,7,4`-Trimethoxyflavone (5,7,4`-TMF)	SCC9	5	BrdU incorporation	24	G1 phase cell cycle arrest		[107]
Acacetin	HSC3	87.95	MTT	24	Apoptosis through extrinsic and mitochondrial pathways, activation of MAPKs (ERK and p38)		[63]
Apigenin	Ca9-22	>40	MTT	24			[93]
	CAL27	>40	MTT	24			[93]
	CAL27	36.4±1.0	SRB	48	Apoptosis		[108]
	FaDu	~50	WST-1	24			[130]
		~25	WST-1	48			
	FaDu	41.6±1.1	SRB	48	Apoptosis, inhibition of EGFR/ErbB2 signaling		[108]
	FaDu	19.5±1.2	WST-1	48			[109]

Compound	Cell line	Cell growth inhibition IC_{50}, μM[1]	Assay method[2]	Assay time, h	Action mechanism[3]	Effect in human normal oral cells[4]	Ref.
	PCI-13	~25	WST-1	48			[109]
	PCI-15B	~25	WST-1	48			[109]
	SAS	>40	MTT	24			[93]
	SCC4	>100	MTT	24	Inhibition of invasion and migration, decrease in MMP-2 and u-PA activity, decrease in TIMP-2 and PAI-1 levels		[110]
	SCC9	40	BrdU incorporation	24	G2/M phase cell cycle arrest		[107]
	SCC15	38.3±1.0	SRB	48	Apoptosis		[108]
	SCC25		MTT	24, 48	Apoptosis, G0/G1 and G2/M phase cell cycle arrest		[55]
	SCC25	11.20	MTT	72	Apoptosis, G2/M phase cell cycle arrest, reduction in clonogenicity, formation of intracellular ROS, depletion of GSH, upregulation of TNF-R and TRAIL-R signaling		[74]
Baicalein	CAL27	~100	MTT		Apoptosis and autophagy, generation of ROS		[50]
	HSC3	>200	MTT	24	G1 phase cell cycle arrest through hypophosphorylation of Rb, promotion of GSK-3ß pathway and activation of AhR		[112]
		~200	MTT	48			
	KB	61.66					[141]
	SCC4	>100	MTT	24	Inhibition of invasion and migration, decrease in MMP-2 and u-PA activity, decrease in TIMP-2 and PAI-1 levels		[110]

Table 4.3. (Continued)

Compound	Cell line	Cell growth inhibition IC_{50}, μM[1]	Assay method[2]	Assay time, h	Action mechanism[3]	Effect in human normal oral cells[4]	Ref.
	SCC4				Apoptosis through Ca^{2+}-associated mitochondrial pathway, generation of ROS		[111]
	SCC25	70	MTT	72	Slight increase in apoptosis, G2/M phase cell cycle arrest, decrease in PCNA-positive cells		[22]
Chrysin	FaDu	10	BrdU incorporation	24			[107]
	KB	12.88					[141]
	SCC9	108	BrdU incorporation	24	G2/M phase cell cycle arrest		[107]
		~50-100	CV				
		~50-100	MTT				
Diosmetin	SCC9	>200	MTT	72			[81]
Flavone	SCC9	8.3	BrdU incorporation	24			[107]
Luteolin	686LN				Apoptosis		[104]
	KB	13.18					[141]
	OC2	~80	MTT	24		IC_{50} >100 μM in HGF	[66]
	SCC4	>100	MTT	24	Apoptosis through mitochondrial pathway, G1 phase cell cycle arrest		
		~60	MTT	48			
		~20	MTT	72			

Compound	Cell line	Cell growth inhibition IC_{50}, μM[1]	Assay method[2]	Assay time, h	Action mechanism[3]	Effect in human normal oral cells[4]	Ref.
	OCSC from GNM		MTT	24	Inhibition of self-renewal ability of cancer stem cells, inactivation of IL-6/STAT3 signaling	Limited effect (IC_{50} >>40 μM) S-G	[64]
	OCSC from SAS		MTT	24	Inhibition of self-renewal ability of cancer stem cells, inactivation of IL-6/STAT3 signaling		
	SCC25		MTT	72	Inhibition of migration, activation of DNA damage pathway		[65]
	Tu212	6.96	SRB	72	Inhibition of colony formation		[113]
	Tu212				Apoptosis. Synergistic effect with EGCG for apoptosis through death receptor, mitochondria-independent pathway		[104]
	Tu686				Apoptosis		[104]
Morusin	KB	6.92					[141]
Scutellarin	HSC4		MTT	4…24	Inhibition of motility, increase in E-cadherin expression, decrease in integrin αvß6 levels		[52]
	SAS		MTT	4…24	Inhibition of motility, reduction of cell adhesion to fibronectin, increase in E-cadherin expression, decrease in integrin αvß6 levels		[52]
	SAS				Decrease in MMP-2, -9 and integrin αvß6 expression		[57]
Tangeretin	SCC9	19.3	BrdU incorporation	24			[107]

Table 4.3. (Continued)

Compound	Cell line	Cell growth inhibition IC_{50}, μM[1]	Assay method[2]	Assay time, h	Action mechanism[3]	Effect in human normal oral cells[4]	Ref.
Vitexin	OC2	>100	AB	48	Apoptosis through p53-dependent pathway, inhibition of migration and metastasis, increase in PAI-1 expression, decrease in MMP-2 levels		[30]
FLAVONOLS							
Datiscetin	Primary culture of OSCC (stage 4)	54	Thymidine incorporation	72		IC_{50} 54 μM in primary culture of normal oral mucosa	[119]
Fisetin	HSC2	39	MTT	24		IC_{50} 608 μM in HGF, 170 μM in HPC, 641 μM in HPLF	[116]
	HSC3	188	MTT	24			
	HSG	375	MTT	24			
	SCC4	52.8	Counting	48	Apoptosis through mitochondrial, death receptor and ER stress pathways, G2/M phase cell cycle arrest, production of ROS and Ca^{2+}		[8]
Galangin	Primary culture of OSCC (stage 4)	50	Thymidine incorporation	72		IC_{50} 54 μM in primary culture of normal oral mucosa	[119]
Kaempferol	FaDu	~50	WST-1	24			[130]
		~25	WST-1	48			
	FaDu	45.9±5.4	WST-1	48			[109]
	HSC2	743	MTT	24		IC_{50} >1000 μM in HGF, >1000 μM in HPC, >1000 μM in HPLF	[116]
	HSC3	>1000	MTT	24			
	HSG	>1000	MTT	24			
	PCI-13	>100	WST-1	24, 48			[109]

Compound	Cell line	Cell growth inhibition IC_{50}, μM[1]	Assay method[2]	Assay time, h	Action mechanism[3]	Effect in human normal oral cells[4]	Ref.
	PCI-15B	>100	WST-1	24, 48			[109]
	Primary culture of OSCC (stage 4)	35	Thymidine incorporation	72		IC_{50} 37 μM in primary culture of normal oral mucosa	[119]
	SCC4	>100	MTT	24	Inhibition of migration and invasion, decrease in MMP-2 expression and activity, inhibition of ERK1/2 signaling		[67]
	SCC25				Apoptosis		[115]
	SCC1483	~40	CellTiter96	24	Apoptosis		[115]
	SCC-QLL1				Apoptosis		[115]
Morin	Primary culture of OSCC (stage 4)	110	Thymidine incorporation	72	G2/M phase cell cycle arrest, no apoptosis, activation of JNK/p38 signaling, inhibition of PK-B/AKT pathway	IC_{50} 180 μM in primary culture of normal oral mucosa (G2/M arrest)	[119]
Myricetin	SCC25	~150	MTT	48	G0/G1 phase cell cycle arrest, no apoptosis, inhibition of migration and invasiveness, inhibition of ERK1/2 and AKT signaling		[3]
Quercetin	ACC-2	77	MTT	24	Apoptosis through mitochondrial pathway, suppression of PI3K/Akt/IKK-α/NF-κB signaling		[9]
		49	MTT	48			
		30	MTT	72			
	ACC-M	59	MTT	24	Apoptosis through mitochondrial pathway, suppression of PI3K/Akt/IKK-α/NF-κB signaling		[9]
		37	MTT	48			
		28	MTT	72			

Table 4.3. (Continued)

Compound	Cell line	Cell growth inhibition IC_{50}, μM[1]	Assay method[2]	Assay time, h	Action mechanism[3]	Effect in human normal oral cells[4]	Ref.
	HSC2	35	MTT	24		IC_{50} 750 μM in HGF, 909 μM in HPC, >1000 μM in HPLF	[116]
	HSC3	250	MTT	24			
	HSG	745	MTT	24			
	HSC3	20	Counting	24	Apoptosis through extrinsic pathway, G2/M phase cell cycle arrest, inhibition of EGFR/Akt signaling, stimulation of FOXO1	IC_{50} >50 μM in HGF	[117]
	TW206	45	Counting	24	Apoptosis, G2/M phase cell cycle arrest		
	KB	81.28					[141]
	KB/VCR	75	MTT	48	Apoptosis through mitochondrial pathway, G1 phase cell cycle arrest, inhibition of invasion and migration		[118]
	Primary culture of OSCC (stage 4)	30	Thymidine incorporation	72		IC_{50} 31 μM in primary culture of normal oral mucosa (G2/M arrest)	[119]
	SAS		Flow cytometry	24	Inhibition of invasion and migration, decrease in MMP-2, -9 activities, inhibition of MAPK and PI3K/AKT signaling		[33]
	SCC9	>200	MTT	24	Necrosis, apoptosis, S phase cell cycle arrest		[76]
		94	MTT	48			
		50	MTT	72			

Compound	Cell line	Cell growth inhibition IC_{50}, μM[1]	Assay method[2]	Assay time, h	Action mechanism[3]	Effect in human normal oral cells[4]	Ref.
	SCC9	>5	MTT	48			[82]
	SCC9	>10	MTT	72			[81]
	SCC25	50	MTT	24	Apoptosis, G1 phase cell cycle arrest, inhibition of migration and invasion	NA up to 100 μM in HGF	[51]
	SCC25				Apoptosis		[115]
	SCC1483	~40	CellTiter96	24	Apoptosis		[115]
	SCC-QLL1				Apoptosis		[115]
Quercitrin	SCC9	>200	MTT	72	Minimal cellular uptake, no hydrolysis to quercetin		[81]
Rutin	Ca9-22	>40	MTT	24			[93]
	CAL27	>40	MTT	24			[93]
	SAS	>40	MTT	24			[93]
Spiraeoside	SCC9	>100	MTT	72	Effect associated with hydrolysis to quercetin		[81]
ISOFLAVONES							
Biochanin A	SCC15		MTT	48	Decrease in ERK expression		[121]
	SCC25	~50	MTT	48	Decrease in Akt expression		[121]
Daidzein	HSC2	91	MTT	24		IC_{50} 524 μM in HGF, 150 μM in HPC, 346 μM in HPLF	[140]
	HSG	508	MTT	24			
	SCC15		MTT	48	No change in ERK expression		[121]
	SCC25	>100	MTT	48	No change in Akt expression		[121]
Formononetin	KB	37.15					[141]
Genistein	HN4	5	TB	96	Apoptosis, G2/M and S phases cell cycle arrest		[54]
	HSC2	156	MTT	24		IC_{50} 533 μM in HGF, 444 μM in HPC, 481 μM in HPLF	[140]
	HSG	259	MTT	24			

Table 4.3. (Continued)

Compound	Cell line	Cell growth inhibition IC_{50}, μM[1]	Assay method[2]	Assay time, h	Action mechanism[3]	Effect in human normal oral cells[4]	Ref.
	HSC3	10…20	SRB	48	Apoptosis		[120]
	KB	20…40	SRB	48	Apoptosis		[120]
	SCC9	>10	MTT	48			[82]
	SCC9	>10	MTT	72			[81]
	SCC15		MTT	48	Decrease in ERK expression		[121]
	SCC25	~50	MTT	48	Decrease in Akt expression		[121]
	SCC25	~200	MTT	72	G2/M phase cell cycle arrest, no apoptosis, suppression of PCNA, decrease in PGE2 level via inhibition of COX-2 activity		[22]
Genistin	SCC9	>100	MTT	72	Rapid cellular uptake, hydrolysis to genistein		[81]
PRENYLATED FLAVONOIDS							
Cudraflavone B	HN4	18.3±2.2	MTT	24	Apoptosis through mitochondrial pathway, activation of MAPK/NF-κB and SIRT1 signaling	NA at 15 μM in NHOK (72h)	[124]
		12.6±1.9	MTT	48			
		10.9±1.1	MTT	72			
	HN12	19.5±1.7	MTT	24	Apoptosis through mitochondrial pathway, activation of MAPK/NF-κB and SIRT1 signaling		
		12.0±1.5	MTT	48			
		10.7±1.7	MTT	72			
Icaritin	CAL27	15.99	CCK-8	24	Apoptosis, autophagy, inhibition of STAT3 signaling		[18]
		11.87	CCK-8	48			
		9.93	CCK-8	72			
	KB	>10	CCK-8	24, 48	Apoptosis through mitochondrial pathway, inhibition of Sp1/DNMT1 signaling		[123]

Compound	Cell line	Cell growth inhibition IC_{50}, μM[1]	Assay method[2]	Assay time, h	Action mechanism[3]	Effect in human normal oral cells[4]	Ref.
	SCC9	10.36	CCK-8	24	Apoptosis, autophagy, inhibition of STAT3 signaling		[18]
		6.11	CCK-8	48			
		3.95	CCK-8	72			
	SCC9	≥10	CCK-8	24, 48	Apoptosis through mitochondrial pathway, inhibition of Sp1/DNMT1 signaling		[123]
ROTENOIDS							
Deguelin	A253		MTT	72	Inhibition of Akt signaling disrupting its association with Hsp-90		[77]
	Ca9-22	0.4±0.05	MTS	72	Inhibition of invasion, no effect on migration	IC_{50} 0.8±0.2 μM in OMF	[2]
	CAL27	0.3±0.05	MTS	72			
	HSC3	>10	MTS	72	Inhibition of invasion and migration		
	SAS	0.3±0.02	MTS	72	Inhibition of invasion and migration, decrease in MMP-2 level, inhibition of TNF-α-induced NF-κB signaling		
	SCC4	>10	MTS	72	Inhibition of migration		
	TW1.5	>10	MTS	72			
	TW2.6	>10	MTS	72			
	SCC9		MTT	72	Inhibition of Akt signaling disrupting its association with Hsp-90		[77]

Table 4.3. (Continued)

Compound	Cell line	Cell growth inhibition IC_{50}, μM[1]	Assay method[2]	Assay time, h	Action mechanism[3]	Effect in human normal oral cells[4]	Ref.
Rotenone	CAL27	0.11±0.02	MTT	48	Apoptosis	IC_{50} 31.43±2.31 μM in OMF	[125]
	SAS	0.95±0.05	MTT	48	Apoptosis through mitochondrial and death receptor pathways, G2/M phase cell cycle arrest, production of ROS		
	Tw2.6	1.24±0.11	MTT	48	Apoptosis		

[1]NA, not active.

[2]AB, Alamar Blue; BrdU, bromodeoxyuridine; CCK-8, cell counting kit-8; CV, crystal violet; MTS, 3-(4,5-dimethylthiazol-2-yl)-5-(3-carboxymethoxtphenyl)-2-(4-sulfophenyl)-2H-tetrazolium; MTT, 3-(4,5-dimethylthiazol-2-yl)-2,5-diphenyltetrazolium bromide; NR, neutral red; SRB, sulforhodamine B; RTCA, real-time cell analyzer; TB, trypan blue exclusion; WST-1, water soluble tetrazolium-1.

[3]AhR, aryl hydrocarbon receptor; Akt, protein kinase B; APP, amyloid precursor protein; BaP, benzo[α]pyrene; BTG2, B-cell translocation gene 2; CCL5, chemotactic cytokine protein RANTES; CCR5, chemokine receptor type 5; COX2, cyclooxygenase-2; CTTN, cortactin; DNMT1, DNA (cytosine-5-)-methyltransferase 1; DSB, double strand brake; CYP, cytochrome P450; ECM, extracellular matrix; EGFR, epidermal growth factor receptor; ER, endoplasmic reticulum; ERK, extracellular signal-regulated kinase; FAK, focal adhesion kinase; FasL, factor associated suicide ligand; FOXO1, fork-head box O1; GSH, reduced glutathione; GSK-3ß, glycogen synthase kinase-3ß; H_2O_2, hydrogen peroxide; HER-2, human epidermal growth factor receptor-2; HGF, hepatocyte growth factor, HK2, hexokinase 2; HO-1, heme oxygenase-1; Hsp-90, heat shock protein-90; IDO, indoleamine 2,3-dioxygenase; γ-IFN, γ-interferon; IKK, IκB kinase; IL, interleukin; JAK, Janus kinase; JNK, c-Jun amino-terminal kinase; MAPK, mitogen-activated protein kinase; Mcl-1, myeloid cell leukemia-1; MDR, multidrug resistance; MMP, matrix metalloproteinase; MT1-MMP, membrane type 1-MMP; NF-κB, nuclear factor κB; Nrf2, nuclear factor E2-related factor 2; PAI, plasminogen activator inhibitor; PCNA, proliferating cell nuclear antigen; PGE2, prostaglandin E2; PI3K, phosphatidylinositol 3-kinase; PKC, protein kinase C; PMA, phorbol-12-myristate-13-acetate; RhoA, Ras homolog gene family, member A; Rb, retinoblastoma; RECK, reversion-inducing cysteine-rich protein with Kazal motifs; ROS, reactive oxygen species; SIRT1, silent information regulator 1; SIRT3, sirtuin 3; Sp1, specificity protein 1; Src, proto-oncogene tyrosine-protein kinase; STAT, signal transducer and activator of transcription; TIMP, tissue inhibitor of metalloproteinase; TNF-α, tumor necrosis factor-α; TNF-R, tumor necrosis factor receptor; TRAIL-R, TNF-related apoptosis-inducing ligand receptor; u-PA, urokinase-type plasminogen activator.

[4]GN56, gingival fibroblasts; HGF, human gingival fibroblasts; HPC, human pulp cells; HPLF, human periodontal ligament fibroblasts; MSK Leuk1, premalignant cell line established from a dysplastic leukoplakia lesion; MSK Leuk1a, more progressed cell line than MSK Leuk1; NHOK, normal human oral keratinocytes; NOE, normal oral epithelial cells derived from anterior tonsillar pillar mucosa; NS-SV-AC, immortalized human normal salivary gland cell line; NS-SV-DC, immortalized human normal salivary gland cell line; OF, oral fibroblasts; OMF, oral mucosa fibroblasts; S-G, immortalized human gingival epithelial cell line; Smulow-Glickman cells, human normal gingival epithelial cell line.

protein in SCC25 cells while increasing SIRT3 activity in human gingival fibroblasts, leading to differences in downstream stress responses [83]. Differential oxidative stress in response to EGCG was described also in other studies as this flavanol induced oxidative stress in OSC2 and OSC4 oral cancer cells, but reduced the production of ROS to background levels in normal oral cells, showing that EGCG promotes tumor cell destruction while contributing to normal cell survival [73, 97]. Selective inhibitory activity of EGCG against oral cancer cells was observed also in CAL27, HSC2 and HSG1 lines as compared to normal human gingival fibroblasts, whereas EGCG behaved as a prooxidant agent in malignant cells [98]. At that, a cyclin-dependent kinase p57 might play an important role in differential action of EGCG being substantially increased in normal cells but remaining consistent in cancerous cells following to treatment with EGCG and exerting probably a protection to normal cells [99].

A therapeutically important and attractive effect of EGCG on oral cancer cells involves its antimetastatic action. At that, this flavanol can suppress the metastasis through multiple mechanisms. EGCG was demonstrated to inhibit invasion, migration and motility of SCC9 cells through suppression of expression and activity of matrix metalloproteinase-2 (MMP-2) and urokinase plasminogen activator (u-PA). Moreover, EGCG also revealed the anti-epithelial-to-mesenchymal-transition (anti-EMT) action reducing cell-matrix interactions as characterized by a decrease in the expression of EMT-related proteins, including p-focal adhesion kinase (FAK), p-Src, snail-1 and vimentin [49]. In several oral cancer cell lines (HSC3, HSC4, SCC9, SCC25), EGCG was found to partially reverse the hypermethylation status and enhance the expression of a novel tumor suppressor gene, the reversion-inducing cysteine-rich protein with Kazal motifs (RECK) that negatively regulates matrix metalloproteinases (MMPs) and inhibits tumor invasion and metastasis. As a result of this demethylation, treatment with EGCG led to a suppressed expression and activity of MMP-2 and MMP-9 in HSC3 and SCC9 cells and inhibited the activity of MMP-2 in HSC4 and SCC25 cells producing a significant decrease in invasive ability of malignant oral cells [100]. In KB cells, EGCG significantly inhibited hepatocyte growth factor (HGF)-promoted tumor cell proliferation, migration and invasion, as well as expression of MMP-2 and MMP-9 [4]. HGF is a stromal-derived mediator in tumor-stromal interactions stimulating the progression and metastasis of oral cancer cells. Inhibition of HGF-induced proliferation, motility and invasiveness by EGCG was also observed in FaDu cells by suppressing the expression and activity of MMP-9 and u-PA [56]. In OC2 cells, exposure to EGCG also led to inhibition of invasive and migratory abilities by reducing the expression and activity of MMP-2, MMP-9 and u-PA [101]. In YD-10B cells, EGCG disturbed functional invadopodia formation and eliminated extracellular matrix (ECM) degradation by inhibition of Src, FAK, cortactin (CTTN), RhoA, MMP-2, MMP-9 and membrane type 1 (MT1)-MMP activities, leading to significantly suppressed migration and invasion of cancer cells [102]. Similar decrease in migration and invasive capabilities was also detected in HSC3 cells [102]. In SGT cells, exposure to EGCG led to suppression of invasion and migration, primarily via ß1 integrin-mediated signaling and decrease in MMP-2 and MMP-9 expression and activity [70]. Treatment of OEC-M1 cells with EGCG substantially reduced the expression and activity of MMP-13 (collagenase-3) that digests collagen and other ECM components and helps malignant cells to survive in adverse environment, and suppressed the proteolytic activity of MMP-2 and MMP-9 [90]. Increased expression of MMPs has been reported to be related to local invasion or regional lymph node metastasis of oral cancer, whereas MMP-13 was even proposed to serve as a marker for oral malignancies. A significant increase in the expression of MMP-13 was indeed found in oral squamous cell carcinoma cells as compared to normal oral keratinocytes [90].

In addition, molecular evidences were provided that EGCG might stimulate tumor cell recognition by the immune system since exposure to EGCG decreased the levels of an immunomodulatory protein, indoleamine 2,3-dioxygenase (IDO) that is frequently expressed by tumor cells and related to immune escape in several oral cancer cell lines (Ca9-22, CAL27, HSC3, SAS, SCC4). In HSC3 cells, EGCG suppressed IDO expression and activity by blocking the interferon-γ (IFN-γ)-triggered JAK-PKC-δ-STAT1 signaling pathway revealing that EGCG can be considered as a potential therapeutic agent to potentiate the antitumor immunity and increase sensitivity to conventional chemotherapeutic drugs [103].

Furthermore, combination of EGCG with other dietary agents, such as flavone luteolin led to a synergistic increase in growth inhibition of Tu212 cells and induction of death receptor-mediated (but mitochondria-independent) apoptosis characterized by 3-5-fold more apoptotic effects than the additive level induced by EGCG and luteolin [104]. A synergistic growth inhibition was also observed in 1483 cells by combining EGCG with curcumin at a 1:1 ratio allowing to reduce the effective dose of EGCG for 8.5-fold [58]. Such synergistic activities might circumvent possible toxic effects related to the treatment of cells with high concentrations of single agents.

The above presented and in Table 4.3 summarized *in vitro* results clearly show that EGCG can be considered as a potential chemopreventive and/or chemotherapeutic agent for combating oral cancer, either alone or in combination with other natural compounds. This green tea component might provide a drug candidate for clinical management of oral tumors in the future and its further studies are imperative. A concise overview of the different anticancer effects of EGCG in human oral cancer cell lines is depicted in Figure 4.1.

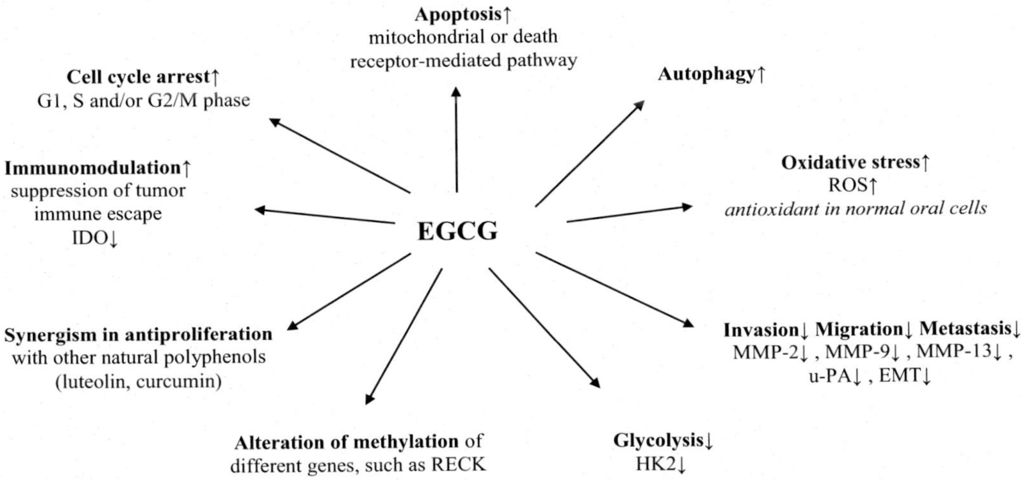

Figure 4.1. Major *in vitro* action mechanisms of EGCG in oral cancer cells (EMT, epithelial to mesenchymal transition; HK2, hexokinase 2; IDO, indoleamine 2,3-dioxygenase; MMP, matrix metalloproteinase; RECK, reversion-inducing cysteine-rich protein with Kazal motifs; ROS, reactive oxygen species; u-PA, urokinase-type plasminogen activator).

Compared to EGCG, the current knowledge about possible activities of other green tea ingredients in oral cancer cells is substantially less. Catechin gallate (CG) was shown to inhibit growth of CAL27 and HSG cells being less sensitive to human normal gingival fibroblasts [105]. Similarly, CG as well as epicatechin gallate (ECG) and epigallocatechin (EGC) induced more effective antiproliferative effects in oral cancer HSC2 cells than in human normal gingival fibroblasts with ECG inducing apoptotic cell death in HSC2 (but not in HGF) cells [106]. At that, ECG and the minor polyphenol constituent of green tea, CG, were as potent as EGCG in HSC2 oral cancer cells while catechin and epicatechin revealed almost no inhibitory activities on human oral cancer cell proliferation [106] (Table 4.3). In addition, ECG was shown to suppress also the growth of murine oral squamous cell carcinoma cells (SCC7) by inducing G1 phase cell cycle arrest and apoptotic cell death through inhibition of ß-catenin signaling [5]. The effects of these green tea polyphenols in different oral cancer cell systems obviously need further studies as well as potential co-action of different flavanols has remained to be clarified.

4.3.1.2. Flavones

Different flavones have been studied regarding to their *in vitro* ability to suppress the proliferation of human oral cancer cells. Majority of these works have focused on apigenin, baicalein and luteolin (Table 4.3).

Although apigenin is a common natural flavone the data about its molecular mechanisms in human oral cancer cells are only scarce to date [36, 55]. This flavone was shown to reduce the viability of SCC25 cells by blocking the cell cycle at G2/M phase and inducing apoptosis [55]. In the same line, apigenin was also able to generate oxidative stress and reduce the cellular clonogenicity, whereas the apoptosis was triggered by upregulation of both tumor necrosis factor receptor (TNF-R) and TNF-related apoptosis-inducing ligand receptor (TRAIL-R) signaling pathways, decrease in Bcl-2 and increase in Bax expression and activation of caspase-3 [74]. Arrest of the cell cycle at G2/M phase following to exposure to apigenin was also observed in SCC9 cells [107]. In other human oral cancer cell lines (CAL27, SCC15, FaDu) apigenin suppressed the survival and induced apoptosis, with a substantial impairment of EGFR/ErbB2 downstream signaling in FaDu cells. Overexpression and a crucial role of these epidermal growth factor receptor (EGFR) family members have been demonstrated in oral cancer genesis and progression [108]. The cell growth inhibitory activities of apigenin were reported also in PCI-13 and PCI-15B oral cancer cells [109]. Furthermore, treatment of SCC4 cells with apigenin led to a substantial inhibition of invasion and migration via decrease in the activity of MMP-2 and u-PA and reduction of the levels of tissue inhibitor of metalloproteinase-2 (TIMP-2) and plasminogen activator inhibitor-1 (PAI-1) [110] (Table 4.3). Such properties of apigenin could be exploited for treatment of oral cancer in the future.

Another natural flavone, baicalein, induced both apoptosis as well as autophagy in CAL27 cells with an important role of ROS generation in autophagic cell death. Pharmacological blocking of autophagic response sensitized the cells to baicalein-triggered apoptosis displaying a cytoprotective role of autophagy in CAL27 cells [50]. Apoptotic death was observed also in SCC4 cells after exposure to baicalein with elevation of Bax, cytochrome c, caspase-3 and caspase-9 levels, increase in cytoplasmic Ca^{2+}, reduction of Bcl-2 and production of ROS [111]. In the same line, baicalein also inhibited invasive and migratory capabilities of cells via suppression of MMP-2 and u-PA activities and decrease in TIMP-2 and PAI-1 levels [110]. In SCC25 cells, baicalein exhibited growth inhibition through the arrest of cell cycle at G2/M phase and a slight increase in apoptotic death [22]. However, in HSC3 cells, baicalein blocked the cell cycle at G1 phase via stimulation of GSK-3ß pathway, activation of aryl hydrocarbon receptor (AhR) and reduction of retinoblastoma (Rb) [112] (Table 4.3). These data indicate that besides the preventive action baicalein could also have some potential to treat oral cancer and should therefore be considered for further clinical trials.

Luteolin reduced the survival of OC2 and SCC4 oral cancer cells with no effect on human normal gingival fibroblasts [66]. In SCC4 cells, this flavone caused the G1 phase cell cycle arrest and induced mitochondrial apoptosis with increase in the level of proapoptotic Bax, decrease in antiapoptotic Bcl-2 and activation of caspases-9 and -3 [66]. Apoptotic death was observed also in several other human oral cancer cell lines treated with luteolin (686LN, Tu686, Tu212), whereas a synergistic effect with EGCG was reported in Tu212 cells [104]. Moreover, in Tu212 cells, luteolin suppressed also colony formation [113]. In SCC25 line, luteolin exhibited cytotoxic activity through a DNA damage pathway and induced a significant reduction of migratory potential [65]. Furthermore, luteolin was able to effectively suppress the proliferation and self-renewal ability as well as stemness markers of oral cancer stem cells derived from SAS and GNM oral squamous carcinoma cells by inactivating IL-6/STAT3 signaling pathway [64] (Table 4.3). Thus, luteolin can represent an important candidate for clinical application by eliminating the potential roots of oral cancer.

Current data about the cytotoxicity of other flavones toward different human oral cancer systems are even scantier. 7-Hydroxyflavone was shown to inhibit invasion and migration of SCC4 cells by reducing MMP-2 and u-PA activities and increasing expression of TIMP-2 and PAI-1 [110]. 7,8-Dihydroxyflavone was described as an antiproliferative and proapoptotic compound in HN22 and HSC4 cells [68]. In SCC9 cells, chrysin was about tenfold less potent antiproliferative agent than its methylated analog, 5,7-dimethoxyflavone, and the situation was similar considering apigenin and its methylated analog, 4`,5,7-trimethoxyflavone. While these fully methylated flavones caused G1 phase cell cycle arrest, their unmethylated derivatives induced G2/M phase blockage in SCC9 cells [107]. However, in FaDu cells, chrysin and 5,7-dimethoxyflavone suppressed the proliferation with rather similar efficacies [107]. Fully methylated chrysin and apigenin appeared to be the two most potent methylated flavones in SCC9 cells as other analogs (7-methoxyflavone, 7,8-dimethoxyflavone, 3`,5-dimethoxyflavone, 4`,5-dimethoxyflavone, 4`,7-dimethoxyflavone and polymethoxylated tangeretin) revealed lower antiproliferative effects [107]. Moreover, chemoprotectant action of 5,7-dimethoxyflavone can come also from its direct inhibition of cytochrome P450 enzymes (CYP1B1/1A1) in SCC9 cells leading to a decreased formation of tobacco-related carcinogen-DNA adducts and prevention of oral carcinogenesis [114]. In addition, a monomethylated flavone, acacetin (4`-methoxy-5,7-dihydroxyflavone or 4`-methoxyapigenin) was found to induce apoptosis in HSC3 cells with activation of caspases-3, -8 and -9, increase in the Bax:Bcl-2 ratio and release of cytochrome c into the cytoplasm, mediated by MAPK signaling cascades [63] (Table 4.3). However, the role of methylation on chemopreventive action of flavonoids is still rather poorly understood so far and certainly needs further and more thorough studies.

Likewise, the effect of glycosylated moiety of flavonoids on their anticancer activities against oral tumorigenesis is not well known. Scutellarin, containing a glucuronide group in its 7-position, significantly inhibited HSC4 and SAS cells proliferation and motility. The expression of E-cadherin and desmoplakin increased while the level of αvß6 integrin decreased after treatment with scutellarin, regulating cell adhesion and resulting in a reduced metastatic potential [52]. In SAS cells, scutellarin also decreased MMP-2 and MMP-9 expression [57]. Vitexin, an analog of apigenin with a glycoside moiety in its 8-position, caused a p53-dependent apoptosis and exerted antimetastatic potential in OC2 cells, accompanied with increased PAI-1 and decreased MMP-2 levels [30] (Table 4.3). These data demonstrate that several intact flavones but also their methylated or glycosylated derivatives could possess some potential as chemopreventive and/or therapeutic agents against oral cancer.

4.3.1.3. Flavonols

The most widely studied flavonol regarding to the antitumor properties against oral cancer is quercetin. This natural agent targets different proteins involved in major cancer hallmarks, such as proliferation, apoptosis and metastasis [31]. Quercetin efficiently decreased cell viability and induced mitochondria-dependent apoptosis through downregulation of the PI3K/Akt/IKK-α/NF-κB signaling pathway in both low metastatic salivary adenoid cystic carcinoma line ACC-2 as well as high metastatic line ACC-M. The respective cell death was described by increased Bax:Bcl-2 ratio, stimulation of caspases-3 and -9, cleavage of poly(ADP-ribose) polymerase (PARP) and release of cytochrome c [9]. Increase in the apoptotic cell population in response to quercetin treatment was observed also in other oral cancer cell lines, such as SCC25, SCC1483 and SCC-QLL1 [115]. Moreover, in HSC2, HSC3 and HSG oral cancer cells, the cytotoxic effect of quercetin was significantly more potent than in the normal human cells (gingival fibroblasts, pulp cells, and periodontal ligament fibroblasts) [116]. In EGFR-overexpressing HSC3 cells, quercetin induced factor associated suicide ligand (FasL)-mediated extrinsic apoptosis through inhibition of EGFR/PI3K/Akt signaling pathway with concomitant activation of fork-head box O1 (FOXO1). It is well known that invasive oral squamous cell carcinomas commonly overexpress EGFR being related to unfavorable clinical outcome and poor prognosis. Moreover, exposure to quercetin was followed by arrest of HSC3 cells in G2/M phase of cell cycle [117]. Apoptosis and elevation of cell numbers in G2/M phase upon treatment with quercetin was found also in TW206 cells [117]. On the contrary, in SCC25 cells, quercetin induced cell cycle blockage at G1 phase, caused mitochondria-mediated apoptosis characterized by increased Bax:Bcl-2 ratio, altered PARP activity and stimulation of caspase-3, and effectively suppressed migratory and invasive abilities [51]. The G1 phase cell cycle arrest was found also in KB/VCR cells in response to quercetin, besides induction of mitochondria-mediated apoptosis and suppression of invasion and migration [118]. The antimetastatic action of quercetin was more thoroughly investigated in SAS cells revealing that this flavonol downregulated MAPK and PI3K/AKT signaling pathways resulting in suppression of the expression and activity of MMP-2 and MMP-9 [33]. However, in SCC9 cells, necrotic shock was initially reported (at 24 and 48 h) followed by later apoptotic cell death (72 h) after treatment with quercetin, whereas the cell cycle was mainly blocked in S phase [76] (Table 4.3). All these *in vitro* data show that specific antitumor effects of quercetin in oral cancer cells considerably depend on certain cell line and may vary to a rather large extent. Nevertheless, quercetin can be considered as a potential agent for prevention and treatment of different oral cancers. Furthermore, some growth inhibitory activity was shown also for quercetin glycosides. The antiproliferative activity of spiraeoside, the 4`-glycoside of quercetin, was probably related to its hydrolysis to quercetin in SCC9 cells. On the contrary, quercitrin, the 3-rhamnoside of quercetin, displayed only minimal cellular uptake and almost no hydrolysis, exerting no effect on growth of SCC9 cells. Thus, the nature of glycoside moiety may have an important role in anticancer effects of flavonoid sugar conjugates [81].

Another common natural flavonol kaempferol was also shown to exert some inhibitory effects on survival of different human oral cancer cells, such as FaDu [109], and induced apoptosis in SCC25, SCC1483 and SCC-QLL1 cells [115]. In SCC4 cells, this flavonol exhibited no cytotoxicity but suppressed migration and invasion by downregulating expression and enzymatic activity of MMP-2 through inhibition of the ERK1/2 signaling pathway [67] (Table 4.3).

Current knowledge about the potential antitumor activity of other flavonols against oral cancer is still rather limited. Fisetin was shown to promote ROS production and Ca^{2+} release, induce cell death through endoplasmic reticulum (ER) stress, mitochondria-dependent and death receptor-mediated pathways, and arrest cell cycle at G2/M phase in SCC4 cells [8]. Morin blocked cell cycle at G2/M phase but displayed no evidence of apoptosis in primary culture of human oral squamous cell carcinoma (stage 4). These growth inhibitory effects of morin revealed tumor selectivity being significantly less effective in normal oral mucosa cells. Interestingly, other flavonols, such as quercetin, galangin and datiscetin exhibited no tumor selectivity in this *in vitro* cellular system [119]. No signs of apoptosis were detected also after treatment of SCC25 cells with myricetin, whereas the inhibitory effect on cell proliferation was predominantly resulted from the blockage of cell cycle at G0/G1 phase. In addition, myricetin suppressed invasive and motile ability of SCC25 cells [3] (Table 4.3). Altogether, these results indicate that several flavonols could have some therapeutic potential for clinical management of oral malignancies in the future.

4.3.1.4. Isoflavones

Isoflavones are widely occurring in soybeans with genistein as the most studied compound against development of oral cancer. This isoflavone was found to exert significant antiproliferative action in different human oral cancer cell lines, including SSC9, HSC3 and KB [82, 120]. In HSC3 and KB cells, genistein induced apoptotic death

[120]. In HN4 cells, genistein inhibited proliferation by triggering apoptosis and arresting the cell cycle at S and G2/M phases [54]. In SCC25 cells, genistein suppressed cell growth via G2/M phase cell cycle arrest, but did not trigger apoptosis; decrease in expression of proliferating cell nuclear antigen (PCNA) as an indicator of cellular proliferative potential and inhibition of prostaglandin E2 (PGE2) synthesis by direct suppression of cyclooxygenase-2 (COX-2) activity were observed. Oral cancer patients have been shown to possess higher serum levels of PGE2 than healthy volunteers facilitating the growth of malignant cells [22]. Both genistein and another isoflavone, biochanin A suppressed cell growth in SCC15 and SCC25 lines; however, the effect of daidzein was significantly less potent [121]. Moreover, the antiproliferative activity detected with treatment of SCC9 cells with genistin, the 7-glycoside of genistein, could be practically attributed to genistein due to rapid cellular uptake of this glycoside and efficient hydrolysis to parent genistein [81] (Table 4.3). These preliminary evidences demonstrate that genistein can be a potential agent for prevention of oral carcinogenesis.

4.3.1.5. Other Flavonoids and Related Compounds

There are only few *in vitro* reports associating other natural flavonoids or flavonoid-related agents with oral cancer development (Table 4.3). A citrus flavanone naringenin was demonstrated to suppress SCC25 oral cancer cell growth related to blockage of cell cycle at G0/G1 phase but triggering no apoptosis. Naringenin reduced also cellular motility and invasiveness highlighting its potential as an antimetastatic agent [3].

Silibinin is the most well known flavanolignan and one of the active ingredients of silymarin. This compound was found to markedly decrease invasion and motility of SCC4 cells without any considerable cytotoxicity. The antimigratory effect was regulated by suppression of expression and activity of MMP-2 and u-PA and enhancement of TIMP-2 and PAI-1 levels through inhibition of ERK1/2 signaling pathway [10]. In KB cells, this flavanolignan behaved as a cytotoxic agent with triggering of intracellular ROS production and induction of apoptotic cell death [122].

Prenylflavone icaritin served as an antiproliferative agent in KB and SCC9 cells and induced mitochondria-dependent apoptosis by increasing the expression of miR-124 through Sp1/DNMT1 signaling [123]. In SCC9 and CAL27 cells, icaritin activated besides apoptosis also autophagic cell death via inhibition of JAK2/STAT3 signaling pathway [18]. Another prenylated flavone, cudraflavone B, suppressed proliferation of both primary (HN4) as well as metastatic oral squamous cell carcinoma cells (HN12) by induction of apoptosis in mitochondria-dependent manner through activation of MAPK/NF-κB and SIRT1 pathways. At that, no cytotoxicity was observed in normal human oral keratinocytes [124]. Thus, prenylflavones possess several features for successful development to chemotherapeutic agents against oral cancer.

Rotenoids can also serve as potentially attractive leads for oral cancer treatment. Low-doses of deguelin significantly suppressed metastasis-related abilities of several oral cancer cell lines, such as Ca9-22, HSC3 and SAS. In SAS cells, suppression of invasion and migration was mediated by inhibition of TNF-α-induced NF-κB signaling and downregulation of MMP-2 expression. It is well known that elevated serum and salivary levels of TNF-α are associated with higher histological grading of oral cancer [2]. Therefore, low doses of deguelin could be clinically applied for prevention of metastasis of oral tumors. Deguelin can also inhibit Akt signaling in oral cancer cells (A253, SCC9) leading to depletion of heat shock protein-90 (Hsp-90) client proteins [77]. Another rotenoid, rotenone, suppressed the proliferation of CAL27, SAS and Tw2.6 oral cancer cells revealing significantly less potency towards normal oral mucosal fibroblasts. Such differential action in malignant and normal oral cells can be due to an elevated oxidative stress in cancer cells as compared to the normal counterparts. In SAS cells, exposure to rotenone resulted in production of intracellular ROS, induction of mitochondrial and death receptor-mediated apoptosis and blockage of cell cycle at G2/M phase [125] (Table 4.3). As rotenoids inhibited the survival of different oral cancer cells already at low micromolar doses revealing also some tumor selectivity, these compounds might be potential leads for the development of further candidates for oral cancer treatment.

Chalcones are important precursors of flavonoids but current knowledge about their potential antitumor effects against human oral carcinogenesis is still rather poor. Two chalcones, isoliquiritigenin and butein, displayed significantly higher cytotoxic activities against several human oral cancer cell lines (HSC2, HSC3, HSG) than human normal cells (gingival fibroblasts, pulp cells, periodontal ligament fibroblasts) [116]. Isoliquiritigenin, occurring mostly in roots of licorice, was shown to induce G2/M phase cell cycle arrest, apoptosis and DNA damage in different oral cancer cells (Ca9-22, HSC3, OEC-M1, SAS and SCC4), while normal human oral cells were much less sensitive to this natural compound, probably because of the weak DNA repair system in oral malignant cells. Furthermore, isoliquiritigenin suppressed migration, inhibited MMP-2 and MMP-9 expression and reduced colony formation in highly malignant HSC3 and SAS cells [13]. In Tca8113 cells, isoliquiritigenin suppressed cellular proliferation, adhesion, migration and invasion through inhibition of MMP-2 and MMP-9 activities, and promoted apoptosis by upregulating Bax and downregulating Bcl-2 protein level. At lower doses (up to about 50 μM), this licorice chalcone induced cell accumulation in S phase while at higher doses it arrested the cell cycle at G1 phase [126]. Isoliquiritigenin 2`-methyl ether behaved also as an antiproliferative agent in HN4 and HN12 cells and induced apoptosis through heme oxygenase-1 (HO-1)-associated pathway involving MAP kinases, NF-κB and Nrf2 [127]. Another chalcone, licochalcone A inhibited the growth and induced apoptosis in HN22 and HSC4 oral cancer cells via suppression of specificity protein 1 (Sp1) expression and regulation of its downstream proteins [14]. Apoptosis was shown to be the cell death form also in KB cells upon treatment with licochalcone A, whereas an extrinsic pathway through upregulation of FasL and activation of caspases-3 and -8 was triggered. At that, no effect on normal human oral keratinocytes was observed [45]. In SCC25 cells, licochalcone A induced apoptosis through both death receptor-mediated and mitochondrial pathways and blocked the cell cycle progression at S and G2/M phases [37]. Moreover, treatment with licochalcone A led to a significant inhibition of adhesion, migration and invasion of SCC25 cells by reduction of MMP-2 activity, elevation of TIMP-2 level, increase in epithelial marker E-cadherin and decrease in mesenchymal marker N-cadherin, showing a clear antimetastatic potential of this natural compound [48]. Also, licochalcone B triggered both mitochondrial and death receptor-dependent apoptosis in HN22 and HSC4 cells and promoted ROS production, but arrested the cell cycle at G1 phase [12]. Sappanchalcone decreased the growth of primary (HN4) and metastatic oral squamous cell carcinoma cells (HN12) due to rapid induction of mitochondrial apoptosis through activation of p38, ERK, JNK and NF-κB signaling [78]. The exposure of HSC3 cells to flavokawain B resulted in blockage of cell cycle in G2/M phase and induction of apoptosis accompanied by mitochondrial dysfunction, death receptor activation and ER stress. Production of intracellular ROS, disruption of PI3K/Akt and p38 MAPK signaling pathways played an important role in these mechanisms [24]. Moreover, treatment of KB cells with flavokawain B also triggered apoptosis through both intrinsic and extrinsic pathways and arrested the cell cycle at G2/M phase, while exhibiting some cytotoxicity only at remarkably higher doses in normal human gingival fibroblasts. In KB cells, this chalcone decreased also MMP-9 activity and u-PA expression indicating a potential antimetastatic potential [128]. One more chalcone derivative studied so far in human oral cancer cell lines (HSC3, SAS7, SCC4), 3,3`,4`,5`-tetramethoxychalcone, behaved as an antiproliferative agent, induced G2/M phase cell cycle arrest and induced mitochondria-dependent apoptosis through generation of DNA double-strand breaks (DSBs) in SCC4 cells. No remarkable inhibition was detected in the growth of normal oral fibroblasts. In addition, this chalcone derivative also decreased the migration potential of SCC4 cells by impairing the C-C chemokine ligand 5 (CCL5)/ C-C chemokine receptor type 5 (CCR5) axis [46] (Table 4.3). Thus, chalcones exert multiple anticancer effects in different human oral cancer cell lines and are good candidates for development as chemopreventive and/or chemotherapeutic agents against oral cancer.

Altogether, based on numerous *in vitro* experiments, flavonoids have various functions in human oral cancer cells by inducing apoptosis via different cellular pathways and/or arresting the cell cycle progression at different phases. It is well known that the major cause of cancer-related death is due to the spread of malignant cells from primary site to locoregional and distant organs. Inhibition of migration, adhesion and invasion of oral tumor cells by several flavonoids can therefore be crucial in suppression of metastasis and improvement of prognosis of patients. At that, the molecular action of certain flavonoids depends largely not only on their structure but also on the specific oral cancer cell line. On one hand, this indicates that flavonoids possess multiple targets in malignant oral cells. On the other hand, variety of molecular alterations following to treatment with these polyphenols also suggests heterogeneity of this tumor. Furthermore, a lot of published data reveal that differently from chemotherapeutic drugs currently in clinical use, flavonoids can act as potent anticancer agents with considerably less toxicity or even some protective effects to normal healthy cells. It is clear that flavonoids possess potential use as chemopreventive and/or chemotherapeutic compounds for treating different human oral cancer types in the future. However, further studies on molecular mechanisms as well as tumor cell specificity are critical before testing their possible clinical application. At that, it would be interesting to clarify the role of HPV infection in oral cancer cells as one of the major risk factor for oral tumorigenesis to determine whether the antitumor action of flavonoids can depend also on certain HPV subtypes possibly providing further HPV subtype-specific tools for prevention or treatment of oral malignancies.

4.3.2. *In Vivo* Results

As the *in vitro* experiments do not consider homeostatic situation characteristic for animals, it is very important to verify the promising *in vitro* results observed with various flavonoids also in *in vivo* conditions. However, compared to *in vitro* studies only limited number of *in vivo* experiments has been performed to date, carried out by using different animal models of oral tumorigenesis. The concise overview of these works is presented in Table 4.4.

Table 4.4. *In vivo* effects of flavonoids in animal models of oral cancer

Compound	Biological system[1]	Administration route and amount flavonoids	Assay time	Biological effects[2]	Ref.
CHALCONES					
2-Hydroxychalcone	4NQO-induced rat oral tumorigenesis	In diet, 500 ppm	32 weeks	Decrease in incidence of tongue carcinoma, preneoplastic lesions and frequency of severe dysplasia in initiation and postinitiation phases, reduction of proliferation markers (BrdUrd labeling index and polyamine levels)	[38]
Chalcone	4NQO-induced rat oral tumorigenesis	In diet, 500 ppm	32 weeks	Decrease in incidence of tongue carcinoma, preneoplastic lesions and frequency of severe dysplasia in initiation and postinitiation phases, reduction of proliferation markers (BrdUrd labeling index and polyamine levels)	[38]
Flavokawain B	KB cells injected into right flank of mice	Intraperitoneal injection, 0.75 mg/kg	27 days	Decrease in tumor volume and incidence, inhibition of proliferation and angiogenesis, increase in apoptosis. No signs of toxicity	[128]
Isoliquiritigenin	HSC3 cells injected into mice	Oral gavage, 5 mg/kg	26 days	Decrease in tumor size and metastasizing	[13]
FLAVANOLS					
Catechin (C)	DMN-OAC-induced tumors in right cheek pouch of hamsters	Drinking water, 1.7 mg/animal/day/15 ml	13 months	Decrease in tumor incidence and delay in tumor formation, increase in number of survivors	[129]
Epigallocatechin gallate (EGCG)	SCC9 cells injected into right axilla of mice	Oral gavage, 10 and 20 mg/kg/day	45 days	Inhibition of tumor growth, reduction of tumor size and weight. No signs of toxicity	[49]
	SCC VII/SF cells injected into flank of mice	Intraperitoneal injection, 25, 50 or 75 mg/kg/day	22 days	Inhibition of tumor growth, increase in apoptosis	[4]
	Tu212 cells injected into right flank of mice	Oral gavage, 125 mg/kg	47 days	Minimal effect on tumor growth as single agent. Inhibition of tumor development in combination with luteolin (10 mg/kg)	[104]
	Tu212 cells injected into right flank of mice	Oral gavage, 125 mg/kg	8 weeks	Decrease in tumor growth	[137]
	YD-10B cells injected into tongue of mice	Intraperitoneal injection, 20 mg/kg	4 weeks	Inhibition of tumor volume and regional infiltration, suppression of CTTN, FAK, MT1-MMP, MMP-2, -9 expression	[102]

Compound	Biological system[1]	Administration route and amount flavonoids	Assay time	Biological effects[2]	Ref.
FLAVANONES					
Hesperetin	DMBA-induced hamster oral tumorigenesis	Topic application in 2.5% solution		Insignificant reduction of tumor burden	[85]
	DMBA-induced hamster oral tumorigenesis	Oral gavage, 40 mg/kg	16 weeks	Decrease in metabolic activity	[61]
Hesperidin	4NQO-induced oral tumorigenesis in rats	By diet, 0.5 g/kg, i.e., 500 ppm	32 weeks	Decrease in incidence of tongue carcinoma in initiation, not postinitiation phase, decrease in hyperplasia and dysplasia in initiation and postinitiation phases, inhibition of proliferation biomarkers (BrdUrd-labeling index and polyamine levels). No signs of toxicity	[29]
	4NQO-induced oral tumorigenesis in rats	By diet, 1000 ppm	32 weeks	Decrease in incidence of tongue neoplasms and preneoplasms in initiation and postinitiation phases, inhibition of proliferation biomarkers (BrdUrd-labeling index and polyamine levels). No synergism in combination with diosmin (900 ppm diosmin and 100 ppm hesperidin). No signs of toxicity	[86]
Naringenin	DMBA-induced hamster oral tumorigenesis	Topic application in 2.5% solution		Reduction of tumor number	[85]
	DMBA-induced hamster oral tumorigenesis	Per-oral, 50 mg/kg/day	16 weeks	Decrease in tumor incidence and severity of preneoplastic lesions, restoration of lipid peroxidation and antioxidants status, decrease in PCNA and p53 expression	[27]
	DMBA-induced hamster buccal pouch tumorigenesis	Per-oral, 50 mg/kg	16 weeks	Prevention of cancer formation, increase in phospholipids, decrease in tryptophan, phenylalanine and nucleic acid contents	[28]
	DMBA-induced hamster buccal pouch tumorigenesis	Per-oral, 50 mg/kg/day	16 weeks	Increase in lipids and glycogen content, decrease in protein levels and nucleic acid contents	[25]
Naringin	DMBA-induced hamster oral tumorigenesis	Topic application in 2.5% solution		Reduction of tumor number, 70% reduction of tumor burden	[85]

Table 4.4. (Continued)

Compound	Biological system[1]	Administration route and amount flavonoids	Assay time	Biological effects[2]	Ref.
Neohesperidin	DMBA-induced hamster oral tumorigenesis	Topic application in 2.5% solution		Insignificant reduction of tumor burden	[85]
FLAVONES					
Apigenin	DMBA-induced hamster oral tumorigenesis	Subcutaneously, 0.16 mg/day (potassium apigenin)	11 weeks	Inhibition of tumor initiation, incidence and volume, decrease in aggressive behavior	[41]
	DMBA-induced hamster oral tumorigenesis	Per-oral, 2.5 mg/kg/day	16 weeks	Reduction of tumor incidence, tumor volume and burden, reversion of the status of lipid peroxidation, antioxidants, phase I and II detoxification agents to near normal range	[26]
	DMBA-induced hamster oral tumorigenesis	Per-oral, 2.5 mg/kg/day	16 weeks	Reduction of tumor incidence, tumor volume and burden, decrease in p53, PCNA, VEGF, COX-2, NF-κB expression, increase in caspases-3 and -9	[17]
	FaDu cells injected into left flank of mice	By gavage, 50 μg/day	28 days	**Increase in tumor burden and growth**	[130]
	FaDu cells injected into left flank of mice	By gavage, 50 μg/day	28 days	**Increase in tumor size, volume and weight**	[109]
Diosmin	DMBA-induced hamster buccal pouch tumorigenesis	Per-oral, 100 mg/kg	14 weeks	Decrease in neoplastic changes, suppression of development of carcinomas, antioxidant activity	[32]
	DMBA-induced hamster buccal pouch tumorigenesis	Per-oral, 100 mg/kg	14 weeks	Reduction of tumor incidence, inhibition of JAK-1/STAT3 signaling, induction of apoptosis, prevention of abnormal expression of VEGF and HIF-1α	[131]
	4NQO-induced oral tumorigenesis in rats	By diet, 1000 ppm	32 weeks	Decrease in incidence of tongue neoplasms, hyperplasia and dysplasia in initiation and postinitiation phases, reduction of proliferation markers (BrdUrd labeling index and polyamine levels). No synergism in combination with hesperidin (900 ppm diosmin and 100 ppm hesperidin)	[86]

Compound	Biological system[1]	Administration route and amount flavonoids	Assay time	Biological effects[2]	Ref.
Luteolin	SCC4 cells injected into right axilla of mice	Intraperitoneal injection, 3 or 5 mg/kg	44 days	Reduction in tumor growth	[66]
	Tu212 cells injected into right flank of mice	Intraperitoneal injection, 3.3 mg/kg	30 days	No significant effect on tumor growth	[113]
	Tu212 cells injected into right flank of mice	Oral gavage, 10 mg/kg	47 days	Minimal effect on tumor growth as single agent. Inhibition of tumor development in combination with EGCG (125 mg/kg)	[104]
Nobiletin	DMBA-induced hamster oral tumorigenesis	Topic application in 2% solution		Insignificant reduction of tumor burden	[85]
Scutellarin	SAS cells injected into left flank of mice	Oral gavage, 5, 10 or 20 mg/kg	3 weeks	Inhibition of tumor growth, reduction of proliferative activity, induction of apoptosis, suppression of invasion and metastasis by decrease in MMP-2 and -9 levels, decrease in integrin αvß6 expression. No signs of toxicity	[57]
Tangeretin	DMBA-induced hamster oral tumorigenesis	Topic application in 2% solution		Insignificant reduction of tumor burden	[85]
FLAVONOLS					
Kaempferol	FaDu cells injected into left flank of mice	By gavage, 50 µg/day	28 days	**Increase in tumor burden and growth**	[130]
	FaDu cells injected into left flank of mice	By gavage, 50 µg/day	28 days	**Increase in tumor size, insignificant increase in tumor weight**	[109]
Morin	4NQO-induced tongue tumorigenesis in rats	Per-oral, 100 or 500 ppm	32 weeks	Suppression of carcinogenesis, reduction of incidence of tongue carcinoma, decrease in quinone reductase activity in the posterior tongue, increase in GST activity, suppression of cell proliferation. No signs of toxicity	[134]

Table 4.4. (Continued)

Compound	Biological system[1]	Administration route and amount flavonoids	Assay time	Biological effects[2]	Ref.
Quercetin	DMBA-induced hamster buccal pouch tumorigenesis	Per-oral, 2.0%	16 weeks	Decrease in papilloma and tumor incidence	[132]
	DMBA-induced hamster buccal pouch tumorigenesis	Intragastric, 25 mg/kg	18 weeks	Decrease in tumor burden and incidence of preneoplastic and neoplastic lesions in simultaneous treatment, delay of tumor growth in posttreatment, modulation of proliferation, cell survival, apoptosis, invasion and angiogenesis markers	[84]
	DMBA-induced hamster buccal pouch tumorigenesis	Intagastric, 25 mg/kg	18 weeks	Decrease in tumor burden and incidence of preneoplastic and neoplastic lesions in simultaneous treatment, attenuation of CYP-mediated ROS generation, abrogation of ROS-induced activation of oncogenic NF-κB signaling circuit	[133]
	4NQO-induced oral tumorigenesis in rats	In diet, 500 ppm	32 weeks	Decrease in incidence of tongue carcinoma, preneoplastic lesions and frequency of severe dysplasia in initiation and postinitiation phases, reduction of proliferation markers (BrdUrd labeling index and polyamine levels)	[38]
	ACC-M cells injected into mice	Per-oral, 1 g/kg/day	32 days	Inhibition of tumor growth, induction of apoptosis, suppression of Akt and IKK-α activation	[9]
	HSC3 cells injected into flank of mice	By gavage, 3 mg/kg/day	35 days	Decrease in tumor size and weight, inhibition of tumor growth, induction of FOXO1 expression	[117]

Compound	Biological system[1]	Administration route and amount flavonoids	Assay time	Biological effects[2]	Ref.
ISOFLAVONES					
Genistein	DMBA-induced hamster buccal pouch tumorigenesis	By gavage, 10 mg/kg/day	18 weeks	**No significant effect on tumor incidence, volume, multiplicity and latency, dysplastic lesions or vessel density. Promoting oral submucosa stroma tumorigenesis, induction of poorly-differentiated fibrosarcomas**	[60]
	HSC3 cells injected into right flank of mice	Intraperitoneal injection, 1 mg/kg	30 days	No significant growth inhibition, some decrease in microvessel density	[120]
	KB cells injected into right flank of mice	Intraperitoneal injection, 1 mg/kg	30 days	No significant growth inhibition, some decrease in microvessel density	[120]
PRENYLATED FLAVONOIDS					
Icaritin	4NQO-induced oral tumorigenesis in mice	Intraperitoneal injection, 10 mg/kg	8 weeks	Decrease in size of lesions, reduction of the number of malignant tumors and number of tumor lesions. No signs of toxicity	[18]
ROTENOIDS					
Deguelin	SAS cells injected into anterior tongue of mice	0.8 or 4 mg/kg	12 days	Decrease in tumor size, reduction of lymph node metastatic tumors, increase in survival rate	[2]

[1]see Tables 4.1 and 4.2.

[2]Akt, protein kinase B; BrdUrd, bromodeoxyuridine; COX2, cyclooxygenase-2; CTTN, cortactin; CYP, cytochrome P450; FAK, focal adhesion kinase; FOXO1, fork-head box O1; GST, glutathione S-transferase; HIF-1α, hypoxia-inducible factor-1α; IKK, IκB kinase; JAK, Janus kinase; MMP, matrix metalloproteinase; MT1-MMP, membrane type 1-MMP; NF-κB, nuclear factor κB; PCNA, proliferating cell nuclear antigen; ROS, reactive oxygen species; STAT3, signal transducer and activator of transcription 3; VEGF, vascular endothelial growth factor.

4.3.2.1. Flavanols

The role of EGCG in oral carcinogenesis was investigated in xenograft nude mice subcutaneously injected with different human oral cancer cells (SCC9, Tu212, YD-10B). This green tea flavanol inhibited tumor growth and reduced tumor weight in SCC9 cells injected mice without displaying any obvious signs of toxicity [49]. In animals inoculated with YD-10B cells, intraperitoneal injection of EGCG led to a significant decrease in tumor volume and suppressed regional infiltration of tumor cells into nearby stroma by inhibiting functional invadopodia formation. These antiinvasive properties were associated with suppression of expression of Src, CTTN, FAK, MT1-MMP, MMP-2 and MMP-9 proteins in tumor tissue [102]. Although in Tu212-bearing nude mice, EGCG itself exerted a minimal effect on tumor growth, it significantly inhibited carcinogenesis in combination with luteolin without any notable evidence of toxicity [104]. Administration of EGCG to syngeneic C3H/HeJ mice injected with mice oral cancer cells SCC VII/SF also resulted in significant suppression of tumor growth and decrease in PCNA-positive cells. Moreover, the elevation of TUNEL-positive cells was an indicator for increased induction of apoptosis [4].

To date, the only *in vivo* work focused on potential effect of catechin on oral tumorigenesis was performed in methyl-(acetoxymethyl)-nitrosamine (DMN-OAC)-induced oral cancer model of Syrian golden hamsters. Administration of catechin in drinking water to hamsters significantly suppressed DMN-OAC-induced tumor burden and incidence in the cheek pouches of animals and delayed also the tumor formation [129] (Table 4.4).

4.3.2.2. Flavanones

Compared to a very small amount of *in vitro* anticancer studies performed with flavanones in human oral cancer cell lines, somewhat more results are available about their role in experimental animal systems. Most of these works have been performed with DMBA-induced hamster buccal pouch tumorigenesis model. Orally administrated naringenin was shown to prevent the tumor formation with occurrence of mild to moderate preneoplastic lesions; the status of antioxidants and lipid peroxidation in the buccal mucosa tissues was restored to near normal range. Also, the expression of PCNA and p53 as markers of promoted tumor cell growth decreased significantly in response to naringenin treatment [27]. Administration of naringenin improved the biochemical constituents to normal range by elevating lipids and their order, increasing glycogen contents and reducing levels of proteins and nucleic acid contents in DMBA-initiated tumor tissues [25]. In more detail, the levels of tryptophan and phenylalanine were suppressed, exerting altogether a significant chemopreventive effect of this citrus flavonoid [28]. However, whereas topic application of naringenin and its 7-neohesperidoside, naringin, reduced the tumor number in DMBA-painted hamster cheek pouch model, other flavanones, such as hesperetin and neohesperidin, caused no significant decrease in tumor burden [85]; although a reduction of metabolic activity upon oral treatment of hamsters with hesperetin was observed [61]. Feeding 500 ppm of hesperidin, the 7-rutinoside of hesperetin, to rats during initiation phase of 4NQO-induced oral tumorigenesis resulted in a significant decrease in the incidence of carcinoma, squamous cell hyperplasia and dysplasia in the tongue, probably due to suppression of cell proliferation as determined by inhibition of cell proliferation biomarker expression [29, 86]. Hesperidin at 1000 ppm suppressed tumorigenesis also when administrated during postinitiation phase, suggesting the presence of a possible threshold effect of this citrus flavonoid on oral cancer development. At that, no toxicity to the liver, kidney or lung of rats appeared [86] (Table 4.4).

4.3.2.3. Flavones

In DMBA-induced hamster buccal pouch tumorigenesis model, orally administrated apigenin (2.5 mg/kg body weight) significantly prevented tumor incidence and burden, probably due to suppression of abnormal cell proliferation. Moreover, apigenin also improved the status of antioxidant defense system and modulated the activities of phase I and phase II detoxification pathways toward elevated elimination of toxic DMBA metabolites, dihydrodiol epoxides [26]. In addition, apigenin exerted antiproliferative, antiinflammatory, antiangiogenic and proapoptotic potential in DMBA-treated hamsters as determined by downregulation of the expression of PCNA, c-fos, p53, Bcl-2, VEGF, COX-2, NF-κB and upregulation of Bax, caspases-3 and -9 [17]. Subcutaneous administration of water-soluble potassium apigenin along with DMBA to hamsters led to a significant suppression of tumor initiation, incidence and volume as compared to animals treated with DMBA alone [41]. These data clearly indicate that apigenin could possess chemopreventive potential against oral tumorigenesis induced by DMBA in hamsters. However, contrary to the suppression of chemical-initiated carcinogenesis in hamster buccal pouch and also to the *in vitro* decrease in the viability of FaDu cells, apigenin treatment significantly increased the tumor volume and weight in xenograft athymic mice injected with FaDu cells [109, 130]. Thus, *in vitro* results might not be always predictive of *in vivo* effects and must be considered with caution. Under certain conditions, tumor growth may be promoted rather than suppressed and the molecular mechanisms behind these events certainly need further unraveling.

In xenograft mice model subcutaneously injected with Tu212 cells, administration of luteolin revealed no significant effect on tumor growth [113]. However, combination of luteolin with EGCG led to a substantial inhibition of tumor growth [104]. Also, continuous administration of luteolin reduced the tumor growth in xenografted mice bearing SCC4 cells [66].

Polymethoxylated flavones tangeretin and nobiletin were almost inactive in DMBA-induced hamster buccal pouch tumorigenesis model [85]. However, flavone glycosides still revealed anticancer effects against oral tumorigenesis. Diosmin, the 7-rutinoside of diosmetin, reduced the neoplastic changes and formation of carcinomas in DMBA-painted hamsters while exerting antioxidant potential [32]. This dietary flavone inhibited IL-6/STAT3 signaling during DMBA-induced hamster buccal pouch tumorigenesis leading to induction of apoptotic cell death and suppression of angiogenesis by preventing the expression of VEGF and hypoxia-inducible factor-1α (HIF-1α) [131]. In 4NQO-initiated oral tumorigenesis model in rats, diosmin caused a substantial decrease in the frequency of tongue carcinoma, hyperplasia and dysplasia with a remarkable decrease in the expression of cell proliferation biomarkers in the tongue epithelium [86]. Another flavone glycoside, scutellarin, inhibited tumor growth in xenograft nude mice bearing SAS cells by reducing tumor cell proliferation, triggering apoptosis and downregulating invasion and metastasis. Indeed, scutellarin decreased the expression of MMP-2, MMP-9 and integrin αvß6 in tumors and increased the number of collagen fibers in the tumor microenvironment [57] (Table 4.4).

4.3.2.4. Flavonols

Consistent with several *in vitro* results, *in vivo* studies with experimental animals have also confirmed the chemopreventive and chemotherapeutic potential of quercetin against oral tumorigenesis. This abundantly occurring flavonol suppressed the incidence of papillomas and tumors in DMBA-induced hamster buccal pouch carcinogenesis [132]. The tumor burden and incidence of preneoplastic lesions were decreased with simultaneous administration of quercetin and DMBA, while posttreatment of quercetin led to a substantial tumor growth delay. Administration of quercetin induced apoptosis and suppressed angiogenesis and invasion with modulation of a plethora of markers involved in proliferation, cell survival, apoptosis, angiogenesis and invasion [84]. Moreover, quercetin attenuated CYP-mediated ROS production and abolished ROS-triggered activation of NF-κB signaling in DMBA-painted hamsters, thereby promoting defense against ROS-induced DNA and protein damage [133]. Treatment of 4NQO-induced oral tumorigenesis in rats during initiation or postinitiation phases with quercetin resulted in a significant decrease in the frequency of tongue carcinoma and preneoplastic lesions with reduction of cell proliferation markers, such as bromodeoxyuridine (BrdUrd) labeling index, in tongue epithelium. At that, no pathological alterations to other organs, such as lung, liver or kidney, were found [38]. Quercetin prevented tumor growth also in xenograft nude mice injected with ACC-M cells, associating with the induction of apoptosis and downregulation of Akt and IKK-α activation [9]. In addition, this dietary flavonol inhibited tumor growth, tumor size and weight in mice xenograft model with HSC3 cells [117]. These data clearly show that dietary quercetin can provide promise as an attractive candidate for oral cancer prevention and therapy.

In a rat model of 4NQO-induced carcinogenesis, another natural flavonol, morin, caused a significant decrease in the incidence of tongue carcinoma. Morin treatment modified detoxifying enzyme activities, including quinone reductase and glutathione S-transferase (GST) that provide defense against tumorigenesis events. As 4NQO requires metabolic activation by quinone reductase to reveal its carcinogenic potential, suppression of activity of this phase II enzyme in posterior tongue by morin could lead to a substantial decrease in carcinoma incidence. In addition, morin also suppressed cellular proliferation without any clinical signs of toxicity in major organs, such as liver, lungs, kidneys and heart [134].

However, treatment of FaDu engineered xenograft mice with kaempferol revealed quite opposite effect by increasing the tumor size and growth [109, 130]. Although kaempferol decreased FaDu cell viability in *in vitro* conditions, this action was not verified in *in vivo* assays, indicating the requirement of caution when transferring the *in vitro* results to *in vivo* systems. Under certain conditions, tumor growth may be promoted instead of expected suppression of malignant processes (Table 4.4).

4.3.2.5. Other Flavonoids and Related Compounds

Despite several *in vitro* studies showing antiproliferative activities of the most well known isoflavone, genistein, in human oral cancer cell lines no significant growth inhibition was found for this soy isoflavone in HSC3- and KB-tumor-bearing mice [120]. Likewise, genistein induced no significant effect on tumor incidence, volume, multiplicity, latency or vascular density also in DMBA-initiated hamster buccal pouch carcinogenesis. On the contrary, treatment with genistein promoted oral submucosa stroma tumorigenesis by inducing poorly differentiated fibrosarcomas [60]. Thus, differently from the compelling results obtained with cell line studies, *in vivo* assays indicate that soy foods with their major isoflavone component, genistein, may not be helpful in suppressing the cancerous transformations in oral cavity. Quite the contrary, caution must be taken before administration of genistein to the individuals diagnosed with oral premalignant lesions or those being under high pressure of risk factors, including exposure to tobacco products. However, it is still needed to count that there are many important differences between tumors in rodents and humans.

The prenylflavone icaritin suppressed 4NQO-induced oral carcinogenesis in mice by reducing the number and size of tumors and premalignant lesions while displaying no signs of adverse reactions [18]. Rotenoid deguelin also efficiently suppressed tumor size in SAS engineered xenograft mice, reduced lymph node metastatic tumors and prolonged survival of experimental animals [2]. Suppression of oral tumorigenesis *in vivo* has been described also for several chalcone derivatives. Isoliquiritigenin was found to reduce tumor size and metastasizing potential in xenograft mice injected with HSC3 cells [13]. Administration of flavokawain B significantly decreased tumor volume and incidence in KB cell-derived tumor xenografts in nude mice by reducing proliferation, inducing apoptosis and inhibiting angiogenesis with no signs of toxicity [128]. Chalcone and 2-hydroxychalcone suppressed 4NQO-initiated carcinogenesis in rats, decreased the frequency of tongue neoplasms and premalignant lesions while feeding the animals during either initiation or postinitiation phases [38] (Table 4.4).

In summary, the molecular mechanisms underlying anticancer action of flavonoids in experimental animal models are probably complex, whereas initiated signaling cascades might also be interconnected. Several animal studies have shown suppressing effects of dietary flavonoids and flavonoid related compounds in different oral tumorigenesis models of rodents. However, caution must still be taken when transferring *in vitro* results to *in vivo* conditions. So far, no *in vitro* studies have been published concerning promotion of malignant oral cell growth by flavonoids. On the contrary, some *in vivo* works can still be found reporting favoring effects of flavone apigenin, flavonol kaempferol and isoflavone genistein on oral tumorigenesis. It is possible that such activities of certain flavonoids might be specific to tumor subtypes, host organisms (mice, rats or hamsters) or transplanted cell lines. For instance, no tumorigenesis promoting effects of flavonoids have been described in either xenograft models or chemical-induced oral tumorigenesis in rats. The mechanisms underlying tumor growth favoring instead impairing activities certainly need further clarification. However, based on our current knowledge about suppressing action of flavonoids on oral tumorigenesis, these natural compounds might provide important novel chemopreventive and/or chemotherapeutic tools for further fighting against oral malignancies. At that, both the treatment time (initiation or postinitiation phase in the case of chemical-induced carcinogenesis) as well as administration route (topic, subcutaneous, oral, intraperitoneal or intragastric) might be critical. Considering no or very few adverse effects observed with flavonoids, high efficacy, capability for oral administration and low cost, these phytochemicals remain compelling candidates for development of future drugs highlighting the necessity of further clinical trials.

4.3.3. Co-Effects of Flavonoids and Chemotherapeutic Drugs on Oral Tumorigenesis

Dietary flavonoids can exert interesting preclinical effects if applied in combination with standard cancer treatment modalities. Indeed, these natural polyphenols have been shown to increase the action of conventional chemotherapeutic drugs and also to relieve

Table 4.5. Effect of flavonoids on conventional treatment modalities of oral cancer

Compound	Drug, treatment	Biological system[1]	Assay conditions[2]	Co-effect[3]	Ref.
FLAVANOLS					
Epigallocatechin gallate (EGCG)	5-Fluorouracil (5-FU)	YCU-H891 cells	WST-1, 48 h	Synergistic enhancement of cell growth inhibition, 45-fold increase in antiproliferative effects of 5-FU even at 0.22 μM EGCG, suppression of colony formation	[72]
	Cisplatin (CDDP)	OSC2 cells	MTT, 24 h EGCG followed by 48 h drug treatment	**Significant decrease in 5 μM CDDP-induced cytotoxicity by pretreatment with 25 and 50 μM EGCG (no additional effect at 100 or 200 μM EGCG), enhancement of p21/WAF1 levels**	[73]
	Cisplatin (CDDP)	OSC4 cells	MTT, 24 h EGCG followed by 48 h drug treatment	Additive decrease in cell viability by pretreatment with 200 μM EGCG	[73]
	Doxorubicin (DOX)	KB-A-1 cells	MTT, 72 h	Increase in growth inhibition of DOX-resistant cells, reversal of MDR, 2.2-fold enhancement of intracellular concentration of DOX	[135]
	Erlotinib	38 cells	SRB, 72 h	Synergistic inhibition of cell growth	[137]
	Erlotinib	1986LN cells		Synergistic growth inhibition, suppression of erlotinib-induced p21 and p27 expression	[59]
	Erlotinib	SQCCY1 cells	SRB, 72 h	Synergistic growth inhibition	[137]
	Erlotinib	SQCCY1 cells		Synergistic growth inhibition, suppression of erlotinib-induced p21 and p27 expression	[59]
	Erlotinib	Tu212 cells	SRB, 72 h	Synergistic growth inhibition, additive G0/G1 phase cell cycle arrest, increase in mitochondrial apoptosis, targeting EGFR/AKT pathway	[137]
	Erlotinib	Tu212 cells injected into right flank of mice	Oral gavage, 125 mg/kg EGCG, 50 mg/kg drug	Greater suppression of tumor growth compared to either agent alone, increase in inhibition of proliferation, induction of apoptosis	[137]
	Erlotinib	Tu686 cells		Synergistic growth inhibition, suppression of erlotinib-induced p21 and p27 expression, increase in drug-initiated apoptosis in wt p53 cells	[59]

Table 4.5. (Continued)

Compound	Drug, treatment	Biological system[1]	Assay conditions[2]	Co-effect[3]	Ref.
	Gefitinib	CAL27 cells		Synergistic inhibition of invasion and migration mediated through MAPK signaling, attenuation of MMP-2 expression and activity	[136]
	Taxol (paclitaxel)	YCU-H891 cells	WST-1, 72 h	Increase in inhibitory effect of 1.5 ng/ml Taxol even at 0.22 μM EGCG, no further enhancement of drug effect at higher EGCG doses (2.2 and 10.9 μM)	[91]
	γ-irradiation	OSC2 cells	MTT, 24 h EGCG followed by 48 h irradiation	**Significant increase in cell viability irradiated at 5 or 10 Gy with 12.5-50 μM EGCG pretreatment, enhancement of γ-ray-induced p21/WAF1 expression**	[73]
FLAVONES					
Apigenin	5-Fluorouracil (5-FU)	SCC25 cells	MTT, 72 h	Synergistic effect of 5-FU cytotoxicity with 5 μM apigenin	[74]
	Cisplatin (CDDP)	SCC25 cells	MTT, 72 h	Synergistic effect of CDDP cytotoxicity with 5 μM apigenin	[74]
Luteolin	Cisplatin (CDDP)	SCC25 cells	MTT, 72 h	Improvement of 5 and 10 μM CDDP cytotoxicity by 10 μM luteolin	[65]
	Paclitaxel	SCC4 cells	MTT, 4, 7 days	Enhancement of paclitaxel cytotoxicity at 5 and 10 μM luteolin	[66]
	Paclitaxel	SCC4 cells injected into right axilla of mice	Intraperitoneal injection, 3 mg/kg luteolin, 1 mg/kg drug	Synergistic reduction of tumor growth	[66]
	Radiation	OCSC from GNM	MTT, 24 h	Enhancement of cancer stem cells radiosensitivity, synergistic inhibition of invasion and colony formation, inactivation of IL-6/STAT3 signaling	[64]
	Radiation	OCSC from SAS	MTT, 24 h	Enhancement of cancer stem cells radiosensitivity, synergistic inhibition of invasion and colony formation, inactivation of IL-6/STAT3 signaling	[64]

Compound	Drug, treatment	Biological system[1]	Assay conditions[2]	Co-effect[3]	Ref.
Scutellarin	Low-intensity ultrasound	SAS cells	MTT, various times	Enhancement of scutellarin-caused cell damage, inhibition of cell growth and motility, induction of apoptosis	[71]
	Low-intensity ultrasound	SAS cells injected into back flank of mice	Oral gavage, 10 mg/kg	Decrease in tumor volume compared to scutellarin alone, increase in apoptosis, reduction of proliferation index, decrease in MMP-2, -9 expression	[71]
FLAVONOLS					
Myricetin	Doxorubicin (DOX)	SCC84 cells	MTT, 4 h myricetin followed by 16 h drug treatment	Resistance of HPV- cells to 2 μM DOX in absence or presence of myricetin (up to 12.5 μM)	[42]
	Doxorubicin (DOX)	UMSCC29 cells	MTT, 4 h myricetin followed by 16 h drug treatment	Resistance of HPV- cells to 2 μM DOX in absence or presence of myricetin (up to 12.5 μM)	[42]
	Doxorubicin (DOX)	UMSCC47 cells and UMSCC47-TC-Clone3	MTT, 4 h myricetin followed by 16 h drug treatment	Sensitization of HPV+ cells to DOX, increase in 2 μM DOX-induced cell death by myricetin	[42]
	Doxorubicin (DOX)	UPCI-SCC90-UP-Clone35	MTT, 4 h myricetin followed by 16 h drug treatment	Sensitization of HPV+ cells to DOX, increase in 2 μM DOX-induced cell death by myricetin	[42]
Quercetin	Cisplatin (CDDP)	DRSPs from SCC25	MTT, 48 h	Chemosensitization of DRSPs to 10 μM CDDP and reduction of growth by 100 μM quercetin, inhibition of invasion and migration, reversal of EMT-associated marker expression	[75]
	Cisplatin (CDDP)	SCC25 cells and DRSPs injected into mice		Increase in inhibition of tumor growth compared to CDDP alone, regulation of EMT-associated marker expression	[75]
	Vincristine (VCR)	KB/VCR cells	Click-rT Edu, 24 h	Enhancement of cell sensitivity to VCR, decrease in proliferation rate, increase in apoptosis, inhibition of invasion and migration, reversal of resistance by suppressing P-gp expression	[118]

Table 4.5. (Continued)

Compound	Drug, treatment	Biological system[1]	Assay conditions[2]	Co-effect[3]	Ref.
ISOFLAVONES					
Genistein	Cetuximab	HSC3 cells	SRB, 48 h	Increase in growth inhibition compared to single agents alone at 25 μg/ml drug and 5 μM genistein, reduction of EGFR expression, increase in apoptotic index	[120]
	Cetuximab	HSC3 cells injected into right flank of mice	Intraperitoneal injection, 1 mg/kg genistein, 1 mg/kg drug	Increase in tumor growth inhibition and substantial growth delay, decrease in proliferating cell number and microvessel density	[120]
	Cetuximab	KB cells	SRB, 48 h	Increase in growth inhibition compared to single agents alone at 50 and 100 μg/ml drug and 10 and 20 μM genistein, increase in apoptotic index	[120]
	Cetuximab	KB cells injected into right flank of mice	Intraperitoneal injection, 1 mg/kg genistein, 1 mg/kg drug	Increase in tumor growth inhibition and substantial growth delay, decrease in proliferating cell number and microvessel density	[120]

[1]see Tables 4.1 and 4.2.

[2]MTT, 3-(4,5-dimethylthiazol-2-yl)-2,5-diphenyltetrazolium bromide; SRB, sulforhodamine B; WST-1, water soluble tetrazolium-1.

[3]Akt, protein kinase B; EGFR, epidermal growth factor receptor; EMT, epithelial-mesenchymal transition; HPV, human papillomavirus, IL, interleukin; MAPK, mitogen-activated protein kinase; MDR, multidrug resistance; MMP, matrix metalloproteinase; P-gp, P-glycoprotein; STAT, signal transducer and activator of transcription; wt, wild type.

their adverse side effects. However, to date, only rather few studies have been performed focusing on the role of a limited number of most commonly occurring flavonoids in the efficacy of a small selection of therapeutic agents. The results of these studies are summarized in Table 4.5.

In most combination studies, co-effect of standard chemotherapeutics with EGCG has been investigated. This green tea flavanol potentiated the growth inhibitory action of 5-fluorouracil (5-FU) in YCU-H891 cells, even at low micromolar doses achievable in plasma after drinking of several cups of green tea or administering of an oral dose of EGCG. In fact, the half maximal inhibitory concentration (IC_{50}) of 5-FU decreased for 45-fold in the presence of 0.1 μg/ml of EGCG [72]. In a similar dose range (0.1-1.0 μg/ml), EGCG substantially increased the sensitivity of YCU-H891 cells to growth inhibitory effect of paclitaxel [91]. In doxorubicin (DOX)-resistant KB cells (KB-A-1), treatment with 10 μg/ml EGCG reduced the IC_{50} value of DOX by 1.9-fold, increased cellular uptake of this chemotherapeutic drug by 2.2-fold and reversed the multidrug resistance (MDR) [135]. In CAL27 cells, co-treatment of EGCG with gefitinib, an EGFR inhibitor, resulted in a synergistic suppression of the metastatic potential by targeting MAPK signaling and attenuating enzymatic activity and protein level of MMP-2. More precisely, 25 μM EGCG enhanced 10 μM gefitinib-initiated invasion and migration abilities of CAL27 cells by 2.12- and 1.56-fold, respectively [136]. Synergistic cell growth inhibition was observed also when EGCG was combined with erlotinib, an EGFR tyrosine kinase inhibitor. Such potentiated antitumor action was found in different human oral cancer cell lines, such as SQCCY1, erlotinib-sensitive Tu212 and chemoresistant 38 cells [137], as well as 1986LN and Tu686 cells [59]. In 1986LN, SQCCY1 and Tu686 cells, combination of EGCG and erlotinib induced synergistic growth inhibition and suppressed drug-induced expression of p21 and p27, although EGCG alone revealed almost no effect on the expression of these cell cycle regulatory proteins [59]. In addition, EGCG significantly potentiated erlotinib-triggered apoptotic cell death in p53 wild type Tu686 cells [59]. In Tu212 cells, combination of erlotinib with 30 μM EGCG resulted in an additive arrest of cell cycle progression at G0/G1 phase and increased markedly apoptosis in mitochondria-dependent manner by synergistic downregulation of EGFR/AKT pathway. Furthermore, treatment of experimental xenograft mice subcutaneously injected with Tu212 cells with a combination of EGCG and erlotinib led to a synergistic enhancement of tumor growth suppression and delay of tumor progression as compared to either single agents, resulted mainly from a more effective inhibition of cell proliferation and induction of apoptosis, with no considerable side effects [137]. Pretreatment of OSC4 cells with high concentration of EGCG (200 μM) followed by addition of cisplatin (CDDP) resulted in an additive decrease in cell viability [73]. However, pretreatment of OSC2 cells with 25 μM and 50 μM EGCG significantly reduced the 5 μM CDDP-induced cytotoxicity and inhibition of DNA synthesis. Likewise, although γ-irradiation led to a decrease in viability of OSC2 cells, pretreatment with 12.5-50 μM EGCG increased the viability of γ-irradiated OSC2 cells as compared to non-pretreated γ-irradiated cells [73]. Thus, application of EGCG can even exert protective effects to tumor cells against cancer therapy, whereas the specific action of this flavanol largely depends on its concentration as well as certain oral cancer cell lines (Table 4.5). Consequently, addition of EGCG to the current chemotherapeutic schemes used to treat oral cancer might provide a promising strategy to significantly enhance therapeutic outcome and improve the survival of patients, certainly requiring further clinical trials. However, it is crucial to bear in mind that under certain conditions, consumption of this natural flavanol can possibly also impair the effectiveness of cancer therapy by protecting the tumor cells and resulting in the quite opposite effects as expected.

Antitumor action of several standard chemotherapeutic drugs has been found to be modulated also by dietary flavones. Administration of 5 μM apigenin in combination with 5-FU or CDDP induced synergistic cytotoxicity in SCC25 cells and this pattern could be further exploited to reverse unresponsiveness of oral cancer cells to 5-FU or CDDP [74]. Growth inhibitory effect of CDDP was improved also by luteolin in SCC25 cells [65]. Similarly, luteolin significantly increased the cytotoxic activity of paclitaxel in SCC4 cells and combination of luteolin and paclitaxel synergistically suppressed tumor growth in xenograft mice subcutaneously inoculated with SCC4 cells [66]. Moreover, in oral cancer stem cells (OCSCs) derived from GNM or SAS cells, luteolin was able to restore radiosensitivity, and enhanced synergistically invasiveness and colony-forming potential when applied in combination with radiation. Such increase in tumorigenicity was probably mediated through downregulation of IL-6/STAT3 signaling [64]. It has been shown in several studies that low-intensity ultrasound can affect the cell membrane permeability and therefore improve the efficacy of drug delivery. Combining low-intensity ultrasound with administration of scutellarin led to irregularities in shape of malignant cells, inhibited cell growth, decreased migration and invasion, and caused apoptosis in SAS cells. In SAS engineered xenograft mice, such combined treatment significantly delayed tumor growth, inhibited tumor volume, reduced proliferation, suppressed angiogenesis and lymphangiogenesis, triggered apoptosis and inhibited MMP-2 and MMP-9 expression levels. Promotion of the permeability of scutellarin into tumor cells by ultrasound was most probably implicated in these effects [71].

Flavonol quercetin, at a dose range of 25-100 μM, reversed P-glycoprotein (P-gp)-mediated MDR in vincristine (VCR)-resistant KB cells (KB/VCR) by suppressing the expression of P-gp and sensitizing the cells to VCR. This effect was accompanied by reduction of proliferation rate, enhancement of apoptosis and inhibition of invasion and migration [118]. Combination of 100 μM quercetin with 10 μM CDDP resulted in a considerable growth inhibition and sensitization of drug-resistant spheres (DRSPs) induced from SCC25 cells to CDDP by targeting p38 MAPK-Hsp27 axis. In addition, the migration and invasion abilities of DRSPs were substantially suppressed with reversal of EMT-associated markers, i.e., upregulation of E-cadherin and downregulation of vimentin and Twist-1. Furthermore, combination of quercetin and CDDP led to a significant inhibition of the size and volume of tumors in xenograft mice injected with DRSPs [75]. Another natural flavonol, myricetin, sensitized HPV-positive oral cancer cells (UMSCC47, UPCI-SCC90), but not HPV-negative cells (SCC84, UMSCC29), to 2 μM DOX-initiated apoptotic cell death showing that infection with certain HPV subtypes might play an important role in the action of flavonoids in oral cancer cells [42].

Combined treatment of oral cancer cells with the most abundant soy isoflavone, genistein, and an anti-EGFR monoclonal antibody, cetuximab, led to a reduced expression of downstream proteins of EGFR signaling in HSC3 cells, but not in KB cells, whereas both cells displayed a significant decrease in proliferation and an additive increase in apoptosis. At that, sensitivity to genistein was somewhat higher in HSC3 than KB cell line and *in vitro* combination effect was observed only at low doses of both compounds. Furthermore, combination of genistein and cetuximab also caused a delay and suppression of tumor growth and reduced neovascularization in two xenograft mice models injected with either HSC3 or KB cells [120] (Table 4.5).

On the whole, combined treatment of natural flavonoids and current anticancer modalities might provide an effective novel strategy to combat oral malignancies and improve prognosis of patients. Several studies have indeed demonstrated sensitizing potential of certain flavonoids toward conventional chemotherapeutic agents or irradiation. Considering that resistance of malignant cells to chemotherapeutics is a common clinical problem and a major cause of treatment failure, and drug-associated toxic adverse effects have remained an important bottleneck in cancer treatment, co-administration with flavonoids could lower the effective doses and minimize toxicities, representing thus a compelling treatment approach for the future. However, it is still crucial to be aware of potential threats in treating cancer patients under standard treatment regimens with flavonoids as negative co-effects in some cellular models have also been described. Therefore, with no doubt, further studies both *in vitro* cell cultures as well as *in vivo* animal systems are highly needed before any recommendations can be given to oral cancer patients.

Table 4.6. Effect of nanoencapsulated flavonoids on oral cancer models

Parent compound	Flavonoid nanoparticles[1]	Biological system[2]	Assay conditions[3]	Comparative biological effects	Ref.
FLAVANOLIGNANS					
Silibinin	Eudragit-encapsulated silibinin	KB cells	MTT, 48 h	Cytotoxicity: IC_{50} 31.09 μM for silibinin nanoparticles, 78.66 μM for silibinin	[122]
FLAVANONES					
Hesperetin	Eudragit-encapsulated hesperetin	DMBA-induced hamster oral tumorigenesis	Per-oral, 40 hesperetin mg/kg/day	Decrease in metabolic activity: treatment with nanoparticulate hesperetin more effective than native hesperetin	[61]
Naringenin	Eudragit-encapsulated naringenin	DMBA-induced hamster buccal pouch tumorigenesis	Per-oral, 50 mg/kg/day	Complete prevention of tumor formation with naringenin-loaded nanoparticles as compared to free naringenin, reduction of degree of histological lesions. More potent antiproliferative, antilipid peroxidative and antioxidative potentials of naringenin-loaded nanoparticles than free naringenin	[27]
	Eudragit-encapsulated naringenin	DMBA-induced hamster buccal pouch tumorigenesis	Per-oral, 50 mg/kg/day	Increase in lipids and glycogen content, decrease in protein levels and nucleic acid contents: more potent effect of naringenin-loaded nanoparticles than free naringenin	[25]
	Eudragit-encapsulated naringenin	DMBA-induced hamster buccal pouch tumorigenesis	Per-oral, 50 mg/kg/day	Prevention of cancer formation, increase in phospholipids, decrease in tryptophan, phenylalanine and nucleic acid contents: more potent effect of naringenin-loaded nanoparticles than free naringenin	[28]
FLAVONES					
Luteolin	PLA-PEG polymer encapsulated luteolin	Tu212 cells	SRB, 72 h	Cell growth inhibition: IC_{50} 4.13 μM for nano-luteolin, 6.96 μM for luteolin. More effective colony formation inhibition by nano-luteolin than free luteolin	[113]
	PLA-PEG polymer encapsulated luteolin	Tu212 cells injected into right flank of mice	Intraperitoneal injection, 3.3 mg/kg	Inhibition of tumor growth by nano-luteolin, no significant effect with luteolin. No signs of toxicity in nano-luteolin-treated mice	[113]

[1]PLA-PEG, polylactic acid-polyethylene glycol.
[2]see Tables 4.1 and 4.2.
[3]MTT, 3-(4,5-dimethylthiazol-2-yl)-2,5-diphenyltetrazolium bromide; SRB, sulforhodamine B.

4.4. FURTHER OPPORTUNITIES AND FUTURE PERSPECTIVES

While numerous *in vitro* and *in vivo* studies have demonstrated an important antitumor role of natural flavonoids against oral cancer, the major bottleneck in practical application of these phytochemicals is their poor bioavailability, i.e., low solubility, poor absorption rate and an extensive metabolism. Therefore, the achievement of pharmacologically relevant concentrations in the human body is still a big challenge. New technologies enabling to overcome these drawbacks are highly needed, whereas nanotechnological approach has already been successfully exploited to promote the *in vivo* efficiency and targeted delivery of several flavonoids. This technological development can improve both the outcome of chemoprevention as well as cancer treatment. On the other hand, local exposure of these dietary polyphenols to oral cavity is also crucial, whereas buccal drug delivery systems represent another compelling approach. Transmucosal oral drug delivery can present several advantages over gastrointestinal route by offering rapid transport to systemic circulation, avoiding metabolic conversion by the liver enzymes and elimination within digestive tract [1]. It is also not less important to emphasize that drug administration through oral transmucosal route is non-invasive and comfortable for patients. However, only few studies have still been performed on antitumor effects of topically applied flavonoids in oral cavity.

Experiments with luteolin, hesperetin, naringenin and silibinin have clearly shown the important benefits of nanotechnological methods. In KB human oral cancer cells, silibinin-loaded nanoparticles were found to display significantly stronger cytotoxic efficacy as compared to the free silibinin, while the levels of intracellular ROS were considerably higher in silibinin-loaded nanoparticles-treated cells leading to an apoptotic cell death [122]. Like luteolin also nano-luteolin suppressed the growth of Tu212 cells with somewhat stronger inhibitory action on colony formation with nanoparticulate flavone. Moreover, while nano-luteolin significantly inhibited tumor growth in xenograft mice injected with Tu212 cells, the free luteolin revealed no significant effect [113]. In DMBA-induced hamster buccal pouch tumorigenesis model, treatment with hesperetin-loaded nanoparticles exhibited considerably stronger decrease in metabolic activity than hesperetin [61]. Also, administration of nanoparticulate naringenin to DMBA-painted hamsters completely blocked the tumor incidence and formation as compared to the free naringenin with a substantial decrease also in the degree of histological lesions. In addition, naringenin-loaded nanoparticles revealed more potent antiproliferative, antilipid peroxidative and antioxidative abilities [27]. Furthermore, nanoparticulate naringenin was more effective in improving the biochemical constituents of the DMBA-exposured hamster buccal mucosa to a normal range than free naringenin by increasing the levels of lipids and their order and glycogen contents and decreasing proteins and nucleic acid contents [25]. At that, the level of tryptophan and phenylalanine was significantly reduced [28] (Table 4.6). Protection of flavonoids from the metabolic conversion by nanoparticulate carriers might improve their oral bioavailability resulting in a greater accumulation of these phytochemicals inside the tumor tissue and exhibiting stronger anticancer activities. Altogether, these data clearly show that flavonoids-loaded nanoparticles might be further applied as an effective drug delivery system to develop more efficient tools for the prevention and/or treatment of oral cancer.

REFERENCES

[1] Iriti M, Varoni EM. Chemopreventive potential of flavonoids in oral squamous cell carcinoma in human studies. *Nutrients* 2013; 5: 2564-76.

[2] Liu YP, Lee JJ, Lai TC, Lee CH, Hsiao YW, Chen PS, Liu WT, Hong CY, Lin SK, Ping Kuo MT, Lu PJ, Hsiao M. Suppressive function of low-dose deguelin on the invasion of oral cancer cells by downregulating tumor necrosis factor alpha-induced nuclear factor-kappa B signaling. *Head Neck* 2016; 38 Suppl 1: E524-34.

[3] Maggioni D, Nicolini G, Rigolio R, Biffi L, Pignataro L, Gaini R, Garavello W. Myricetin and naringenin inhibit human squamous cell carcinoma proliferation and migration in vitro. *Nutr Cancer* 2014; 66: 1257-67.

[4] Koh YW, Choi EC, Kang SU, Hwang HS, Lee MH, Pyun J, Park R, Lee Y, Kim CH. Green tea (-)-epigallocatechin-3-gallate inhibits HGF-induced progression in oral cavity cancer through suppression of HGF/c-Met. *J Nutr Biochem* 2011; 22: 1074-83.

[5] Lim YC, Lee SH, Song MH, Yamaguchi K, Yoon JH, Choi EC, Baek SJ. Growth inhibition and apoptosis by (-)-epicatechin gallate are mediated by cyclin D1 suppression in head and neck squamous carcinoma cells. *Eur J Cancer* 2006; 42: 3260-6.

[6] Godsey J, Grundmann O. Review of various herbal supplements as complementary treatments for oral cancer. *J Diet Suppl* 2016; 13: 538-50.

[7] Gavin A, Pham JT, Wang D, Brownlow B, Elbayoumi TA. Layered nanoemulsions as mucoadhesive buccal systems for controlled delivery of oral cancer therapeutics. *Int J Nanomedicine* 2015; 10: 1569-84.

[8] Su CH, Kuo CL, Lu KW, Yu FS, Ma YS, Yang JL, Chu YL, Chueh FS, Liu KC, Chung JG. Fisetin-induced apoptosis of human oral cancer SCC-4 cells through reactive oxygen species production, endoplasmic reticulum stress, caspase-, and mitochondria-dependent signaling pathways. *Environ Toxicol* 2017; 32: 1725-41.

[9] Sun ZJ, Chen G, Hu X, Zhang W, Liu Y, Zhu LX, Zhou Q, Zhao YF. Activation of PI3K/Akt/IKK-alpha/NF-kappaB signaling pathway is required for the apoptosis-evasion in human salivary adenoid cystic carcinoma: its inhibition by quercetin. *Apoptosis* 2010; 15: 850-63.

[10] Chen PN, Hsieh YS, Chiang CL, Chiou HL, Yang SF, Chu SC. Silibinin inhibits invasion of oral cancer cells by suppressing the MAPK pathway. *J Dent Res* 2006; 85: 220-5.

[11] Gao F, Li M, Liu WB, Zhou ZS, Zhang R, Li JL, Zhou KC. Epigallocatechin gallate inhibits human tongue carcinoma cells via HK2-mediated glycolysis. *Oncol Rep* 2015; 33: 1533-9.

[12] Oh H, Yoon G, Shin JC, Park SM, Cho SS, Cho JH, Lee MH, Liu K, Cho YS, Chae JI, Shim JH. Licochalcone B induces apoptosis of human oral squamous cell carcinoma through the extrinsic- and intrinsic-signaling pathways. *Int J Oncol* 2016; 48: 1749-57.

[13] Hsia SM, Yu CC, Shih YH, Yuanchien Chen M, Wang TH, Huang YT, Shieh TM. Isoliquiritigenin as a cause of DNA damage and inhibitor of ataxia-telangiectasia mutated expression leading to G2/M phase arrest and apoptosis in oral squamous cell carcinoma. *Head Neck* 2016; 38 Suppl 1: E360-71.

[14] Cho JJ, Chae JI, Yoon G, Kim KH, Cho JH, Cho SS, Cho YS, Shim JH. Licochalcone A, a natural chalconoid isolated from Glycyrrhiza inflata root, induces apoptosis via Sp1 and Sp1 regulatory proteins in oral squamous cell carcinoma. *Int J Oncol* 2014; 45: 667-74.

[15] Yuan CH, Horng CT, Lee CF, Chiang NN, Tsai FJ, Lu CC, Chiang JH, Hsu YM, Yang JS, Chen FA. Epigallocatechin gallate sensitizes cisplatin-resistant oral cancer CAR cell apoptosis and autophagy through stimulating AKT/STAT3 pathway and suppressing multidrug resistance 1 signaling. *Environ Toxicol* 2017; 32: 845-55.

[16] Irimie AI, Braicu C, Zanoaga O, Pileczki V, Gherman C, Berindan-Neagoe I, Campian RS. Epigallocatechin-3-gallate suppresses cell proliferation and promotes apoptosis and autophagy in oral cancer SCC-4 cells. *Onco Targets Ther* 2015; 8: 461-70.

[17] Silvan S, Manoharan S. Apigenin prevents deregulation in the expression pattern of cell-proliferative, apoptotic, inflammatory and angiogenic markers during 7,12-dimethylbenz[a]anthracene-induced hamster buccal pouch carcinogenesis. *Arch Oral Biol* 2013; 58: 94-101.

[18] Yang JG, Lu R, Ye XJ, Zhang J, Tan YQ, Zhou G. Icaritin reduces oral squamous cell carcinoma progression via the inhibition of STAT3 signaling. *Int J Mol Sci* 2017; 18: E132.

[19] Ferlay J, Soerjomataram I, Ervik M, Dikshit R, Eser S, Mathers C, Rebelo M, Parkin DM, Forman D, Bray F. *GLOBOCAN 2012 v1.1, Cancer Incidence and Mortality Worldwide: IARC CancerBase* No. 11[Internet]. Lyon, France: International Agency for Research on Cancer; 2014. Available from: http://globocan.iarc.fr.

[20] Ding Y, Yao H, Yao Y, Fai LY, Zhang Z. Protection of dietary polyphenols against oral cancer. *Nutrients* 2013; 5: 2173-91.

[21] Garavello W, Lucenteforte E, Bosetti C, La Vecchia C. The role of foods and nutrients on oral and pharyngeal cancer risk. *Minerva Stomatol* 2009; 58: 25-34.

[22] Ye F, Wu J, Dunn T, Yi J, Tong X, Zhang D. Inhibition of cyclooxygenase-2 activity in head and neck cancer cells by genistein. *Cancer Lett* 2004; 211: 39-46.

[23] Bhavana SM, Lakshmi CR. Oral oncoprevention by phytochemicals – a systematic review disclosing the therapeutic dilemma. *Adv Pharm Bull* 2014; 4 Suppl 1: 413-20.

[24] Hseu YC, Lee MS, Wu CR, Cho HJ, Lin KY, Lai GH, Wang SY, Kuo YH, Kumar KJ, Yang HL. The chalcone flavokawain B induces G2/M cell-cycle arrest and apoptosis in human oral carcinoma HSC-3 cells through the intracellular ROS generation and downregulation of the Akt/p38 MAPK signaling pathway. *J Agric Food Chem* 2012; 60: 2385-97.

[25] Krishnakumar N, Sulfikkarali NK, Manoharan S, Nirmal RM. Screening of chemopreventive effect of naringenin-loaded nanoparticles in DMBA-induced hamster buccal pouch carcinogenesis by FT-IR spectroscopy. *Mol Cell Biochem* 2013; 382: 27-36.

[26] Silvan S, Manoharan S, Baskaran N, Anusuya C, Karthikeyan S, Prabhakar MM. Chemopreventive potential of apigenin in 7,12-dimethylbenz(a)anthracene induced experimental oral carcinogenesis. *Eur J Pharmacol* 2011; 670: 571-7.

[27] Sulfikkarali N, Krishnakumar N, Manoharan S, Nirmal RM. Chemopreventive efficacy of naringenin-loaded nanoparticles in 7,12-dimethylbenz(a)anthracene induced experimental oral carcinogenesis. *Pathol Oncol Res* 2013; 19: 287-96.

[28] Krishnakumar N, Sulfikkarali NK, Manoharan S, Venkatachalam P. Raman spectroscopic investigation of the chemopreventive response of naringenin and its nanoparticles in DMBA-induced oral carcinogenesis. *Spectrochim Acta A Mol Biomol Spectrosc* 2013; 115: 648-53.

[29] Tanaka T, Makita H, Ohnishi M, Hirose Y, Wang A, Mori H, Satoh K, Hara A, Ogawa H. Chemoprevention of 4-nitroquinoline 1-oxide-induced oral carcinogenesis by dietary curcumin and hesperidin: comparison with the protective effect of beta-carotene. *Cancer Res* 1994; 54: 4653-9.

[30] Yang SH, Liao PH, Pan YF, Chen SL, Chou SS, Chou MY. The novel p53-dependent metastatic and apoptotic pathway induced by vitexin in human oral cancer OC2 cells. *Phytother Res* 2013; 27: 1154-61.

[31] Bundela S, Sharma A, Bisen PS. Potential compounds for oral cancer treatment: resveratrol, nimbolide, lovastatin, bortezomib, vorinostat, berberine, pterosrilbene, deguelin, andrographolide, and colchicine. *PLoS One* 2015; 10: e0141719.

[32] Suresh K, Rajasekar M, Kumar RA, Sivakumar K. Dose tolerance study of diosmin against 7,12-dimethylbenz(a)anthracene (DMBA) induced hamster buccal pouch carcinogenesis. *Int J Pharm Biol Arch* 2014; 5: 82-9.

[33] Lai WW, Hsu SC, Chueh FS, Chen YY, Yang JS, Lin JP, Lien JC, Tsai CH, Chung JG. Quercetin inhibits migration and invasion of SAS human oral cancer cells through inhibition of NF-κB and matrix metalloproteinase-2/-9 signaling pathways. *Anticancer Res* 2013; 33: 1941-50.

[34] Messadi DV, Sato K. Oral cancer chemoprevention: current status and future direction. *J Calif Dent Assoc* 2016; 44: 101-11.

[35] Lee UL, Choi SW. The chemopreventive properties and therapeutic modulation of green tea polyphenols in oral squamous cell carcinoma. *ISRN Oncol* 2011; 2011: 403707.

[36] Maggioni D, Biffi L, Nicolini G, Garavello W. Flavonoids in oral cancer prevention and therapy. *Eur J Cancer Prev* 2015; 24: 517-28.

[37] Zeng G, Shen H, Yang Y, Cai X, Xun W. Licochalcone A as a potent antitumor agent suppresses growth of human oral cancer SCC-25 cells in vitro via caspase-3 dependent pathways. *Tumour Biol* 2014; 35: 6549-55.

[38] Makita H, Tanaka T, Fujitsuka H, Tatematsu N, Satoh K, Hara A, Mori H. Chemoprevention of 4-nitroquinoline 1-oxide-induced rat oral carcinogenesis by the dietary flavonoids chalcone, 2-hydroxychalcone, and quercetin. *Cancer Res* 1996; 56: 4904-9.

[39] Varoni EM, Lodi G, Sardella A, Carrassi A, Iriti M. Plant polyphenols and oral health: old phytochemicals for new fields. *Curr Med Chem* 2012; 19: 1706-20.

[40] King M, Chatelain K, Farris D, Jensen D, Pickup J, Swapp A, O´Malley S, Kingsley K. Oral squamous cell carcinoma proliferative phenotype is modulated by proanthocyanidins: a potential prevention and treatment alternative for oral cancer. *BMC Complement Altern Med* 2007; 7: 22.

[41] Gomez-Garcia FJ, Lopez-Jornet MP, Alvarez-Sanchez N, Castillp-Sanchez J, Benavente-Garcia O, Vicente Ortega V. Effect of the phenolic compounds apigenin and carnosic acid on oral carcinogenesis in hamster induced by DMBA. *Oral Dis* 2013; 19: 279-86.

[42] Yuan CH, Filippova M, Krstenansky JL, Duerksen-Hughes PJ. Flavonol and imidazole derivatives block HPV16 E6 activities and reactivate apoptotic pathways in HPV$^+$ cells. *Cell Death Dis* 2016; 7: 2060.

[43] Lee JC, Chung LC, Chen YJ, Feng TH, Chen WT, Juang HH. Upregulation of B-cell translocation gene 2 by epigallocatechin-3-gallate via p38 and ERK signaling blocks cell proliferation in human oral squamous cell carcinoma cells. *Cancer Lett* 2015; 360: 310-8.

[44] Narotzki B, Reznick AZ, Aizenbud D, Levy Y. Green tea: a promising natural product in oral health. *Arch Oral Biol* 2012; 57: 429-35.

[45] Kim JS, Park MR, Lee SY, Kim DK, Moon SM, Kim CS, Cho SS, Yoon G, Im HJ, You JS, Oh JS, Kim SG. Licochalcone A induces apoptosis in KB human oral cancer cells via a caspase-dependent FasL signaling pathway. *Oncol Rep* 2014; 31: 755-62.

[46] Lai CK, Rao YK, Chang KR, Lin CW, Su HL, Chang CS, Lai CH, Tzeng YM. 3,3`,4`,5`-Tetramethoxychalcone inhibits human oral cancer cell proliferation and migration via p53-mediated mitochondrial-dependent apoptosis. *Anticancer Res* 2014; 34: 1811-9.

[47] de Moura CF, Noguti J, de Jesus GP, Ribeiro FA, Garcia FA, Gollucke AP, Aguiar O Jr, Ribeiro DA. Polyphenols as a chemopreventive agent in oral carcinogenesis: putative mechanisms of action using in-vitro and in-vivo test systems. *Eur J Cancer Prev* 2013; 22: 467-72.

[48] Shen H, Zeng G, Tang G, Cai X, Bi L, Huang C, Yang Y. Antimetastatic effects of licochalcone A on oral cancer via regulating metastasis-associated proteases. *Tumour Biol* 2014; 35: 7467-74.

[49] Chen PN, Chu SC, Kuo WH, Chou MY, Lin JK, Hsieh YS. Epigallocatechin-3 gallate inhibits invasion, epithelial-mesenchymal transition, and tumor growth in oral cancer cells. *J Agric Food Chem* 2011; 59: 3836-44.

[50] Li B, Lu M, Jiang XX, Pan MX, Mao JW, Chen M. Inhibiting reactive oxygen species-dependent autophagy enhanced baicalein-induced apoptosis in oral squamous cell carcinoma. *J Nat Med* 2017; 71: 433-41.

[51] Chen SF, Nien S, Wu CH, Liu CL, Chang YC, Lin YS. Reappraisal of the anticancer efficacy of quercetin in oral cancer cells. *J Chin Med Assoc* 2013; 76: 146-52.

[52] Li H, Huang D, Gao Z, Lv Y, Zhang L, Cui H, Zheng J. Scutellarin inhibits cell migration by regulating production of αvß6 integrin and E-cadherin in human tongue cancer cells. *Oncol Rep* 2010; 24: 1153-60.

[53] Zlotogorski A, Dayan A, Dayan D, Chaushu G, Salo T, Vered M. Nutraceuticals as new treatment approaches for oral cancer: II. Green tea extracts and resveratrol. *Oral Oncol* 2013; 49: 502-6.

[54] Alhasan SA, Pietrasczkiwicz H, Alonso MD, Ensley J, Sarkar FH. Genistein-induced cell cycle arrest and apoptosis in a head and neck squamous cell carcinoma cell line. *Nutr Cancer* 1999; 34: 12-9.

[55] Maggioni D, Garavello W, Rigolio R, Pignataro L, Gaini R, Nicolini G. Apigenin impairs oral squamous cell carcinoma growth in vitro inducing cell cycle arrest and apoptosis. *Int J Oncol* 2013; 43: 1675-82.

[56] Lim YC, Park HY, Hwang HS, Kang SU, Pyun JH, Lee MH, Choi EC, Kim CH. (-)-Epigallocatechin-3-gallate (EGCG) inhibits HGF-induced invasion and metastasis in hypopharyngeal carcinoma cells. *Cancer Lett* 2008; 271: 140-52.

[57] Li H, Huang D, Gao Z, Chen Y, Zhang L, Zheng J. Scutellarin inhibits the growth and invasion of human tongue squamous carcinoma through the inhibition of matrix metalloproteinase-2 and -9 and αvß6 integrin. *Int J Oncol* 2013; 42: 1674-81.

[58] Khafif A, Schantz SP, Chou TC, Edelstein D, Sacks PG. Quantitation of chemopreventive synergism between (-)-epigallocatechin-3-gallate and curcumin in normal, premalignant and malignant human oral epithelial cells. *Carcinogenesis* 1998; 19: 419-24.

[59] Amin AR, Khuri FR, Chen ZG, Shin DM. Synergistic growth inhibition of squamous cell carcinoma of the head and neck by erlotinib and epigallocatechin-3-gallate: the role of p53-dependent inhibition of nuclear factor-kappaB. *Cancer Prev Res* (Phila) 2009; 2: 538-45.

[60] Yang Y, Zhou ZT, Ge JP. Effect of genistein on DMBA-induced oral carcinogenesis in hamster. *Carcinogenesis* 2006; 27: 578-83.

[61] Gurushankar K, Nazeer SS, Jayasree RS, Krishnakumar N. Evaluation of antitumor activity of hesperetin-loaded nanoparticles against DMBA-induced oral carcinogenesis based on tissue autofluorescence spectroscopy and multivariate analysis. *J Fluoresc* 2015; 24: 931-9.

[62] Kim YS, Kim CH. Chemopreventive role of green tea in head and neck cancers. *Integr Med Res* 2014; 3: 11-5.

[63] Kim CD, Cha JD, Li S, Cha IH. The mechanism of acacetin-induced apoptosis on oral squamous cell carcinoma. *Arch Oral Biol* 2015; 60: 1283-98.

[64] Tu DG, Lin WT, Yu CC, Lee SS, Peng CY, Lin T, Yu CH. Chemotherapeutic effects of luteolin on radio-sensitivity enhancement and interleukin-6/signal transducer and activator of transcription 3 signaling repression of oral cancer stem cells. *J Fromos Med Assoc* 2016; 115: 1032-8.

[65] Tjioe KC, Tostes Oliveira D, Gavard J. Luteolin impacts on the DNA damage pathway in oral squamous cell carcinoma. *Nutr Cancer* 2016; 68: 838-47.

[66] Yang SF, Yang WE, Chang HR, Chu SC, Hsieh YS. Luteolin induces apoptosis in oral squamous cancer cells. *J Dent Res* 2008; 87: 401-6.

[67] Lin CW, Chen PN, Chen MK, Yang WE, Tang CH, Yang SF, Hsieh YS. Kaempferol reduces matrix metalloproteinase-2 expression by down-regulating ERK1/2 and the activator protein-1 signaling pathways in oral cancer cells. *PLoS One* 2013; 8: e80883.

[68] Lee RH, Shin JC, Kim KH, Choi YH, Chae JI, Shim JH. Apoptotic effects of 7,8-dihydroxyflavone in human oral squamous cancer cells through suppression of Sp1. *Oncol Rep* 2015; 33: 631-8.

[69] Katz J, Blake E, Medrano TA, Sun Y, Shiverick KT. Isoflavones and gamma irradiation inhibit cell growth in human salivary gland cells. *Cancer Lett* 2008; 270: 87-94.

[70] Park JH, Yoon JH, Kim SA, Ahn SG, Yoon JH. (-)-Epigallocatechin-3-gallate inhibits invasion and migration of salivary gland adenocarcinoma cells. *Oncol Rep* 2010; 23: 585-90.

[71] Li H, Fan H, Wang Z, Zheng J, Cao W. Potentiation of scutellarin on human tongue carcinoma xenograft by low-intensity ultrasound. *PLoS One* 2013; 8: e59473.

[72] Masuda M, Suzui M, Weinstein IB. Effects of epigallocatechin-3-gallate on growth, epidermal growth factor receptor signaling pathways, gene expression, and chemosensitivity in human head and neck squamous cell carcinoma cell lines. *Clin Cancer Res* 2001; 7: 4220-9.

[73] Yamamoto T, Staples J, Wataha J, Lewis J, Lockwood P, Schoenlein P, Rao S, Osaki T, Dickinson D, Kamatani T, Schuster G, Hsu S. Protective effects of EGCG on salivary gland cells treated with gamma-radiation or cis-platinum(II)diammine dichloride. *Anticancer Res* 2004; 24: 3065-73.

[74] Chan LP, Chou TH, Ding HY, Chen PR, Chiang FY, Kuo PL, Liang CH. Apigenin induces apoptosis via tumor necrosis factor receptor- and Bcl-2-mediated pathway and enhances susceptibility of head and neck squamous cell carcinoma to 5-fluorouracil and cisplatin. *Biochim Biophys Acta* 2012; 1820: 1081-91.

[75] Chen SF, Nieh S, Jao SW, Liu CL, Wu CH, Chang YC, Yang CY, Lin YS. Quercetin suppresses drug-resistant spheres via the p38 MAPK-Hsp27 apoptotic pathway in oral cancer cells. *PLoS One* 2012; 7: e49275.

[76] Haghiac M, Walle T. Quercetin induces necrosis and apoptosis in SCC-9 oral cancer cells. *Nutr Cancer* 2005; 53: 220-31.

[77] Yang YL, Ji C, Bi ZG, Lu CC, Wang R, Gu B, Cheng L. Deguelin induces both apoptosis and autophagy in cultured head and neck squamous cell carcinoma cells. *PLoS One* 2013; 8: e54736.

[78] Lee YM, Kim YC, Choi BJ, Lee DW, Yoon JH, Kim EC. Mechanism of sappanchalcone-induced growth inhibition and apoptosis in human oral cancer cells. *Toxicol In Vitro* 2011; 25: 1782-8.

[79] Lucenteforte E, Garavello W, Bosetti C, La Vecchia C. Dietary factors and oral and pharyngeal cancer risk. *Oral Oncol* 2009; 45: 461-7.

[80] Rossi M, Garavello W, Talamini R, Negri E, Bosetti C, Dal Maso L, Lagiou P, Tavani A, Polesel J, Barzan L, Ramazzotti V, Franceschi S, La Vecchia C. Flavonoids and the risk of oral and pharyngeal cancer: a case-control study from Italy. *Cancer Epidemiol Biomarkers Prev* 2007; 16: 1621-5.

[81] Browning AM, Walle UK, Walle T. Flavonoid glycosides inhibit oral cancer cell proliferation – role of cellular uptake and hydrolysis to the aglycones. *J Pharm Pharmacol* 2005; 57: 1037-42.

[82] Walle T, Browning AM, Steed LL, Reed SG, Walle UK. Flavonoid glucosides are hydrolyzed and thus activated in the oral cavity in humans. *J Nutr* 2005; 135: 48-52.

[83] Tao L, Park JY, Lambert JD. Differential prooxidative effects of the green tea polyphenol, (-)-epigallocatechin-3-gallate, in normal and oral cancer cells are related to differences in sirtuin 3 signaling. *Mol Nutr Food Res* 2015; 59: 203-11.

[84] Priyadarsini RV, Vinothini G, Murugan RS, Manikandan P, Nagini S. The flavonoid quercetin modulates the hallmark capabilities of hamster buccal pouch tumors. *Nutr Cancer* 2011; 63: 218-26.

[85] Miller EG, Peacock JJ, Bourland TC, Taylor SE, Wright JM, Patil BS, Miller EG. Inhibition of oral carcinogenesis by citrus flavonoids. *Nutr Cancer* 2008; 60: 69-74.

[86] Tanaka T, Makita H, Ohnishi M, Mori H, Satoh K, Hara A, Sumida T, Fukutani K, Tanaka T, Ogawa H. Chemoprevention of 4-nitroquinoline 1-oxide-induced oral carcinogenesis in rats by flavonoids diosmin and hesperidin, each alone and in combination. *Cancer Res* 1997; 57: 246-52.

[87] Hsu S, Lewis J, Singh B, Schoenlein P, Osaki T, Athar M, Porter AG, Schuster G. Green tea polyphenol targets the mitochondria in tumor cells inducing caspase 3-dependent apoptosis. *Anticancer Res* 2003; 23: 1533-9.

[88] Liu X, Zhang DY, Zhang W, Zhao X, Yuan C, Ye F. The effect of green tea extract and EGCG on the signaling network in squamous cell carcinoma. *Nutr Cancer* 2011; 63: 466-75.

[89] Chen LL, Han WF, Geng Y, Su JS. A genome-wide study of DNA methylation modified by epigallocatechin-3-gallate in the CAL-27 cell line. *Mol Med Rep* 2015; 12: 5886-90.

[90] Chiang WC, Wong YK, Lin SC, Chang KW, Liu CJ. Increase of MMP-13 expression in multi-stage oral carcinogenesis and epigallocatechin-3-gallate suppress MMP-13 expression. *Oral Dis* 2006; 12: 27-33.

[91] Masuda M, Suzui M, Lim JT, Weinstein IB. Epigallocatechin-3-gallate inhibits activation of HER-2/neu and downstream signaling pathways in human head and neck and breast carcinoma cells. *Clin Cancer Res* 2003; 9: 3486-91.

[92] Tao L, Forester SC, Lambert JD. The role of the mitochondrial oxidative stress in the cytotoxic effects of the green tea catechin, (-)-epigallocatechin-3-gallate, in oral cells. *Mol Nutr Food Res* 2014; 58: 665-76.

[93] Lin HY, Hou SC, Chen SC, Kao MC, Yu CC, Funayama S, Ho CT, Way TD. (-)-Epigallocatechin gallate induces Fas/CD95-mediated apoptosis through inhibiting constitutive and IL-6-induced JAK/STAT3 signaling in head and neck squamous cell carcinoma cells. *J Agric Food Chem* 2012; 60: 2480-9.

[94] Khafif A, Schantz SP, al-Riwi M, Edelstein D, Sacks PG. Green tea regulates cell cycle progression in oral leukoplakia. *Head Neck* 1998; 20: 528-34.

[95] Yamamoto T, Digumarthi H, Aranbayeva Z, Wataha J, Lewis J, Messer R, Qin H, Dickinson D, Osaki T, Schuster GS, Hsu S. EGCG-targeted p57/KIP2 reduces tumorigenicity of oral carcinoma cells: role of c-Jun N-terminal kinase. *Toxicol Appl Pharmacol* 2007; 224: 318-25.

[96] Ko SY, Chang KW, Lin SC, Hsu HC, Liu TY. The repressive effect of green tea ingredients on amyloid precursor protein (APP) expression in oral carcinoma cells in vitro and in vivo. *Cancer Lett* 2007; 245(1-2): 81-9.

[97] Yamamoto T, Hsu S, Lewis J, Wataha J, Dickinson D, Singh B, Bollag WB, Lockwood P, Ueta E, Osaki T, Schuster G. Green tea polyphenol causes differential oxidative environments in tumor versus normal epithelial cells. *J Pharmacol Exp Ther* 2003; 307: 230-6.

[98] Weisburg JH, Weissman DB, Sedaghat T, Babich H. In vitro cytotoxicity of epigallocatechin gallate and tea extracts to cancerous and normal cells from the human oral cavity. *Basic Clin Pharmacol Toxicol* 2004; 95: 191-200.

[99] Hsu S, Lewis JB, Borke JL, Singh B, Dickinson DP, Caughman GB, Athar M, Drake L, Aiken AC, Huynh CT, Das BR, Osaki T, Schuster GS. Chemopreventive effects of green tea polyphenols correlate with reversible induction of p57 expression. *Anticancer Res* 2001; 21: 3743-8.

[100] Kato K, Long NK, Makita H, Toida M, Yamashita T, Hatakeyama D, Hara A, Mori H, Shibata T. Effects of green tea polyphenol on methylation status of RECK gene and cancer cell invasion in oral squamous cell carcinoma cells. *Br J Cancer* 2008; 99: 647-54.

[101] Ho YC, Yang SF, Peng CY, Chou MY, Chang YC. Epigallocatechin-3-gallate inhibits the invasion of human oral cancer cells and decreases the productions of matrix metalloproteinases and urokinase-plasminogen activator. *J Oral Pathol Med* 2007; 36: 588-93.

[102] Hwang YS, Park KK, Chung WY. Epigallocatechin-3 gallate inhibits cancer invasion by repressing functional invadopodia formation in oral squamous cell carcinoma. *Eur J Pharmacol* 2013; 715: 286-95.

[103] Cheng CW, Shieh PC, Lin YC, Chen YJ, Lin YH, Kuo DH, Liu JY, Kao JY, Kao MC, Way TD. Indoleamine 2,3-dioxygenase, an immunomodulatory protein, is suppressed by (-)-epigallocatechin-3-gallate via blocking of gamma-interferon-induced JAK-PKC-delta-STAT1 signaling in human oral cancer cells. *J Agric Food Chem* 2010; 58: 887-94.

[104] Amin AR, Wang D, Zhang H, Peng S, Shin HJ, Brandes JC, Tighiouart M, Khuri FR, Chen ZG, Shin DM. Enhanced anti-tumor activity by the combination of the natural compounds (-)-epigallocatechin-3-gallate and luteolin: potential role of p53. *J Biol Chem* 2010; 285: 34557-65.

[105] Babich H, Zuckerbraun HL, Weinerman SM. In vitro cytotoxicity of (-)-catechin gallate, a minor polyphenol in green tea. *Toxicol Lett* 2007; 171: 171-80.

[106] Babich H, Krupka ME, Nissim HA, Zuckerbraun HL. Differential in vitro cytotoxicity of (-)-epicatechin gallate (ECG) to cancer and normal cells from the human oral cavity. *Toxicol In Vitro* 2005; 19: 231-42.

[107] Walle T, Ta N, Kawamori T, Wen X, Tsuji PA, Walle UK. Cancer chemopreventive properties of orally bioavailable flavonoids – methylated versus unmethylated flavones. *Biochem Pharmacol* 2007; 73: 1288-96.

[108] Masuelli L, Marzocchella L, Quaranta A, Palumbo C, Pompa G, Izzi V, Canini A, Modesti A, Galvano F, Bei R. Apigenin induces apoptosis and impairs head and neck carcinomas EGFR/ErbB2 signaling. *Front Biosci* (Landmark Ed) 2011; 16: 1060-8.

[109] Swanson HI, Choi EY, Helton WB, Gairola CG, Valentino J. Impact of apigenin and kaempferol on human head and neck squamous cell carcinoma. *Oral Surg Oral Med Oral Pathol Oral Radiol* 2014; 117: 214-20.

[110] Yang SF, Yang WE, Kuo WH, Chang HR, Chu SC, Hsieh YS. Antimetastatic potentials of flavones on oral cancer cell via an inhibition of matrix-degrading proteases. *Arch Oral Biol* 2008; 53: 287-94.

[111] Lin YT, Yang JS, Lin HJ, Tan TW, Tang NY, Chaing JH, Chang YH, Lu HF, Chung JG. Baicalein induces apoptosis in SCC-4 human tongue cancer cells via a Ca2+-dependent mitochondrial pathway. *In Vivo* 2007; 21: 1053-8.

[112] Cheng YH, Li LA, Lin P, Cheng LC, Hung CH, Chang NW, Lin C. Baicalein induces G1 arrest in oral cancer cells by enhancing the degradation of cyclin D1 and activating AhR to decrease Rb phosphorylation. *Toxicol Appl Pharmacol* 2012; 263: 360-7.

[113] Majumdar D, Jung KH, Zhang H, Nannapaneni S, Wang X, Amin AR, Chen Z, Chen ZG, Shin DM. Luteolin nanoparticle in chemoprevention: in vitro and in vivo anticancer activity. *Cancer Prev Res* (Phila) 2014; 7: 65-73.

[114] Wen X, Walle T. Preferential induction of CYP1B1 by benzo[a]pyrene in human oral epithelial cells: impact on DNA adduct formation and prevention by polyphenols. *Carcinogenesis* 2005; 26: 1774-81.

[115] Kang JW, Kim JH, Song K, Kim SH, Yoon JH, Kim KS. Kaempferol and quercetin, components of Ginkgo biloba extract (EGb 761), induce caspase-3-dependent apoptosis in oral cavity cancer cells. *Phytother Res* 2010; 24 Suppl 1: S77-82.

[116] Chowdhury SA, Kishino K, Satoh R, Hashimoto K, Kikuchi H, Nishikawa H, Shirataki Y, Sakagami H. Tumor-specificity and apoptosis-inducing activity of stilbenes and flavonoids. *Anticancer Res* 2005; 25: 2055-63.

[117] Huang CY, Chan CY, Chou IT, Lien CH, Hung HC, Lee MF. Quercetin induces growth arrest through activation of FOXO1 transcription factor in EGFR-overexpressing oral cancer cells. *J Nutr Biochem* 2013; 24: 1596-603.

[118] Yuan Z, Wang H, Hu Z, Huang Y, Yao F, Sun S, Wu B. Quercetin inhibits proliferation and drug resistance in KB/VCR oral cancer cells and enhances its sensitivity to vincristine. *Nutr Cancer* 2015; 67: 126-36.

[119] Brown J, O′Prey J, Harrison PR. Enhanced sensitivity of human oral tumours to the flavonol, morin, during cancer progression: involvement of the Akt and stress kinase pathways. *Carcinogenesis* 2003; 24: 171-7.

[120] Park SJ, Kim MJ, Kim YK, Kim SM, Park JY, Myoung H. Combined cetuximab and genistein treatment shows additive anti-cancer effect on oral squamous cell carcinoma. *Cancer Lett* 2010; 292: 54-63.

[121] Johnson TL, Lai MB, Lai JC, Bhushan A. Inhibition of cell proliferation and MAP kinase and Akt pathways in oral squamous cell carcinoma by genistein and biochanin A. *Evid Based Complement Alternat Med* 2010; 7: 351-8.

[122] Gohulkumar M, Gurushankar K, Rajendra Prasad N, Krishnakumar N. Enhanced cytotoxicity and apoptosis-induced anticancer effect of silibinin-loaded nanoparticles in oral carcinoma (KB) cells. *Mater Sci Eng C Mater Biol Appl* 2014; 41: 274-82.

[123] Jin L, Miao J, Liu Y, Li X, Jie Y, Niu Q, Han X. Icaritin induces mitochondrial apoptosis by up-regulating miR-124 in human oral squamous cell carcinoma cells. *Biomed Pharmacother* 2017; 85: 287-95.

[124] Lee HJ, Auh QS, Lee YM, Kang SK, Chang SW, Lee DS, Kim YC, Kim EC. Growth inhibition and apoptosis-inducing effects of cudraflavone B in human oral cancer cells via MAPK, NF-κB, and SIRT1 signaling pathway. *Planta Med* 2013; 79: 1298-306.

[125] Lee J, Huang MS, Yang IC, Lai TC, Wang JL, Pang VF, Hsiao M, Kuo MY. Essential roles of caspases and their upstream regulators in rotenone-induced apoptosis. *Biochem Biophys Res Commun* 2008; 371: 33-8.

[126] Hou C, Li W, Li Z, Gao J, Chen Z, Zhao X, Yang Y, Zhang X, Song Y. Synthetic isoliquiritigenin inhibits human tongue squamous carcinoma cells through its antioxidant mechanism. *Oxid Med Cell Longev* 2017; 2017: 1379430.

[127] Lee YM, Jeong GS, Lim HD, An RB, Kim YC, Kim EC. Isoliquiritigenin 2`-methyl ether induces growth inhibition and apoptosis in oral cancer cells via heme oxygenase-1. *Toxicol In Vitro* 2010; 24: 776-82.

[128] Lin E, Lin WH, Wang SY, Chen CS, Liao JW, Chang HW, Chen SC, Lin KY, Wang L, Yang HL, Hseu YC. Flavokawain B inhibits growth of human squamous carcinoma cells: involvement of apoptosis and cell cycle dysregulation in vitro and in vivo. *J Nutr Biochem* 2012; 23: 368-78.

[129] Azuine MA, Bhide SV. Adjuvant chemoprevention of experimental cancer: catechin and dietary turmeric in forestomach and oral cancer models. *J Ethnopharmacol* 1994; 44: 211-7.

[130] Helton WB, Valentino J, Choi EY, Gairola CG, Swanson H. Apigenin and kaempferol effects on oral cancer FaDu cells. *Otolaryncol Head Neck Surg* 2008; 139 2 Suppl: P180.

[131] Rajasekar M, Suresh K, Sivakumar K. Diosmin induce apoptosis through modulation of STAT-3 signaling in 7,12 dimethylbenz(a)anthracene induced hamster buccal pouch carcinogenesis. *Biomed Pharmacother* 2016; 83: 1064-70.

[132] Balasubramanian S, Govindasamy S. Inhibitory effect of dietary flavonol quercetin on 7,12-dimethylbenz[a]anthracene-induced hamster buccal pouch carcinogenesis. *Carcinogenesis* 1996; 17: 877-9.

[133] Priyadarsini RV, Nagini S. Quercetin suppresses cytochrome P450 mediated ROS generation and NFκB activation to inhibit the development of 7,12-dimethylbenz[a]anthracene (DMBA) induced hamster buccal pouch carcinomas. *Free Radic Res* 2012; 46: 41-9.

[134] Kawabata K, Tanaka T, Honjo S, Kakumoto M, Hara A, Makita H, Tatematsu N, Ushida J, Tsuda H, Mori H. Chemopreventive effect of dietary flavonoid morin on chemically induced rat tongue carcinogenesis. *Int J Cancer* 1999; 83: 381-6.

[135] Wei D, Mei Y, Liu J. Quantification of doxorubicin and validation of reversal effect of tea polyphenols on multidrug resistance in human carcinoma cells. *Biotechnol Lett* 2003; 25: 291-4.

[136] Chang CM, Chang PY, Tu MG, Lu CC, Kuo SC, Amagaya S, Lee CY, Jao HY, Chen MY, Yang JS. Epigallocatechin gallate sensitizes CAL-27 human oral squamous cell carcinoma cells to the anti-metastatic effects of gefitinib (Iressa) via

synergistic suppression of epidermal growth factor receptor and matrix metalloproteinase-2. *Oncol Rep* 2012; 28: 1799-807.

[137] Zhang X, Zhang H, Tighiouart M, Lee JE, Shin HJ, Khuri FR, Yang CS, Chen Z, Shin DM. Synergistic inhibition of head and neck tumor growth by green tea (-)-epigallocatechin-3-gallate and EGFR tyrosine kinase inhibitor. *Int J Cancer* 2008; 123: 1005-14.

[138] Su CH, Chen LJ, Liao JF, Cheng JT. Increase of phosphatase and tensin homolog by silymarin to inhibit human pharynx squamous cancer. *J Med Food* 2013; 16: 778-84.

[139] Ohno H, Araho D, Uesawa Y, Kagaya H, Ishihara M, Sakagami H, Yamamoto M. Evaluation of cytotoxicity and tumor-specificity of licorice flavonoids based on chemical structure. *Anticancer Res* 2013; 33: 3061-8.

[140] Shirataki Y, Wakae M, Yamamoto Y, Hashimoto K, Satoh K, Ishihara M, Kikuchi H, Nishikawa H, Minagawa K, Motohashi N, Sakagami H. Cytotoxicity and radical modulating activity of isoflavones and isoflavanones from Sophora species. *Anticancer Res* 2004; 24: 1481-8.

[141] Gunda SK, Kongaleti SF, Shaik M. Natural flavonoid derivatives as oral human epidermoid carcinoma cell inhibitors. *Int J Comput Biol Drug Des* 2015; 8: 19-39.

Chapter 5

ANTICANCER ACTION OF FLAVONOIDS IN ESOPHAGEAL CANCER MODELS

5.1. INTRODUCTION TO ESOPHAGEAL CANCER

Esophageal cancer is a malignant neoplasm evolving in the esophagus. To date, it is one of the least explored cancers in the whole word, being at the same time one of the deadliest and most aggressive tumors [1-5]. Esophageal cancer ranks the eighth most common cancer and the sixth leading cause of cancer-related death in the world, the fifth in men and the eighth in women [1-3, 5, 6-16]. While the global incidence and mortality has been increasing in the recent years, cancer of esophagus constitutes about 7% of all gastrointestinal tumors [4, 17, 18]. Regarding to incidence rates, this malignancy varies worldwide more than any other cancer, revealing more than 50-fold difference in its geographic distribution, i.e., between high- and low-rate regions [6, 15, 19-22]. Whereas in 2008, the age-standardized rate was 0.8 and 2.7 cases per 100000 in Greece and Canada, respectively, this parameter was 16.7 per 100000 in China, reaching even 30.2 per 100000 in Xinjiang Uyghur Autonomous Region in the north-west China [15, 22]. Indeed, about 70% of all esophageal cancer cases occur in China and in certain parts of north China it is the most common malignancy considering both incidence and mortality rates [3, 14, 19, 23]. The incidence is higher also in Central Asia, south Iran, India, and South Africa [7, 10, 17].

Based on the histological differences, there are two principal forms of esophageal malignancy possessing also distinct risk factors and pathological characteristics, esophageal squamous cell carcinoma (ESCC) and esophageal adenocarcinoma (EAC) [1, 4, 6, 8, 11, 12, 14, 16, 19, 24]. ESCC is the predominant type of esophageal cancer worldwide, being prevalent in East Asia, particularly in China, Southern and Eastern

Africa; EAC is more common in the industrialized Western countries where its incidence is rapidly and continuously increasing [5, 6, 8-10, 12, 13, 16, 24-32]. The highest incidence of EAC in the world is in the UK with 16 cases per 100000 people [29]. ESCC presents as ulcerating or infiltrating epithelial lesions in the middle or upper third of the esophagus, EAC typically develops in the lower third of the esophagus near the gastroesophageal junction [6, 8, 14, 16, 25, 33]. ESCC is rather uncommon in individuals younger than 40 years of age, after that the incidence elevates with each decade of life [6]. Men have a 3-4-fold higher risk for developing this form of esophageal neoplasm than women [6]. The principal precursor lesion for ESCC is epithelial dysplasia and this malignancy develops as the result of progressive sequence of histopathological changes from esophagitis, atrophy, mild to severe dysplasia, carcinoma in situ, and invasive carcinoma [6, 16, 19]. The main risk factors for this type of malignancy include alcohol and tobacco consumption and deficiencies in certain nutrients occurring in fresh fruits and vegetables, such as vitamins and trace minerals [6, 16, 19, 20, 33]. At that, the use of alcohol can further increase the risk among tobacco smokers [6]. Consumption of salty, pickled and moldy foodstuffs might also be related to the development of ESCC due to the frequent contamination of these items with carcinogenic nitrosamines and fungal toxins [6, 20]. Furthermore, consumption of hot food and beverages, including drinking hot tea, may produce the thermal irritation and promote the ESCC development [6, 19, 34, 35]. In addition, previous exposure to radiation therapy is shown to be associated with an increased risk for ESCC [6]. The potential role of infection with human papillomavirus (HPV) is also suggested but its involvement in the tumorigenesis process is still rather unknown [6].

EAC arises from the glandular cells that are present at the junction of the esophagus and the stomach [19]. The gastroesophageal reflux disease induces chronic inflammation, leading to premalignant Barrett`s esophagus, an intestinal metaplasia that occurs when continued exposure to stomach acid results in the damage to esophagus and replacement of the inner esophageal lining by glandular cells [2, 18, 19]. The patients with Barrett`s esophagus have esophageal cancer incidence rate of 7 cases per 1000 per year which exceeds more than 100-fold the esophageal tumor rate in the general US population, whereas the segment length of Barrett`s esophagus is associated with the EAC risk [18, 19]. The other established risk factors for EAC include also *Helicobacter pylori* infection, tobacco use and obesity (characterized by high body mass index), whereas diet is a potentially modifiable factor [16, 25, 33, 36].

Considering the high geographic variation in the incidence of esophageal cancer, it is obvious that environmental and behavioral factors, including lifestyle and dietary habits, contribute to the multifactorial etiology, besides familial risk factors and genetic factors. Several tobacco carcinogens, including polycyclic aromatic hydrocarbons and aromatic amines as well as certain nitrosamines might be causally associated with both types of esophageal cancer, ESCC and EAC [2, 6, 20, 26]. Alcohol drinking is another established risk factor [19]. On the other hand, esophageal tumorigenesis is shown to be sensitive to dietary influences, whereas consumption of red and processed meat and saturated fat may elevate the cancer risk and high intake of fruits and vegetables rich in dietary fiber, vitamins and diverse phytochemicals can reduce the risk of this usually fatal disease [3, 5, 7, 16, 27, 31].

The treatment modalities of esophageal cancer include surgery, chemotherapy, radiotherapy and their combinations [9, 25]. As the initial symptoms are atypical, this cancer is often diagnosed in locally advanced or distantly metastasized stages, making the complete surgical resection of tumor mass very complicated or unlikely for most patients [3, 6, 8, 9, 14, 17, 20, 25]. Systemic chemotherapy is applied as the standard treatment for metastatic esophageal cancer, however, the general status of these patients is often poor with emerging toxic side effects and drug resistance [3, 9, 17]. Although radiotherapy has been proved to be an efficient option for non-surgical management of esophageal cancer, radiation tolerance is a frequent by-effect, revealing altogether that current treatment possibilities for esophageal cancer offer only poor cure rates [10, 25]. Indeed, esophageal cancer is related to a dismal prognosis and the 5-year overall survival rate is estimated to be only about 5-20% with no substantial improvement in the last 40 years [1, 2, 6, 8, 12, 13, 17, 23, 25, 29, 30, 37]. Besides the high recurrence rate after treatment, also the lack of screening programs leading to the delay in cancer detection as well as late presentation, early metastasis and highly invasive nature of this disease are responsible for such a poor prognosis [1, 9, 28]. Therefore, in order to control the spread of this cancer and improve the survival rate but also to reduce the costs of esophageal cancer treatment, there is a constant need for development of methods for earlier detection and diagnosis, novel treatment modalities and more efficient chemotherapeutic agents [3, 7, 23, 25, 30, 38].

As esophageal cancer has become an important public health problem, there is no doubt that strategies for its prevention also need to be evolved. One approach for this is through changes in lifestyle, especially cessation of alcohol consumption and smoking, and elimination of the high salt and moldy foods with potential carcinogenic contamination [1, 6]. Numerous epidemiological studies have shown that the increased intake of fruits and vegetables might significantly decrease the occurrence of esophageal cancer indicating that some phytochemicals in these products could possess antitumorigenic properties and behave as preventive agents [1, 6, 18, 20, 21, 26, 29, 32, 35, 39]. Among such secondary metabolites, flavonoids exert a spectrum of biological properties that may, at least in part, account for the protection derived from regular consumption of fruits and vegetables [29, 35]. At that, an efficient chemopreventive agent should either block the progression of premalignant lesions or extend the latency period for their progression [6]. Further identification of antitumor agents in plant-based food items is very important contributing to the potential prevention and treatment opportunities in the future. In parallel to this, a more precise understanding of the biological mechanisms underlying esophageal carcinogenesis with establishing novel cellular target molecules for different bioactive flavonoids is equally important [14, 28].

5.2. Model Systems to Study the Effect of Flavonoids in Esophageal Carcinogenesis

The potential antitumor effects of structurally various flavonoids have been studied in different *in vitro* cell line and *in vivo* animal models of esophageal cancer. At that, both the cell lines established from esophageal squamous cell carcinoma (ESCC) as well as esophageal adenocarcinoma (EAC) tissues have been used. The overview of human esophageal cancer cell lines applied in works with flavonoids is presented in Table 5.1, survey of the respective animal models with both chemical-induced as well as xenograft implanted tumors is exhibited in Table 5.2.

As can be seen from the Table 5.1, the majority of human esophageal cancer cell lines used in the experimental assays with flavonoids are ESCC lines. As two main esophageal cancer types, ESCC and EAC, possess distinct etiological factors and pathological characteristics, it is possible that chemopreventive properties of flavonoids might differ in models with different origin of malignancies. Therefore, more works with human EAC lines are certainly needed to clarify this issue in the future.

Animal models present an excellent opportunity to assess chemopreventive regimens against cancer. As experimental rodent models, human esophageal cancer cell lines engineered mice xenografts and chemical-induced rat esophageal tumorigenesis have been used (Table 5.2). The food contaminant N-nitrosomethylbenzylamine (NMBA)-induced carcinogenesis in the rat esophagus has proved to be a valuable tool for investigation of the biology and novel targets for intervention, prevention and therapy of ESCC [6, 7]. This nitrosamine is definitely the most potent inducer of rat esophageal tumors and needs metabolic activation and conversion to produce damage to the esophageal DNA [6, 7]. Premalignant lesions induced by NMBA closely mimic those found in the human disease, including epithelial dysplasia [6]. Another carcinogenic chemical applied for induction of esophageal tumorigenesis (ESCC) in rats is N-methyl-N-amylnitrosamine (MNAN) with neoplasms similar to those of humans and incidence of preneoplastic lesions, including hyperplasia and dysplasia [20].

Table 5.1. Human esophageal cancer cell lines used for *in vitro* studies with flavonoids

Cell line name	Cell line type[1]	Other information about cell line[2]	Ref.
EC109, Eca109	ESCC line		[8, 10-12, 14, 17, 23, 25, 27, 28, 30, 37, 38, 40, 44, 45, 48]
EC9706, Eca9706	ESCC line		[3, 13, 28, 46, 47, 49]
HEC-46R1	ESCC line, well-differentiated		[51]
KYSE30	ESCC line		[13, 47]
KYSE150	ESCC line		[12, 41-43, 50]
KYSE270	ESCC line		[9]
KYSE510	ESCC line	Mutant p53	[4, 42, 50]
OE19	EA line; moderately differentiated	Over-expressing HER-2/neu	[32, 43]
OE33, JROECL33	EA line; poorly differentiated	Mutant p53, constitutive expression of COX-2 protein	[5, 29, 32]
SKGT4	EA line		[2]
T.Tn	ESCC line		[9]
TE1	ESCC line	Mutant p53 (mutations in exon 6 and 8)	[8, 14, 28, 52]
TE2	ESCC line	Wild type p53	[52]
TE8	ESCC line		[2]
TE13	ESCC line		[10]

[1]EA, esophageal adenocarcinoma; ESCC, esophageal squamous cell carcinoma.
[2]COX-2, cyclooxygenase-2; HER-2/neu, human epidermal growth factor receptor 2.

Table 5.2. Animal models used for *in vivo* studies of flavonoids on esophageal cancer

Species	Strain	Gender	Age, weeks	Weight, g	Induction of tumorigenesis[1]		Application route and site	Ref.
					Chemical	Implanted cell line[2]		
Mice	BALB/c	Male, female	4…8	20…25		EC109 (Eca109), EC9706 (Eca9706), KYSE30, KYSE150, KYSE270, SKGT4	Subcutaneous injection of cells into the left or right flank, right axilla area, dorsal scapular region, lateral tail vein or left hind limb	[2, 3, 8, 9, 12-14, 17, 25, 28, 30, 47]
Rats	F344, Wistar	Male	6…10		MNAN, NMBA		Intraperitoneal injection, subcutaneous administration	[7, 20, 34]

[1]MNAN, N-methyl-N-amylnitrosamine; NMBA, N-nitrosomethylbenzylamine.
[2]see Table 5.1.

Table 5.3. ***In vitro*** **antiproliferative effects and respective molecular mechanisms of flavonoids in human esophageal cancer cell lines**

Compound	Cell line	Cell growth inhibition IC_{50}, μM[1]	Assay method[2]	Assay time, h	Action mechanism[3]	Effect in human normal oral cells[4]	Ref.
CHALCONES							
Licochalcone A	Eca109	NA up to 10 μM	MTT	24	Downregulation of Akt protein expression at higher doses (10 μM). Synergism in TRAIL-induced apoptosis through inhibition of Akt and promotion of proteasomal degradation of XIAP		[38]
FLAVANOLS							
Catechin (C)	OE19	NA up to ~200 μM	NR	24, 48	No effect on cell cycle or apoptosis		[32]
	OE33	NA up to ~200 μM	NR	24, 48	No effect on cell cycle or apoptosis		[32]
Epicatechin (EC)	OE19	NA up to ~200 μM	NR	24, 48	No effect on cell cycle or apoptosis		[32]
	OE33	NA up to ~200 μM	NR	24, 48	No effect on cell cycle or apoptosis		[32]
Epigallocatechin gallate (EGCG)	Eca109	256	MTT	48	Apoptosis, G1 phase cell cycle arrest, production of ROS, increase in cleaved caspase-3, decrease in VEGF expression		[8]
	Eca109				Apoptosis through activation of caspases-3 and -9 via MAPK JNK and p38 pathways. Significantly higher apoptotic effect in combination with ascorbic acid		[40]
	KYSE150				Reactivation of methylation-silenced gene RARß		[42]

Table 5.3. (Continued)

Compound	Cell line	Cell growth inhibition IC_{50}, μM[1]	Assay method[2]	Assay time, h	Action mechanism[3]	Effect in human normal oral cells[4]	Ref.
	KYSE150				Inhibition of colony formation. Auto-oxidation-dependent suppression of EGFR phosphorylation and protein level		[43]
	KYSE150	20	MTT	24			[41]
	KYSE510	~50	TB	144	Suppression of colony formation, inhibition of DNMT activity, reversal of hypermethylation and reactivation of methylation-silenced genes ($p16^{INK4\alpha}$, RARß, MGMT, hMLH1)		[42]
	OE19				Auto-oxidation-dependent decrease in HER-2/neu phosphorylation and protein level		[43]
	SKGT4		MTT	24, 72, 120	Reduction of invasion, downregulation of expression of p-Erk1/2, c-Jun, COX-2, up-regulation of activated caspase-3		[2]
	TE1	162	MTT	48	Apoptosis, G1 cell cycle arrest, production of ROS, increase in cleaved caspase-3, decrease in VEGF expression		[8]
	TE8		MTT	24, 72, 120	Reduction of invasion, downregulation of expression of p-Erk1/2, c-Jun, COX-2, up-regulation of activated caspase-3		[2]
Peracetylated epigallocatechin gallate (AcEGCG)	KYSE150	10	MTT	24	Increase in intracellular EGCG level compared to parent EGCG		[41]

Compound	Cell line	Cell growth inhibition IC_{50}, μM[1]	Assay method[2]	Assay time, h	Action mechanism[3]	Effect in human normal oral cells[4]	Ref.
FLAVANOLIGNANS							
Silibinin	KYSE270	133	MTT	96	Apoptosis, downregulation of Bcl-2, increase in Bax protein and cleaved caspase-3. Inhibition of migration and invasion, induction of E-cadherin expression, suppression of N-cadherin. Increase in p-AMPK expression, decrease in phosphorylation of mTOR	No cytotoxic effect up to 200 μM in NE2-hTERT	[9]
	T.Tn	95	MTT	96	Apoptosis, downregulation of Bcl-2, increase in Bax protein and cleaved caspase-3. Inhibition of migration and invasion, induction of E-cadherin expression, suppression of N-cadherin. Increase in p-AMPK expression, decrease in phosphorylation of mTOR		
FLAVANONES							
Hesperetin	Eca109	200	CCK-8	72	Apoptosis, depletion of GSH, accumulation of ROS, increase in caspases-3 and -9, release of cytochrome C, decrease in Bcl-2 and survivin, increase in Bax		[25]
Naringenin	EC109	NA up to 80 μM	MTT	48			[44]
Sakuranetin	EC109		MTT	24, 48, 72	Rise in mitochondrial membrane potential loss, apoptosis		[45]
FLAVONES							
3`,4`-Dimethoxyflavone (3`,4`-DMF)						Inhibition of BaP-induced CYP1B1 mRNA expression in HET-1A	[26]

Table 5.3. (Continued)

Compound	Cell line	Cell growth inhibition IC_{50}, μM[1]	Assay method[2]	Assay time, h	Action mechanism[3]	Effect in human normal oral cells[4]	Ref.
5,7-Dimethoxyflavone (5,7-DMF)						Inhibition of BaP-DNA binding, BaP-induced CYP1B1 mRNA and protein expression in HET-1A	[26]
6-Methoxycapillarisin	EC109		MTT	24, 48, 72	Rise in mitochondrial membrane potential loss, apoptosis		[45]
Apigenin	KYSE510	112	MTT	24	Apoptosis through mitochondrial pathway in a p53-independent manner, G2/M phase cell cycle arrest through increases in p63 and p73 protein translation or stability and regulation of $p21^{waf1}$, cyclin B1 and PIG3. Cleavage of caspases-3 and -9, not -8		[4]
	OE33	151	MTT	24	Apoptosis through mitochondrial pathway in a p53-independent manner, G2/M phase cell cycle arrest through increases in p63 and p73 protein translation or stability and regulation of cyclin B1, PIG3, GADD45ß, 14-3-3σ. Cleavage of caspases-3 and -9, not -8		[5]

Compound	Cell line	Cell growth inhibition IC_{50}, μM[1]	Assay method[2]	Assay time, h	Action mechanism[3]	Effect in human normal oral cells[4]	Ref.
Chrysin	EC109	~80	MTT	48			[44]
	KYSE510	63	MTT	24	Apoptosis through mitochondrial pathway in a p53-independent manner, G2/M phase cell cycle arrest through increases in p63 and p73 protein translation or stability and regulation of $p21^{waf1}$, cyclin B1 and PIG3. Cleavage of caspases-3 and -9, not -8		[4]
	OE33	107	MTT	24	Apoptosis through mitochondrial pathway in a p53-independent manner, G2/M phase cell cycle arrest through increases in p63 and p73 protein translation or stability and regulation of cyclin B1, PIG3, GADD45ß, 14-3-3σ. Cleavage of caspases-3 and -9, not -8		[5]
Luteolin	Eca109	70.68±1.72	MTT	24	Apoptosis, G0/G1 phase cell cycle arrest, decrease in cyclin D1, survivin and c-myc mRNA levels, increase in p53 mRNA and protein expression		[27]
	KYSE510	35	MTT	24	Apoptosis through mitochondrial pathway in a p53-independent manner, G2/M phase cell cycle arrest through increases in p63 and p73 protein translation or stability and regulation of $p21^{waf1}$, cyclin B1 and PIG3. Cleavage of caspases-3 and -9, not -8		[4]

Table 5.3. (Continued)

Compound	Cell line	Cell growth inhibition IC_{50}, μM[1]	Assay method[2]	Assay time, h	Action mechanism[3]	Effect in human normal oral cells[4]	Ref.
	OE33	84	MTT	24	Apoptosis through mitochondrial pathway in a p53-independent manner, G2/M phase cell cycle arrest through increases in p63 and p73 protein translation or stability and regulation of cyclin B1, PIG3, GADD45ß, 14-3-3σ. Cleavage of caspases-3 and -9, not -8		[5]
Oroxylin A	Eca109	333.79	CCK-8	24			[10]
	TE13	285.79	CCK-8	24			[10]
Orientin	EC109	>80	MTT	24	Apoptosis, increase in p53 expression, decrease in Bcl-2		[23]
		44.6% inhibition at 80 μM	MTT	48			
		58.8% inhibition at 80 μM	MTT	72			
Vitexin	EC109	>80	MTT	24	Apoptosis, increase in p53 expression, decrease in Bcl-2		[23]
		>80	MTT	48			
		>80	MTT	72			
FLAVONOLS							
Avicularin	EC109	NA up to 80 μM	MTT	48			[44]
Galangin	EC109		MTT	48	Synergistic growth inhibition in combination with berberine	NA up to 50 μM, no synergistic effect in combination with berberine in HEEC	[28]
	Eca9706		MTT	48	Apoptosis, G2/M phase cell cycle arrest, suppression of cell migration, production of ROS, reduction of Wnt3α and ß-catenin, inhibition of PI3K/JAK2/STAT3 pathway. Synergistic effects with berberine		

Compound	Cell line	Cell growth inhibition IC_{50}, μM[1]	Assay method[2]	Assay time, h	Action mechanism[3]	Effect in human normal oral cells[4]	Ref.
	TE1		MTT	48	Synergistic growth inhibition in combination with berberine		[28]
Isorhamnetin	Eca109		TB	24, 48, 72, 96, 120	Partial inhibition of antiproliferative effects by cellular stress response during early exposure. Decrease in IκBα up to 48 h incubation, further increase in IκBα level and decrease in NF-κB activation. Early increase in COX-2 expression, followed by its decline		[11]
	Eca109	126.5±0.3	MTT	48	Apoptosis, G1 phase cell cycle arrest, increase in Bax, decrease in Bcl-2 expression, reinforcing c-fos and p53 expression, reduction of clone formation, suppression of DNA replication		[37]
Kaempferol	Eca109		CellTiter96	24, 48, 72	G0/G1 cell cycle arrest, increase in p21, decrease in cyclin D1 expression. Inhibition of glycolysis via decrease in HK2. Inhibition of EGFR phosphorylation, suppression of ERK1/2 and Akt pathways		[12]
	EC9706	46.3% inhibition at 105 μM	MTT	24	Increase in intracellular Ca^{2+} level		[46]

Table 5.3. (Continued)

Compound	Cell line	Cell growth inhibition IC_{50}, μM[1]	Assay method[2]	Assay time, h	Action mechanism[3]	Effect in human normal oral cells[4]	Ref.
	KYSE150		CellTiter96	24, 48, 72	G0/G1 cell cycle arrest, increase in p21, decrease in cyclin D1 expression. Inhibition of glycolysis via decrease in HK2. Inhibition of EGFR phosphorylation, suppression of ERK1/2 and Akt pathways phosphorylation, suppression of ERK1/2 and Akt pathways		[12]
	KYSE510	72	MTT	24	Apoptosis through mitochondrial pathway in a p53-independent manner, G2/M phase cell cycle arrest through increases in p63 and p73 protein translation or stability and regulation of $p21^{waf1}$, cyclin B1 and PIG3. Cleavage of caspases-3 and -9, not -8		[4]
	OE33	111	MTT	24	Apoptosis through mitochondrial pathway in a p53-independent manner, G2/M phase cell cycle arrest through increases in p63 and p73 protein translation or stability and regulation of cyclin B1, PIG3, GADD45ß, 14-3-3σ. Cleavage of caspases-3 and -9, not -8		[5]
Kaempferol-zinc(II) complex	EC9706	72% inhibition at 105 μM	MTT	24	Apoptosis, increase in intracellular Ca^{2+} level		[46]

Compound	Cell line	Cell growth inhibition IC_{50}, μM[1]	Assay method[2]	Assay time, h	Action mechanism[3]	Effect in human normal oral cells[4]	Ref.
Myricetin	EC9706		CCK-8	48	Apoptosis, G0/G1 phase cell cycle arrest, decrease in survivin, cyclin D1 and Bcl2 expression, increase in caspase-3 and p53 levels		[3]
	EC9706		CCK-8	24, 48, 72	Apoptosis through RSK2 and Bad, G0/G1 phase cell cycle arrest, suppression of migration and invasion		[47]
	KYSE30		CCK-8	24, 48, 72	Apoptosis through RSK2 and Bad, G0/G1 phase cell cycle arrest, suppression of migration and invasion		[47]
	KYSE510	233	MTT	24	Apoptosis through mitochondrial pathway in a p53-independent manner, G2/M phase cell cycle arrest through increases in p63 and p73 protein translation or stability and regulation of $p21^{waf1}$, cyclin B1 and PIG3. Cleavage of caspases-3 and -9, not -8		[4]
	OE33	205	MTT	24	Apoptosis through mitochondrial pathway in a p53-independent manner, G2/M phase cell cycle arrest through increases in p63 and p73 protein translation or stability and regulation of cyclin B1, PIG3, GADD45ß, 14-3-3σ. Cleavage of caspases-3 and -9, not -8		[5]

Table 5.3. (Continued)

Compound	Cell line	Cell growth inhibition IC_{50}, μM[1]	Assay method[2]	Assay time, h	Action mechanism[3]	Effect in human normal oral cells[4]	Ref.
Quercetin	EC109	>80	MTT	48			[44]
	KYSE510	48	MTT	24	Apoptosis through mitochondrial pathway in a p53-independent manner, G2/M phase cell cycle arrest through increases in p63 and p73 protein translation or stability and regulation of $p21^{waf1}$, cyclin B1 and PIG3. Cleavage of caspases-3 and -9, not -8		[4]
	OE33	>50	NR	24	Apoptosis, S phase cell cycle arrest, decrease in COX-2 mRNA expression, suppression of PGE2 production		[29]
		10…50	NR	48			
		10…50	NR	72			
	OE33	78	MTT	24	Apoptosis through mitochondrial pathway in a p53-independent manner, G2/M phase cell cycle arrest through increases in p63 and p73 protein translation or stability and regulation of cyclin B1, PIG3, GADD45ß, 14-3-3σ. Cleavage of caspases-3 and -9, not -8		[5]
Nanoliposomal quercetin	Eca109/ 9706				Anti-cancer stem cell effects. Apoptosis, down-regulation of NF-κBp65, HDAC1 and cyclin D1, up-regulation of caspase-3, promotion of E-cadherin expression		[48]

Table 5.3. (Continued)

Compound	Cell line	Cell growth inhibition IC_{50}, μM[1]	Assay method[2]	Assay time, h	Action mechanism[3]	Effect in human normal oral cells[4]	Ref.
	Eca9706	>60	MTT	48	Apoptosis, acting as HDI, down-regulation of DNMT1, NF-κBp65, HDAC1 and cyclin D1, up-regulation of caspase-3 and $p16^{INK4\alpha}$ expression, increase in E-cadherin. Stronger inhibition of cell growth in combination with butyrate		[49]
ISOFLAVONES							
Alpinumisoflavone	EC9706	>20	MTT	24, 48, 72	Apoptosis via modulation of miR-370/PIM1 signaling, increase in miR-370 and decrease in PIM1 expression		[13]
	KYSE30	>20	MTT	24, 48, 72	Apoptosis via modulation of miR-370/PIM1 signaling, increase in miR-370 and decrease in PIM1 expression		[13]
Biochanin A	HEC-46R1	35.2	CV	96			[51]
	KYSE510		TB	144	Re-expression of RARß mRNA, inhibition of DNMT activity		[50]
Daidzein	HEC-46R1	>157.3	CV	96			[51]
	KYSE510		TB	144	Re-expression of RARß mRNA, inhibition of DNMT activity		[50]

Table 5.3. (Continued)

Compound	Cell line	Cell growth inhibition IC_{50}, μM[1]	Assay method[2]	Assay time, h	Action mechanism[3]	Effect in human normal oral cells[4]	Ref.
Genistein	HEC-46R1	30.7	CV	96			[51]
	KYSE510	2…20	TB	144	No apoptosis, decrease in colony formation. Reversal of DNA hypermethylation and re-expression of methylation-silenced genes (RARß, p16, MGMT), inhibition of DNMT activity. Synergistic action in combination with trichostatin, 5-aza-dCyd and sulphoraphane		[50]
Genistin	HEC-46R1	NA	CV	96			[51]
Prunectin	HEC-46R1	20.8	CV	96			[51]
Puerarin	Eca109	>1000	CCK-8	48	Apoptosis (at 400 μM)		[17]
	HEC-46R1	>192.1	CV	96			[51]
PRENYLATED FLAVONOIDS							
Icariin	EC109	106.13	CCK-8	12	Apoptosis, suppression of adhesion and migration, increase in caspase-9 activity, generation of ROS. Increase in NADPH oxidase activity, decrease in GSH levels and Bcl2 expression. Activation of ER stress signaling	NA up to 80 μM in HET-1A	[14]
		73.65	CCK-8	24			
		38.59	CCK-8	36			
	TE1	115.29	CCK-8	12	Apoptosis, suppression of adhesion and migration, increase in caspase-9 activity, generation of ROS. Increase in NADPH oxidase activity, decrease in GSH levels and Bcl2 expression. Activation of ER stress signaling		
		76.77	CCK-8	24			
		42.21	CCK-8	36			

Compound	Cell line	Cell growth inhibition IC_{50}, μM[1]	Assay method[2]	Assay time, h	Action mechanism[3]	Effect in human normal oral cells[4]	Ref.
Baohuoside-1	Eca109	48.2	MTT	48	Apoptosis, decrease in expression of ß-catenin, survivin and cyclin D1 mRNA and protein		[30]

[1]NA, not active. [2]CCK-8, cell counting kit-8; CV, crystal violet; MTT, 3-(4,5-dimethylthiazol-2-yl)-2,5-diphenyltetrazolium bromide; NR, neutral red; TB, trypan blue exclusion. [3]5-aza-dCyd, 5-aza-2`-deoxycytidine; Akt, protein kinase B; AMPK, AMP-activated protein kinase; BaP, benzo[α]pyrene; COX2, cyclooxygenase-2; DNMT, DNA methyltransferase; CYP, cytochrome P450; EGFR, epidermal growth factor receptor; ER, endoplasmic reticulum; ERK, extracellular signal-regulated kinase; GSH, reduced glutathione; HDAC1, histone deacetylase 1; HDI, histone deacetylases inhibitor; HER-2/neu, human epidermal growth factor receptor-2; HK2, hexokinase 2; hMLH1, human mutL homologue 1; JAK, Janus kinase; JNK, c-Jun amino-terminal kinase; MAPK, mitogen-activated protein kinase; MGMT, O^6-methylguanine methyltransferase; mTOR, mechanistic target of rapamycin; NADPH, nicotinamide adenine dinucleotide phosphate; NF-κB, nuclear factor κB; PGE2, prostaglandin E2; PI3K, phosphatidylinositol 3-kinase; PIG3, p53-inducible gene 3; PIM1, proto-oncogene serine/threonine-protein kinase; RARß, retinoic acid receptor ß; ROS, reactive oxygen species; RSK2, ribosomal S6 kinase 2; STAT, signal transducer and activator of transcription; TRAIL, tumor necrosis factor-related apoptosis-inducing ligand; VEGF, vascular endothelial growth factor; XIAP, X-linked inhibitor of apoptosis protein. [4]HEEC, normal human esophageal cells; HET-1A, normal but immortalized human esophageal epithelial cell line; NE2-hTERT, immortalized normal esophageal epithelial cells.

In the following subsection 5.3, the effects of various flavonoids in different esophageal cancer models are analyzed by studying only individual compounds and not flavonoid mixtures or their natural extracts on the human esophageal cancer cell lines and experimental animals bearing human esophageal carcinoma xenografts.

5.3. Effect of Different Flavonoids on Esophageal Carcinogenesis

Compared to the oral cancer models (chapter 4) and despite the highly aggressive and deadly nature of the esophageal cancer, the data about anticancer role of flavonoids in esophageal cancer systems is considerably more scarce pointing to a clear and urgent need for further studies.

5.3.1. *In Vitro* Results

Results of *in vitro* studies about the effects of flavonoids on human esophageal cancer cell lines are compiled in Table 5.3. When available, activity data of these natural compounds in normal human esophageal cells (HEEC or HET-1A) are also presented for comparison of their effects on malignant and healthy cells.

5.3.1.1. Flavanols

Only few studies about the potential anticancer role of flavanols against esophageal tumorigenesis have been published to date and most of them are focused on the major polyphenolic component of green tea, EGCG. In human ESCC cell lines Eca109 and TE1, EGCG was shown to suppress proliferation with half-maximal inhibitory constants (IC_{50}) of more than 150 μM, induce G1 phase cell cycle arrest and apoptosis accompanied with ROS generation and activation of caspase-3 [8]. In Eca109 cells, EGCG triggered apoptosis via mitogen-activated protein kinase (MAPK) pathways and this effect was significantly enhanced by ascorbic acid [40]. In TE8 and SKGT4 cells, EGCG inhibited viability and invasion capacity through the downregulation of phosphorylated Erk1/2, c-Jun and cycloxygenase-2 (COX-2) expression; inhibitory effects were observed also in combination with curcumin and lovastatin [2]. Compared to Eca109 and TE1 lines, EGCG could inhibit the growth of other ESCC cells, KYSE150 and KYSE510, at substantially lower doses with IC_{50} values in the range of 20-50 μM [41, 42]. This flavanol inhibited DNA methyltransferase (DNMT) activity and reactivated some methylation-silenced genes, including $p16^{INK4\alpha}$, retinoic acid receptor ß (RARß), O^6-methylguanine methyltransferase (MGMT) and human mutL homologue (hMLH1), in KYSE510 cells [42]. Hypermethylation of DNA is a crucial mechanism for silencing several genes participating in cell cycle regulation, DNA repair process and apoptosis, and suppression of DNMT activity can lead to reversal of hypermethylation and hence, prevention of tumorigenesis [42]. However, this flavanol is rather unstable and is extensively biotransformed. In KYSE150 cells, peracetylated EGCG was about two-fold more potent in inhibition of esophageal cancer cell growth and this effect was related to the increased intracellular levels of EGCG [41]. In the same cell line, KYSE150, EGCG suppressed the growth and induced auto-oxidation-dependent inactivation of epidermal growth factor receptor (EGFR), an effect that was prevented or delayed by the presence of superoxide dismutase as an EGCG stabilizing agent [43]. In EAC cell line OE19 that over-expresses HER-2/neu, treatment with EGCG triggered auto-oxidation-dependent decrease in phosphorylation and protein level of HER-2/neu [43].

Two other common flavanols, catechin and epicatechin, exerted no effects on proliferation of EAC cells OE19 and OE33, induced no significant alterations in cell cycle or in percentage of apoptotic cells [32] (Table 5.3).

5.3.1.2. Flavones

Three plant dietary flavones, apigenin, chrysin and luteolin, were all able to induce cytotoxic responses in two human esophageal cancer cell lines, KYSE510 and OE33 (both harboring mutant and non-functional tumor suppressor gene p53), revealing growth inhibitory potency order of luteolin> chrysin> apigenin [4, 5]. Compared to ESCC line KYSE510, EAC cells OE33 were somewhat more resistant to these flavones. In both lines, flavones-engendered cytotoxicity was mediated by arresting cell cycle progression at G2/M phase and causing p53-independent mitochondrial apoptosis through activation of caspases-3 and -9 [4, 5]. Treatment of KYSE510 cells with these flavones led to upregulation of $p21^{waf1}$ and PIG3, and downregulation of cyclin B1 levels [4]. Treatment of OE33 cells with three flavones resulted in upregulation of GADD45ß, 14-3-3σ and PIG3, and downregulation of cyclin B1 [5]. Such alterations of gene expression levels could be caused by increases in p63 and p73 protein translation or stability in KYSE510 and OE33 cells treated with apigenin, chrysin or luteolin [4, 5]. The doses of chrysin which exerted growth inhibitory effects in KYSE510 cells were rather similar to concentrations cytotoxically active in another ESCC line, EC109 [44]. However, differently from KYSE510 cells, luteolin induced G0/G1 phase cell cycle arrest in Eca109 (EC109) cells with triggering also apoptotic changes, indicating that action of flavonoids in esophageal cancer cells is cell line specific [27]. In addition, treatment of Eca109 cells with luteolin resulted in alteration of gene expression with downregulation of cyclin D1, survivin and c-myc, and upregulation of p53 level [27]. As apoptotic inhibitor survivin has been shown to be over-expressed in esophageal cancer tissues and its expression was proposed to reflect the prognosis of ESCC, luteolin might improve the disease prognosis [27].

There are too few studies published so far to estimate the role of methylation on cytotoxic activity of flavones in esophageal cancer cells. Oroxylin A (6-methoxybaicalein) was reported to inhibit the growth of ESCC lines Eca109 and TE13 with IC_{50} values closed to 300 μM [10]. Treatment of EC109 cells with 6-methoxycapillarisin led to antiproliferative action with apoptotic cell death [45]. 5,7-Dimethoxyflavone was able to downregulate benzo[α]pyrene (BaP)-triggered CYP1B1 mRNA and protein levels, and inhibit BaP-DNA binding in normal immortalized human esophageal HET-1A cells. In the same cellular system, 3`,4`-dimethoxyflavone also suppressed CYP1B1 expression, revealing potential chemopreventive action of these dimethoxyflavones against esophageal tumorigenesis as formation of carcinogen-DNA adducts is thought to play a crucial role in esophageal cancer initiation [26].

Two flavone glycosides, orientin (luteolin 8-C-glycoside) and vitexin (apigenin 8-C-glycoside), were demonstrated to suppress the growth of EC109 cells with induction of morphological apoptotic features and regulation of expression of apoptosis-related genes p53 and bcl-2. At that, somewhat stronger anticancer activities for orientin were shown as compared to vitexin at the same concentrations [23] (Table 5.3).

5.3.1.3. Flavonols

Several dietary flavonols have been tested concerning their *in vitro* activities against esophageal carcinogenesis. Among them, kaempferol suppressed tumor cell proliferation in ESCC lines Eca109 and KYSE150 by inducing G0/G1 phase cell cycle arrest and regulating the expression of cell cycle progression-related proteins (p21, cyclin D1) [12]. Moreover, this plant-derived flavonol inhibited tumor glycolysis with suppression of hexokinase-2 and decline of glucose uptake and lactate secretion [12]. These effects were largely attributed to the inhibition of EGFR activity by kaempferol and suppression of EGFR-mediated downstream pathways (ERK1/2 and Akt), showing that EGFR might be an important target for plant polyphenol kaempferol to exhibit its biological effects against esophageal carcinogenesis and consumption of this flavonol can be especially beneficial for EGFR-positive ESCC patients [12]. In EC9706 cells, it was shown that kaempferol in the complex with zinc ions exerted two-fold stronger cytotoxic effects than the free flavonol by inhibiting proliferation and inducing apoptosis via elevation of intracellular calcium ion level [46]. Treatment of human ESCC cells EC9706 with another common dietary flavonol, myricetin, led to the arrest of cell cycle progression in G0/G1 phase and increased the percentage of apoptotic cells [3, 47]. In both EC9706 and KYSE30 cells, myricetin triggered apoptosis through ribosomal S6 kinase 2 (RSK2) and inhibited also migration and invasion of tumor cells [47]. Quercetin was shown to suppress the growth in both human ESCC cells EC109 [44] and EAC line OE33 [29]. In

OE33 cells, quercetin was able to inhibit the mRNA expression of COX-2, decrease production of prostaglandin E2 (PGE2) and elevate the fraction of apoptotic cells [29]. All these three natural flavonols, kaempferol, myricetin and quercetin, were cytotoxic in two human esophageal cancer cell lines, KYSE510 and OE33, with cytotoxic potency profile of quercetin> kaempferol> myricetin [4, 5]. These effects were mediated by perturbation of the cell cycle progression at G2/M phase and induction of mitochondrial-mediated p53-independent apoptosis [4, 5]. The ability of kaempferol to block the cell cycle in different phases in KYSE150 (G0/G1) and KYSE510 cells (G2/M) was explained by the differences in expression of EGFR in these two lines: whereas KYSE150 cells over-express this receptor, the EGFR level in KYSE510 cells is very low [4, 12]. These results clearly show that effects of flavonoids on human esophageal cancer cells substantially depend on certain cell line and type. Treatment of KYSE510 cells with kaempferol, myricetin and quercetin led to upregulation of $p21^{waf1}$ and PIG3, and downregulation of cyclin B1 levels [4]. In OE33 cells, exposure to three flavonols resulted in upregulation of GADD45ß, 14-3-3σ and PIG3, and downregulation of cyclin B1 levels [5]. Such alterations in gene expression levels could be mediated by increases in p63 and p73 protein translation or stability in these two p53-mutant cell lines, KYSE510 and OE33, leading to cell cycle arrest and apoptosis by kaempferol, myricetin and quercetin [4, 5]. Although cytotoxic and even anti cancer stem cells effects were shown for nanoliposomal quercetin in human ESCC cells Eca109 and Eca9706, its potency and potential biological mechanisms were not compared with those characteristic to the parent quercetin [48, 49].

The current knowledge about the potential esophageal tumor-fighting effects of other flavonols is still rather scanty. Galangin was shown to induce apoptosis, arrest the cell cycle progression at G2/M phase, increase ROS production and suppress migration of Eca9706 cells through inactivation of Wnt3α signaling pathway, and revealed synergistic cytotoxic effects in combination with berberine in different human ESCC cell lines (EC109, Eca9706 and TE1) [28]. In normal human esophageal cells HEEC, galangin induced no significant decrease in cell survival up to 50 μM doses indicating higher sensitivity of malignant esophageal cells to this flavonol [28]. Isorhamnetin (3`-methoxyquercetin) significantly reduced the viability of Eca109 cells, induced G0/G1 phase cell cycle arrest and apoptosis with increasing the ratio of pro-apoptotic to anti-apoptotic proteins Bax/Bcl-2 [37]. The antiproliferative and cytotoxic action of this methoxylated flavonol appeared only after longer exposure (72 hours) as during the early treatment, isorhamnetin caused cellular stress response involving activation of NF-κB pathway and upregulation of COX-2 promoting initial survival of Eca109 cells [11]. The only flavonol glycoside studied so far, avicularin (quercetin 3-O-arabinofuranoside), revealed no cytotoxic activity up to 80 μM doses in EC109 cells, being significantly less potent that quercetin in similar assay conditions [44] (Table 5.3).

5.3.1.4. Isoflavones

The most common soy isoflavone, genistein, suppressed the growth of human ESCC KYSE510 cells and decreased colony formation without engendering apoptosis [50]. This natural isoflavone aglycone reversed DNA hypermethylation, inhibited DNMT activity and reactivated several methylation-silenced genes (RARß, p16$^{INK\alpha}$, MGMT) in KYSE510 cells [50]. In addition, genistein revealed synergistic effects in combination with sulphoraphane as compared to the effects of individual agents alone [50]. However, two other isoflavones, daidzein and biochanin A (4`-methoxygenistein), were somewhat less efficient in suppressing DNMT activity, reactivating RARß and inhibiting tumor cell growth [50]. Genistein and its methoxylated isoflavones, biochanin A (4`-methoxygenistein) and prunectin (7-methoxygenistein), were demonstrated to suppress the growth of human ESCC cells HEC-46R1 with rather similar antiproliferative efficacy, whereas daidzein and two isoflavone glycosides, genistin (genistein 7-glycoside) and puerarin (daidzein 8-glycoside), revealed almost no or only very weak growth inhibitory activity up to 200 μM doses [51]. A pyranoisoflavone, alpinumisoflavone, was reported to trigger apoptosis by targeting miR-370/PIM1 signaling in ESCC lines EC9706 and KYSE30 via upregulation of tumor suppressor miR-370 [13]. Altogether, these data show that only esophageal squamous carcinoma cell lines have been studied so far concerning their sensitivity towards flavonoids, as no data have been published about the potential anticancer effects of isoflavones and their structural analogues in human esophageal adenocarcinoma cell lines (Table 5.3).

5.3.1.5. Other Flavonoids and Related Compounds

There are only few studies available concerning the effects of citrus flavanones against human esophageal tumorigenesis. All these works were performed with EC109 (Eca109) esophageal squamous cell carcinoma line, with no information about the behavior of flavanones in esophageal adenocarcinoma cells. Hesperetin was able to trigger mitochondria-mediated apoptotic pathway, increased intracellular ROS production, decreased glutathione (GSH) level and elevated cleaved caspases-9 and -3 in ESCC line Eca109 [25]. However, these effects revealed at rather high concentrations with IC_{50} value of cell viability inhibition of 200 μM [25]. Naringenin exerted no cytotoxic effects up to 80 μM in EC109 cells [44]. On the contrary, methoxylated naringenin, sakuranetin (7-methoxynaringenin), exerted antiproliferative responses and triggered apoptosis by inducing DNA damage and rise of mitochondrial membrane potential loss already at 25 μM concentration in the same cellular system [45].

Flavanolignan silibinin suppressed proliferation, induced apoptosis and inhibited invasion and migration of human ESCC lines KYSE270 and T.Tn through stimulation of adenosine monophosphate-activated protein kinase (AMPK), while revealing no cytotoxic effects in immortalized normal esophageal epithelial cells [9]. It has been shown that signaling pathway of AMPK is often inactivated in human ESCC and hence, this kinase might be a potential therapeutic target for cancer treatment [9].

Prenylated flavonoid baohuoside-I (icariside II) inhibited the growth and strongly induced apoptosis by downregulating ß-catenin-dependent pathway in Eca109 cells. Targeting this signaling may be pivotal in controlling the progression of esophageal cancer, as it regulates a wide array of genes which are involved in proliferation, differentiation and survival, such as cyclin D1 and survivin [30]. Prenylated flavonol glycoside icariin decreased the viability of human ESCC cells EC109 and TE1 with suppression of cell adhesion and migration, decrease in GSH level, production of ROS, inducing mitochondria-mediated intrinsic apoptosis and elevation of nicotinamide adenine dinucleotide phosphate (NADPH) oxidase activity. Exposure to icariin activated endoplasmic reticulum stress-related signaling and intervention to this pathway might represent a novel treatment option in the fight against esophageal tumorigenesis [14]. At that, icariin did not affect the viability of normal human esophageal epithelial cells HET-1A [14].

Lastly, licochalcone-A belonging to chalcones subclass exhibited synergistic activity in tumor necrosis factor-related apoptosis-inducing ligand (TRAIL)-triggered apoptosis in Eca109 cells via inactivation of protein kinase B (Akt) and decrease in the level of X-linked inhibitor of apoptosis protein (XIAP) [38] (Table 5.3).

5.3.2. *In Vivo* Results

The results of studies performed with animals about the potential role of flavonoids against esophageal carcinogenesis are summarized in Table 5.4. These works include both the human ESCC cells-engineered xenografts in mice as well as chemical-induced esophageal carcinogenesis models in rats. Compared to the works conducted with experimental cell lines, substantially less data are available from *in vivo* assays. At that, some flavonoids have been reported to inhibit different phases (initiation, promotion, progression) of esophageal tumorigenesis in rats and suppress the tumor growth in mice bearing human esophageal cancer xenografts. It is important to emphasize that to date no study has revealed any opposite effects for studied flavonoids, such as increase in tumor burden or promoting tumorigenesis.

5.3.2.1. Flavanols

In a rat model of NMBA-induced esophageal carcinogenesis, oral or intraperitoneal administration of EGCG led to a decrease in the incidence and multiplicity of tumors suppressing both the formation as well as development of carcinoma, accompanied by the reduction of NMBA-upregulated expression of cyclin D1 and COX-2 and the generation of PGE2 [7, 34]. As cyclin D1 has been shown to be a strong prognostic factor for esophageal cancer and the expression level of COX-2 might be related to the tumor size [7], agents that inhibit these molecules might be efficient against carcinogenesis in the esophagus. However, such beneficial effects of EGCG were offset by addition of hot water (with temperature over 55^0C) showing that intake of hot beverages or foods can increase the esophageal cancer risk [34]. In SKGT4 cells-engineered nude mouse xenograft experiments, EGCG suppressed the tumor growth and decreased the expression of COX-2, phosphorylated ERK1/2 and a cellular proliferation biomarker Ki67 [2]. In mouse xenograft assay of Eca109 cells, this flavanol inhibited the tumor growth by elevating caspase-3 cleavage and reducing VEGF protein levels [8]. At that, no obvious toxicity was detected to normal tissues [8]. These data show that regular drinking of green tea rich in EGCG may be preventive against esophageal tumorigenesis; however, consumption of very hot tea abolishes protective effects of EGCG. Furthermore, although EGCG is the most abundant flavanol in green tea, other efficient compounds in this natural agent and their biological mechanisms also need further identification and investigation (Table 5.4).

5.3.2.2. Flavonols

The potential tumor-fighting effects of flavonols against esophageal cancer have been studied only for three compounds (kaempferol, myricetin and galangin) in mouse xenograft models. In KYSE150 cells-inoculated mice, kaempferol restrained tumor growth with the reduction of EGFR activity and expression of hexokinase-2 and Ki67, exerting no apparent toxicity [12]. Myricetin suppressed tumor volume in EC9706-injected mice models [3, 47]. Administration of galangin to Eca9706 cells-injected xenograft tumor mice led to a significant inhibition of tumor growth by promoting apoptotic cell death, without any significant toxicity to normal tissues. Moreover, these anticancer effects were synergistic in combination of galangin and berberine [28]. Based on these results, some flavonols may serve as potent candidates for ESCC management in the future (Table 5.4).

Table 5.4. *In vivo* effects of flavonoids in animal models of esophageal cancer

Compound	Biological system[1]	Administration route and amount of flavonoids	Assay time	Biological effects[2]	Ref.
FLAVANOLS					
Epigallocatechin gallate (EGCG)	NMBA-induced rat esophageal tumorigenesis	Oral or intraperitoneal administration, 4 or 10 mg/kg	24 weeks	Suppression of tumor formation, reduction of tumor volume. Decrease in cyclin D1 and COX-2 expression and PGE2 production	[7]
	NMBA-induced rat esophageal tumorigenesis	Per oral, 10 mg/kg	20 weeks	Suppression of tumorigenesis. Decrease in cyclin D1 and COX-2 expression and PGE2 production. Inhibitory effects abolished by addition of hot water (55 or 65 ^{0}C)	[34]
	SKGT4 cells injected into right flank of mice	Per oral, 50 µg/kg/day, 5 days per week	30 days	Suppression of tumor growth. Inhibition of expression of Ki67, p-Erk1/2 and COX-2 expression	[2]
	Eca109 cells injected into the flank of mice	Intraperitoneal injection, 10 mg/kg/day	14 days	Inhibition of tumor growth. Increase in caspase-3 cleavage and decrease in VEGF protein expression. No toxicity to normal tissues (liver, spleen, kidney)	[8]
FLAVANOLIGNANS					
Silibinin	Luciferase-expressing KYSE150 cells injected through the lateral tail vein to mice	Oral gavage, 100 mg/kg/day	8 weeks	Suppression of metastasis (indicated by the lower bioluminescence in lungs)	[9]
	KYSE270 cells injected into left flank of mice	Oral gavage, 100 mg/kg/day	18 days	Suppression of tumor growth. No signs of toxicity	[9]
FLAVANONES					
Hesperetin	Eca109 cells implanted into lower right flank of mice	Intraperitoneal injection, 30, 60 or 90 mg/kg every 3 days	30 days	Inhibition of tumor growth. Induction of cell apoptosis in the tumor mass. Suppression of metastasis. No signs of toxicity	[25]
Hesperidin	MNAN-induced rat esophageal tumorigenesis	Via diet, 1000 ppm	20 weeks	Reduction of incidence of carcinoma and preneoplastic lesions when administered in initiation phase. Decrease in proliferation biomarkers (BrdU-labeling index and polyamine levels).	[20]

Table 5.4. (Continued)

Compound	Biological system[1]	Administration route and amount of flavonoids	Assay time	Biological effects[2]	Ref.
				No superior effect in combination with diosmin (900 ppm diosmin and 100 ppm hesperidin). No signs of toxicity	
FLAVONES					
Diosmin	MNAN-induced rat esophageal tumorigenesis	Via diet, 1000 ppm	20 weeks	Reduction of incidence of carcinoma and preneoplastic lesions when administered in initiation phase. Decrease in proliferation biomarkers (BrdU-labeling index and polyamine levels). No superior effect in combination with hesperidin (900 ppm diosmin and 100 ppm hesperidin). No signs of toxicity	[20]
FLAVONOLS					
Kaempferol	KYSE150 cells injected into right flank of mice	Intraperitoneal injection, 100 mg/kg every 3 days	25 days	Delay in tumor growth. Decrease in EGFR phosphorylation, decline in HK2 expression, decrease in proliferation marker Ki67. No signs of toxicity	[12]
Galangin	Eca9706 cells injected into right axilla of mice	Gavage, 25 mg/kg/day, 15 days before inoculation	35 days	Inhibition of tumor growth, reduction of the weight and volume of tumors. Induction of apoptosis, increase in p53, decrease in Ki67 and Wnt3α expression. Synergistic effect with berberine. No signs of toxicity	[28]
Myricetin	EC9706 cells injected into dorsal scapular region of mice	25 mg/kg	50 days	Inhibition of tumor growth	[3]
	EC9706 cells injected into dorsal scapular region of mice	Intraperitoneal administration, 5 mg/kg bw/day	28 days	Decrease in tumor volume	[47]

Compound	Biological system[1]	Administration route and amount of flavonoids	Assay time	Biological effects[2]	Ref.
ISOFLAVONES					
Alpinumisoflavone	KYSE cells injected into right side of posterior flank of mice	50 or 100 mg/kg/day	30 days	Suppression of tumor growth. Induction of apoptosis through modulation of miR-370/PIM1 signaling. No signs of toxicity	[13]
Puerarin	Eca109 cells injected into mice	25 mg/kg/day	3 weeks	Inhibition of tumor growth, reduction of tumor volumes and weights. Induction of apoptosis. No signs of toxicity	[17]
PRENYLATED FLAVONOIDS					
Icariin	EC109 cells injected into left hind limb of mice	Intraperitoneal administration, 60 or 120 mg/kg/day	28 days	Inhibition of tumor growth. Up-regulation of ER stress-related molecules and proapoptotic protein PUMA, down-regulation of Bcl-2	[14]
Baohuoside-I	Eca109-luciferase-expresing cells injected into left flank of mice	Intralesional injection, 25 mg/kg/day	21 days	Inhibition of tumor growth, reduction of cellularity in tumor mass. Decrease in ß-catenin, cyclin D1 and survivin expression	[30]

[1]see Tables 5.1 and 5.2.

[2]BrdUrd, bromodeoxyuridine; COX2, cyclooxygenase-2; EGFR, epidermal growth factor receptor; ER, endoplasmic reticulum; ERK1/2, extracellular signal-regulated protein kinase 1 and 2; HK2, hexokinase 2; PGE2, prostaglandin E2; PIM1, proto-oncogene serine/threonine-protein kinase; PUMA, p53 upregulated modulator of apoptosis.

5.3.2.3. Other Flavonoids and Related Compounds

Two flavonoid sugar conjugates, i.e., flavone glycoside diosmin (diosmetin 7-rutinoside) and flavanone glycoside hesperidin (hesperetin 7-rutinoside), significantly reduced the incidence of ESCC in MNAN-induced rat esophageal carcinogenesis by decreasing the multiplicity of carcinoma and papilloma formation, when administered to animals during the initiation phase [20]. Exposure to these glycosides resulted in suppression of cell proliferation biomarkers, including 5-bromodeoxyuridine (BrdU-labeling index, silver-stained nucleolar organizer regions` protein number and blood polyamine concentration [20]. A significant increase in serum polyamines has been reported in esophageal cancer patients suggesting that this value might be a good biochemical marker for tumorigenesis in esophagus [20]. Both glycosides suppressed also the incidence of preneoplastic lesions, including nodular and papillary hyperplasia and mild to severe dysplasia, with no obvious pathological alterations in the liver, kidney or lung of rats [20]. However, dietary feeding of combination of these two sugar-conjugates did not lead to any additive or synergistic effects [20]. The aglycone flavanone hesperetin exerted dose dependent inhibition of tumor growth, triggered apoptosis and suppressed metastasis in human xenograft tumor model in mice transplanted with Eca109 cells [25].

Isoflavone sugar conjugate puerarin (daidzein 8-glycoside), suppressed tumor growth and induced apoptosis in Eca109 cells-engineered xenografts in mice [17]. Pyranoisoflavone alpinumisoflavone exerted its cytotoxic effects in KYSE cells-injected xenografts in nude mice via modulation of miR-370/PIM1 pathway [13].

Flavanolignan silibinin suppressed the growth of tumor xenografts and metastasis in mice transplanted with KYSE150 and KYSE270 cells without any obvious toxicity to animals [9]. In Eca109 cells-implanted xenografts in mice, prenylated flavonoid baohuoside-I (icariside II) inhibited tumor growth and downregulated the expression of ß-catenin and its downstream targets, survivin and cyclin D1 [30]. Treatment of athymic nude mice bearing EC109 cells-based xenografts with prenylated flavonol glycoside icariin led to a significant suppression of tumor growth with increase in pro- and decrease in anti-apoptotic proteins [14] (Table 5.4).

Altogether, these data suggest that natural flavonoids could be useful as potential anticancer agents in combating esophageal cancer (esophageal squamous cell carcinoma) and further studies with these compounds as well as other still unexplored flavonoids are urgently needed.

5.3.3. Combination of Flavonoids with Standard Modalities of Esophageal Cancer Treatment

Although a wide spectrum of anticancer drugs is used in clinical treatment regimens, the adverse side effects on normal healthy cells limit their administration, representing a major problem in the management of esophageal cancer [27, 38]. Combination of conventional treatment modalities with flavonoids can possess several advantages, including lowering the doses of chemotherapeutic drugs and overcoming or delaying the emergence of drug resistance. However, so far, only very few flavonoids have been tested in respect to their potential co-effects with standard chemotherapeutics and on the other hand, only two drugs have been used in these studies, 5-fluorouracil (5-FU) and cisplatin, besides the works focused on the role of flavonoids in irradiation schemes. The results of these *in vitro* and *in vivo* experiments are summarized in Table 5.5. It is very important to mention that to date, no unfavorable responses of flavonoids on the outcomes of standard therapies have been described, showing that further research and development of more efficient treatment strategies comprising the natural plant-based polyphenolic agents are highly needed.

Treatment of human ESCC cells EC9706 with combination of 5-FU and flavonol myricetin exerted more efficient antiproliferative, anticlonogenic, proapoptotic and cell cycle progression perturbing responses than either 5-FU or myricetin alone. These effects were accompanied by decrease in the expression of survivin, cyclin D1 and Bcl-2, and increase in caspase-3 and p53 levels. Moreover, in EC9706 cells-engineered xenografts in mice, combined treatment displayed significantly lower tumor growth speed than individual agents alone, suggesting that myricetin enhanced the chemosensitivity of human ESCC cells toward 5-FU [3].

Cytotoxic responses of 5-FU were synergistically enhanced also by co-treatment of human ESCC cells KYSE270 with low-doses of flavanolignan silibinin. In addition, silibinin boosted the sensitivity of KYSE270 cell-based tumor xenografts in nude mice toward 5-FU, as the combined treatment led to significantly smaller malignant neoplasms than either agents individually [9]. Similar potentiating anticancer effects of KYSE270 cells and xenograft tumors were observed also for silibinin in combination with another traditional chemotherapeutic drug, cisplatin [9]. It is clear, that both ongoing experimental research as well as clinical trials is needed to further substantiate the therapeutic value of co-effects between silibinin and conventional chemotherapeutics in the fight against esophageal carcinogenesis.

Table 5.5. Effect of flavonoids on conventional treatment modalities of esophageal cancer

Compound	Drug, treatment	Biological system[1]	Assay conditions[2]	Co-effect[3]	Ref.
FLAVANOLIGNANS					
Silibinin	Cisplatin	KYSE270 cells	MTT, 96 h	Synergistic suppression of proliferation in combination of low-doses of silibinin (15 μM) and cisplatin (10 μM) that alone had no antiproliferative effect	[9]
	Cisplatin	KYSE270 cells injected into left flank of mice	Oral administration, 25 mg/kg/day silibinin; intraperitoneal injection of drug 2 mg/kg twice weekly	Synergistic suppression of tumor volume compared to single agents alone. Minimal toxicity and side effects	[9]
	5-Fluorouracil (5-FU)	KYSE270 cells	MTT, 96 h	Synergistic suppression of proliferation in combination of low-doses of silibinin (15 μM) and 5-FU (2.5 μM) that alone had no antiproliferative effect	[9]
	5-Fluorouracil (5-FU)	KYSE270 cells injected into left flank of mice	Oral administration, 25 mg/kg/day silibinin; intraperitoneal injection of drug 20 mg/kg twice weekly	Synergistic suppression of tumor volume compared to single agents alone. Minimal toxicity and side effects	[9]
FLAVONES					
Oroxylin A	Radiation	TE13 cells	Clonogenic survival assay, 14 days	Significant radiosensitization by 10 and 50 μM oroxylin A. Synergistic induction of apoptosis, blocking of cells in G2/M phase. Decrease in expression of cyclin B1 and Cdc2	[10]
	Radiation	Eca109 cells	Clonogenic survival assay, 14 days	Significant radiosensitization by 10 and 50 μM oroxylin A. Synergistic induction of apoptosis, blocking of cells in G2/M phase. Decrease in expression of cyclin B1 and Cdc2	[10]
FLAVONOLS					
Myricetin	5-Fluorouracil (5-FU)	EC9706	CCK-8, 48 h	Increase in chemosensitivity of 5-FU, significant elevation of growth inhibition. Increase in G0/G1 phase cell cycle arrest, delay in cell cycle progression, enhancement of apoptosis, promotion of chemosensitivity by regulating the expression of survivin, cyclin D1, Bcl-2, caspase-3 and p53	[3]

Compound	Drug, treatment	Biological system[1]	Assay conditions[2]	Co-effect[3]	Ref.
	5-Fluorouracil (5-FU)	EC9706 cells injected into dorsal scapular region of mice	Myricetin 25 mg/kg, 5-FU 20 mg/kg	Stronger inhibition of tumor growth as compared to single agents alone	[3]
ISOFLAVONES					
Genistein	Radiation	TE1	Clonogenic assay, 14 days	Significant enhancement of radiosensitivity by 30 μM genistein combined with 5 Gy irradiation. Abolishment of radiation-induced activation of p42/p44 ERK and AKT/PKB, decrease in cyclin D1 expression, inhibition of activation of survival signals. No change in apoptotic cells	[52]
	Radiation	TE2	Clonogenic assay, 14 days	Significant enhancement of radiosensitivity by 30 μM genistein combined with 5 Gy irradiation. Abolishment of radiation-induced activation of p42/p44 ERK and AKT/PKB, decrease in expression of cyclin D1 and Bcl-2, increase in Bax expression, promotion of PARP cleavage, increase in percentage of apoptotic cells	[52]
Puerarin	5-Fluorouracil (5-FU)	Eca109	CCK-8, 48 h	Synergistic inhibition of cell proliferation. Significant increase in apoptosis compared to single agents alone	[17]
	5-Fluorouracil (5-FU)	Eca109 cells injected into mice	Puerarin 25 mg/kg/day, 5-FU 12 mg/kg/day	Superior antitumor effect of combined treatment (inhibition of tumor growth, reduction of tumor volume and weight) compared to individual agents. Suppression of cellular viability, increase in apoptosis. No elevation of 5-FU toxicity	[17]

[1]see Tables 5.1 and 5.2.
[2]CCK-8, cell counting kit-8; MTT, 3-(4,5-dimethylthiazol-2-yl)-2,5-diphenyltetrazolium bromide.
[3]Akt, protein kinase B; Cdc2, cell division cycle protein 2; ERK, extracellular signal-regulated kinase; PARP, poly (ADP-ribose) polymerase.

Furthermore, exposure of human ESCC cells Eca109 to the combination of puerarin (daidzein 8-glycoside) and 5-FU resulted in a synergistic inhibition of tumor cell proliferation and promotion of apoptosis [17]. This combination exerted superior inhibition of tumor growth in Eca109 cell xenograft mice model compared to either puerarin or 5-FU alone, with no obvious side effects to animals [17]. 5-FU has been considered as the first-line and most commonly used chemotherapeutic drug in clinical management of esophageal cancer; however, due to its serious adverse effects on normal tissues and gradual appearance of drug resistance, novel agents that could be applied in combination therapy are of growing interest.

The clinical outcome of irradiation treatment of advanced esophageal cancers is still unsatisfactory today. Tumor radioresistance is one of the major causes of failure after radiotherapy and therefore, agents that could affect the radioresistance of tumors are crucial for improving the therapeutic results. Genistein remarkably enhanced the radiosensitivity in two human ESCC lines TE1 and TE2 by inhibiting the radiation-induced stimulation of survival signals, p42/p44 ERK and AKT/PKB, and expression of cyclin D1 [52]. However, the cell death modes were different as after combined treatment with genistein and irradiation (5 Gy), a significant increase in the fraction of apoptotic cells was observed in p53 wild-type TE2 cells, but not in p53-mutant TE1 cells [52]. A methoxylated flavone, oroxylin A (6-methoxybaicalein), behaved as a potent radiosensitizer (0-8 Gy) in human ESCC cells TE13 and Eca109 by decreasing the cell survival of tumor cells, blocking them in G2/M phase and synergistically inducing apoptosis [10]. Thus, genistein and oroxylin A might be potent agents in promoting the therapeutic effects of radiotherapy in clinical management of esophageal malignancies (ESCC).

Altogether, flavonoids can function as promising and powerful chemo- and radiosensitizers for treatment of esophageal cancer (especially, ESCC) and it is highly expected that this potential will be applied in the future clinical treatment schemes in order to improve the prognosis and survival of esophageal cancer patients (Table 5.5).

5.4. Current Limitations and Future Challenges

Based on the preclinical experimental results derived from *in vitro* and *in vivo* assays summarized in this chapter, flavonoids could be considered as novel potential lead compounds for treatment of esophageal cancer, both alone as well as in combination with clinically used traditional drugs. However, there are still several limitations that attenuate current conclusions and offer challenges for future studies.

First, the spectrum of different flavonoids tested so far is very small and only some structural analogues have been assessed, mostly in human esophageal cancer cell lines, the studies with experimental animals are even more sparse. Indeed, only a few representatives of major flavonoid classes have been investigated concerning their potential action against esophageal carcinogenesis, whereas, for instance, rotenoids are still completely unexplored. Taking into account that the 5-year survival rate of esophageal cancer has not been essentially improved over the several past decades [6], identifying novel promising efficient agents for treatment of esophageal tumors is of very high importance. In parallel, cellular targets and molecular mechanisms affected by biologically active flavonoids also need further research and unraveling.

Second, considering the fact that the incidence of esophageal adenocarcinoma is rapidly and continuously rising in industrialized Western societies [32], in the next years, much more attention should be turned to the anticancer studies of various flavonoids in human esophageal adenocarcinoma cell lines and xenograft models. To date, most of the investigations have been performed with human esophageal squamous cell carcinoma lines; however, as these two histologically different esophageal cancer subtypes possess distinct etiological factors and pathological characteristics, it is very probable that flavonoids could behave somewhat differently in these two cellular systems. Today, no data at all can be found about the potential anticancer action of isoflavones or flavanones in human esophageal adenocarcinoma lines. Moreover, knowledge about the potential co-effects of flavonoids with conventional treatment modalities, such as chemotherapeutic drugs or irradiation, is completely lacking for adenocarcinoma cells. Considering that the incidence of esophageal adenocarcinoma is increasing in the Western countries, such combinatorial experiments are urgently important. Current results derived with human esophageal squamous cell carcinoma cell lines and respective xenograft models are very promising in respect to such combinatorial therapies, whereas only potentiating action of flavonoids have been described with no disadvantageous effects.

Third, the advantages of modern nanotechnological approaches are still practically not applied in the studies of anticancer effects of flavonoids on different esophageal cancer models, although it is well known that low stability and poor bioavailability can limit the use of these plant-based polyphenolic compounds in future clinical practice.

Last but not least, learning more about the risk factors and identifying molecular determinants related to the development of esophageal cancer makes it possible to inform populations about the major preventable hazards and possibilities to reduce these risks. At that, dietary consumption of certain plant-based flavonoids and products rich in these agents might be crucial in prevention of this very deadly and often fatal malignancy.

REFERENCES

[1] Zikri NN, Riedl KM, Wang LS, Lechner J, Schwartz SJ, Stoner GD. Black raspberry components inhibit proliferation, induce apoptosis, and modulate gene expression in rat esophageal epithelial cells. *Nut Cancer* 2009; 61: 816-26.

[2] Ye F, Zhang GH, Guan BX, Xu XC. Suppression of esophageal cancer cell growth using curcumin, (-)-epigallocatechin-3-gallate and lovastatin. *World J Gastroenterol* 2012; 18: 126-35.

[3] Wang L, Feng J, Chen X, Guo W, Du Y, Wang Y, Zang W, Zhang S, Zhao G. Myricetin enhance chemosensitivity of 5-fluorouracil on esophageal carcinoma in vitro and in vivo. *Cancer Cell Int* 2014; 14: 71.

[4] Zhang Q, Zhao XH, Wang ZJ. Cytotoxicity of flavones and flavonols to a human esophageal squamous cell carcinoma cell line (KYSE-510) by induction of G2/M arrest and apoptosis. *Toxicol In Vitro* 2009; 23: 797-807.

[5] Zhang Q, Zhao XH, Wang ZJ. Flavones and flavonols exert cytotoxic effects on a human oesophageal adenocarcinoma cell line (OE33) by causing G2/M arrest and inducing apoptosis. *Food Chem Toxicol* 2008; 46: 2042-53.

[6] Stoner GD, Wang LS, Chen T. Chemoprevention of esophageal squamous cell carcinoma. *Toxicol Appl Pharmacol* 2007; 224: 337-49.

[7] Li ZG, Shimada Y, Sato F, Maeda M, Itami A, Kaganoi J, Komoto I, Kawabe A, Imamura M. Inhibitory effects of epigallocatechin-3-gallate on N-nitrosomethylbenzylamine-induced esophageal tumorigenesis in F344 rats. *Int J Oncol* 2002; 21: 1275-83.

[8] Liu L, Hou L, Gu S, Zuo X, Meng D, Luo M, Zhang X, Huang S, Zhao X. Molecular mechanism of epigallocatechin-3-gallate in human esophageal squamous cell carcinoma in vitro and in vivo. *Oncol Rep* 2015; 33: 297-303.

[9] Li J, Li B, Xu WW, Chan KW, Guan XY, Qin YR, Lee NP, Chan KT, Law S, Tsao SW, Cheung AL. Role of AMPK signaling in mediating the anticancer effects of silibinin is esophageal squamous cell carcinoma. *Expert Opin Ther Targets* 2016; 20: 7-18.

[10] Tan C, Qian X, Ge Y, Yang B, Wang F, Guan Z, Cai J. Oroxylin A could be a promising radiosensitizer for esophageal squamous cell carcinoma by inducing G2/M arrest and activating apoptosis. *Pathol Oncol Res* 2017; 23: 323-8.

[11] Shi C, Fan LY, Cai Z, Liu YY, Yang CL. Cellular stress response in Eca-109 cells inhibits apoptosis during early exposure to isorhamnetin. *Neoplasma* 2012; 59: 361-9.

[12] Yao S, Wang X, Li C, Zhao T, Jin H, Fang W. Kaempferol inhibits cell proliferation and glycolysis in esophagus squamous cell carcinoma via targeting EGFR signaling pathway. *Tumour Biol* 2016; 37: 10247-56.

[13] Han Y, Yang X, Zhao N, Peng J, Gao H, Qiu X. Alpinumisoflavone induces apoptosis in esophageal squamous cell carcinoma by modulating miR-370/PIM1 signaling. *Am J Cancer Res* 2016; 6: 2755-71.

[14] Fan C, Yang Y, Liu Y, Jiang S, Di S, Hu W, Ma Z, Li T, Zhu Y, Xin Z, Wu G, Han J, Li X, Yan X. Icariin displays anticancer activity against human esophageal cancer cells via regulating endoplasmic reticulum stress-mediated apoptotic signaling. *Sci Rep* 2016; 6: 21145.

[15] Tang L, Lee AH, Xu F, Zhang T, Lei J, Binns CW. Soya and isoflavone intakes associated with reduced risk of oesophageal cancer in north-west China. *Public Health Nutr* 2015; 18: 130-4.

[16] Vermeulen E, Zamora-Ros R, Duell EJ, Lujan-Barroso L, Boeing H, Aleksandrova K, Bueno-de-Mesquita HB, Scalbert A, Romieu I, Fedirko V, Touillaud M, Fagherazzi G, Perquier F, Molina-Montes E, Chirlaque MD, Vicente Argüelles M, Amiano P, Barricarte A, Pala V, Mattiello A, Saieva C, Tumino R, Ricceri F, Trichopoulou A, Vasilopoulou E, Ziara G, Crowe FL, Khaw KT, Wareham NJ, Lukanova A, Grote VA, Tjonneland A, Halkjaer J, Bredsdorff L, Overvad K, Siersema PD, Peeters PH, May AM, Weiderpass E, Skeie G, Hjartaker A, Landberg R, Johansson I, Sonestedt E, Ericson U, Riboli E, Gonzalez CA. Dietary flavonoid intake and esophageal cancer risk in the European prospective investigation into cancer and nutrition cohort. *Am J Epidemiol* 2013; 178: 570-81.

[17] Wang J, Yang ZR, Guo XF, Song J, Zhang JX, Wang J, Dong WG. Synergistic effects of puerarin combined with 5-fluorouracil on esophageal cancer. *Mol Med Rep* 2014; 10: 2535-41.

[18] Petrick JL, Steck SE, Bradshaw PT, Chow WH, Engel LS, He K, Risch HA, Vaughan TL, Gammon MD. Dietary flavonoid intake and Barrett`s esophagus in western Washington State. *Ann Epidemiol* 2015; 25: 730-5.

[19] Yuan JM. Green tea and prevention of esophageal and lung cancers. *Mol Nutr Food Res* 2011; 55: 886-904.

[20] Tanaka T, Makita H, Kawabata K, Mori H, Kakumoto M, Satoh K, Hara A, Sumida T, Fukutani K, Tanaka T, Ogawa H. Modulation of N-methyl-N-amylnitrosamine-induced rat oesophageal tumourigenesis by dietary feeding of diosmin and hesperidin, both alone and in combination. *Carcinogenesis* 1997; 18: 761-9.

[21] Cui L, Liu X, Tian Y, Xie C, Li Q, Cui H, Sun C. Flavonoids, flavonoid subclasses, and esophageal cancer risk: A meta-analysis of epidemiologic studies. *Nutrients* 2016; 8: E350.

[22] Sun CL, Yuan JM, Lee MJ, Yang CS, Gao YT, Ross RK, Yu MC. Urinary tea polyphenols in relation to gastric and esophageal cancers: A prospective study of men in Shanghai, China. *Carcinogenesis* 2002; 23: 1497-503.

[23] An F, Wang S, Tian Q, Zhu D. Effects of orientin and vitexin from Trollius chinensis on the growth and apoptosis of esophageal cancer EC-109 cells. *Oncol Lett* 2015; 10: 2627-2633.

[24] Petrick JL, Steck SE, Bradshaw PT, Trivers KF, Abrahamson PE, Engel LS, He K, Chow WH, Mayne ST, Risch HA, Vaughan TL, Gammon MD. Dietary intake of flavonoids and oesophageal and gastric cancer: Incidence and survival in the United States of America (USA). *Br J Cancer* 2015; 112: 1291-300.

[25] Wu D, Zhang J, Wang J, Li J, Liao F, Dong W. Hesperetin induces apoptosis of esophageal cancer cells via mitochondrial pathway mediated by the increased intracellular reactive oxygen species. *Tumour Biol* 2016; 37: 3451-9.

[26] Wen X, Walle T. Cytochrome P450 1B1, a novel chemopreventive target for benzo[a]pyrene-initiated human esophageal cancer. *Cancer Lett* 2007; 246: 109-14.

[27] Wang TT, Wang SK, Huang GL, Sun GJ. Luteolin induced-growth inhibition and apoptosis of human esophageal squamous carcinoma cell line Eca109 cells in vitro. *Asian Pac J Cancer Prev* 2012; 13: 5455-61.

[28] Ren K, Zhang W, Wu G, Ren J, Lu H, Li Z, Han X. Synergistic anti-cancer effects of galangin and berberine through apoptosis induction and proliferation inhibition in oesophageal carcinoma cells. *Biomed Pharmacother.* 2016; 84: 1748-59.

[29] Cheong E, Ivory K, Doleman J, Parker ML, Rhodes M, Johnson IT. Synthetic and naturally occurring COX-2 inhibitors suppress proliferation in a human oesophageal adenocarcinoma cell line (OE33) by inducing apoptosis and cell cycle arrest. *Carcinogenesis* 2004; 25: 1945-52.

[30] Wang L, Lu A, Liu X, Sang M, Shan B, Meng F, Cao Q, Ji X. The flavonoid baohuoside-I inhibits cell growth and downregulates survivin and cyclin D1 expression in esophageal carcinoma via ß-catenin-dependent signaling. *Oncol Rep* 2011; 26: 1149-56.

[31] Xie Y, Huang S, Su Y. Dietary flavonols intake and risk of esophageal and gastric cancer: A meta-analysis of epidemiological studies. *Nutrients* 2016; 8: 91.

[32] Pierini R, Kroon PA, Guyot S, Ivory K, Johnson IT, Belshaw NJ. Procyanidin effects on oesophageal adenocarcinoma cells strongly depend on flavan-3-ol degree of polymerization. *Mol Nutr Food Res* 2008; 52: 1399-407.

[33] Sun L, Subar AF, Bosire C, Dawsey SM, Kahle LL, Zimmerman TP, Abnet CC, Heller R, Graubard BI, Cook MB, Petrick JL. Dietary flavonoid intake reduces the risk of head and neck but not esophageal or gastric cancer in US men and women. *J Nutr* 2017; 147: 1729-38.

[34] Li ZG, Shimada Y, Sato F, Maeda M, Itami A, Kaganoi J, Komoto I, Kawabe A, Imamura M. Promotion effects of hot water on N-nitrosomethylbenzylamine-induced esophageal tumorigenesis in F344 rats. *Oncol Rep* 2003; 10: 421-6.

[35] Bobe G, Peterson JJ, Gridley G, Hyer M, Dwyer JT, Brown LM. Flavonoid consumption and esophageal cancer among black and white men in the United States. *Int J Cancer* 2009; 125: 1147-54.

[36] Jiao L, Kramer JR, Rugge M, Parente P, Verstovsek G, Alsarraj A, El-Serag. HB. Dietary intake of vegetables, folate, and antioxidants and the risk of Barrett`s esophagus. *Cancer Causes Control* 2013; 24: 1005-14.

[37] Ma G, Yang C, Qu Y, Wei H, Zhang T, Zhang N. The flavonoid component isorhamnetin in vitro inhibits proliferation and induces apoptosis in Eca-109 cells. *Chem Biol Interact* 2007; 167: 153-60.

[38] Yang P, Tuo L, Wu Q, Cao X. Licochalcone-A sensitizes human esophageal carcinoma cells to TRAIL-mediated apoptosis by proteasomal degradation of XIAP. *Hepatogastroenterology* 2014; 61: 1229-34.

[39] Rossi M, Garavello W, Talamini R, La Vecchia C, Franceschi S, Lagiou P, Zambon P, Dal Maso L, Bosetti C, Negri E. Flavonoids and risk of squamous cell esophageal cancer. *Int J Cancer* 2007; 120: 1560-4.

[40] Gao Y, Li W, Jia L, Li B, Chen YC, Tu Y. Enhancement of (-)-epigallocatechin-3-gallate and theaflavin-3-3`-digallate induced apoptosis by ascorbic acid in human lung adenocarcinoma SPC-A-1 cells and esophageal carcinoma Eca-109 cells via MAPK pathways. *Biochem Biophys Res Commun* 2013; 438: 370-4.

[41] Lambert JD, Sang S, Hong J, Kwon SJ, Lee MJ, Ho CT, Yang CS. Peracetylation as a means of enhancing in vitro bioactivity and bioavailability of epigallocatechin-3-gallate. *Drug Metab Dispos* 2006; 34: 2111-6.

[42] Fang MZ, Wang Y, Ai N, Hou Z, Sun Y, Lu H, Welsh W, Yang CS. Tea polyphenol (-)-epigallocatechin-3-gallate inhibits DNA methyltransferase and reactivates methylation-silenced genes in cancer cell lines. *Cancer Res* 2003; 63: 7563-70.

[43] Hou Z, Sang S, You H, Lee MJ, Hong J, Chin KV, Yang CS. Mechanism of action of (-)-epigallocatechin-3-gallate: Auto-oxidation-dependent inactivation of epidermal growth factor receptor and direct effects on growth inhibition in human esophageal cancer KYSE 150 cells. *Cancer Res* 2005; 65: 8049-56.

[44] Deng X, Zhao X, Lan Z, Jiang J, Yin W, Chen L. Anti-tumor effects of flavonoids from the ethnic medicine Docynia delavayi (Franch.) Schneid. and its possible mechanism. *J Med Food* 2014; 17: 787-94.

[45] Hong L, Ying SH. Ethanol extract and isolated constituents from artemisia dracunculus inhibit esophageal squamous cell carcinoma and induce apoptotic cell death. *Drug Res* (Stuttg) 2015; 65: 101-6.

[46] Tu LY, Pi J, Jin H, Cai JY, Deng SP. Synthesis, characterization and anticancer activity of kaempferol-zinc(II) complex. *Bioorg Med Chem Lett* 2016; 26: 2730-4.

[47] Zang W, Wang T, Wang Y, Li M, Xuan X, Ma Y, Du Y, Liu K, Dong Z, Zhao G. Myricetin exerts anti-proliferative, anti-invasive, and pro-apoptotic effects on esophageal carcinoma EC9706 and KYSE30 cells via RSK2. *Tumour Biol* 2014; 35: 12583-92.

[48] Zheng NG, Mo SJ, Li JP, Wu JL. Anti-CSC effects in human esophageal squamous cell carcinomas and Eca109/9706 cells induced by nanoliposomal quercetin alone or combined with CD 133 antiserum. *Asian Pac J Cancer Prev* 2014; 15: 8679-84.

[49] Zheng NG, Wang JL, Yang SL, Wu JL. Aberrant epigenetic alteration in Eca9706 cells modulated by nanoliposomal quercetin combined with butyrate mediated via epigenetic-NF-κB signaling. *Asian Pac J Cancer Prev* 2014; 15: 4539-43.

[50] Fang MZ, Chen D, Sun Y, Jin Z, Christman JK, Yang CS. Reversal of hypermethylation and reactivation of p16INK4a, RARbeta, and MGMT genes by genistein and other isoflavones from soy. *Clin Cancer Res* 2005; 11: 7033-41.

[51] Yanagihara K, Ito A, Toge T, Numoto M. Antiproliferative effects of isoflavones on human cancer cell lines established from the gastrointestinal tract. *Cancer Res* 1993; 53: 5815-21.

[52] Akimoto T, Nonaka T, Ishikawa H, Sakurai H, Saitoh JI, Takahashi T, Mitsuhashi N. Genistein, a tyrosine kinase inhibitor, enhanced radiosensitivity in human esophageal cancer cell lines in vitro: possible involvement of inhibition of survival signal transduction pathways. *Int J Radiat Oncol Biol Phys* 2001; 50: 195-201.

Chapter 6

Anticancer Action of Flavonoids in Gastric Cancer Models

6.1. Introduction to Gastric Cancer

Gastric cancer is a type of gastrointestinal tract tumors being still a prevalent cause of malignancy-related morbidity and mortality worldwide [1-25]. It remains a serious health problem and a great threat to the health and life of humans around the world. Also, gastric cancer is a considerable economic burden and has an important impact on health care systems. Based on the data of GLOBOCAN 2012, gastric cancer, also called stomach cancer, is the fifth most commonly diagnosed cancer after lung, breast, colorectal and prostate tumors in both sexes worldwide. Moreover, gastric cancer is the third leading cause of malignancy-associated death, just following lung and liver cancers [15, 26-30]. In 2012, a total of 951600 new gastric cancer cases and 723100 deaths are estimated to have occurred globally, accounting for 6.8% of all newly diagnosed malignancies and 8.8% of all cancer deaths [9, 15, 26, 31-39]. Although overall incidence and mortality of gastric cancer have steadily declined in the past few decades [5, 11, 19, 35, 40-45], this disease displays a remarkable variation in geographic incidences and the highest rates are observed in Far East Asia, South America, and Eastern Europe [23, 46-53]. The countries of Eastern Asia, such as China, Japan, and Korea, are the high-risk regions for gastric cancer where its incidence ranks second among all malignant diseases being still an increasing public health problem [4, 8, 21, 23, 24, 27, 42, 43, 45, 52, 54-59]. Indeed, more than 40% of global annual gastric cancer cases occur in China alone [60-62]. Gastric cancer reveals also considerable gender discrepancy occurring throughout the world about 2-3 times more frequently in men than in women [46, 63-65]. The reasons behind this difference are not fully known but it is proposed that biological features associated with sex hormones can be involved [63, 64].

Gastric carcinogenesis can occur in two main anatomical localizations: tumors arising in the cardia (the proximal part of the stomach near the gastroesophageal junction) and those in the non-cardia (the distal part of the stomach) [7, 46, 65, 66]. While the incidence of gastric cancers in non-cardia has declined in most countries, the prevalence of cardia tumors is remained rather stable or even risen in several European countries, Japan, and Northern America [46, 65, 67, 68]. Approximately 90% of gastric cancers are adenocarcinomas which are further classified as either poorly or undifferentiated gastric adenocarcinomas (so-called diffuse-type tumors) accounting for about 54% and characterized by high-degree of malignancy, rapid metastasis and poor prognosis [1, 3, 13, 62, 69, 70], or well-differentiated gastric adenocarcinomas (known also as intestinal-type tumors) possessing slow metastasis and better prognosis [62, 65, 71]. The latter histological subtype is outlined by a corpus-dominated chronic gastritis with gastric atrophy and intestinal metaplasia, whereas the diffuse-type tumors are characterized by gastritis throughout the stomach without atrophy [71, 72]. Gastric cancer development is a multistep and multifactorial process, and chronic inflammation is believed to serve a prerequisite for tumorigenesis in the gastric epithelium [7, 71, 73]. In general, gastric carcinogenesis has a dismal prognosis because of its aggressive and early spread to distant organs and the lack of effective measures to stop this progression [7, 38, 39, 74-76]. Due to the non-specific symptoms or asymptomatology in its early stages but also the absence of validated screening programs, most gastric cancer patients are diagnosed in locally advanced or disseminated stages of the disease [2, 6, 10, 19, 22, 29, 34, 41, 45, 77-80]. Moreover, the recurrence rate of gastric cancer is also high [22, 81]. Thus, the average five-year survival rate for gastric cancer is low, ranging only from 10% to 30% [14, 27, 29, 32-34, 41, 42, 54, 56, 78, 82, 83].

Gastric cancer is a multifactorial process, and its initiation and development are often related to different environmental factors, including diet, infections and lifestyle, but also some genetic alterations [7, 9, 46, 59, 65, 66, 84, 85]. Diet is an important modifiable factor in behavioral pattern of humans. Studies have suggested that a high consumption of smoked, salted, pickled and nitrated foods as well as saturated fats and carbohydrates are linked to a significantly increased risk of gastric cancer [1, 3, 13, 25, 46, 59, 67, 69-71, 86-89]. Indeed, dietary components, such as N-nitrosamines and heme iron from cured red meats and high salt content from processed foods are gastric carcinogens or cancer promoters being typically more abundant in Japanese diets than in US diets [25, 46, 47, 73, 90]. On the other hand, high and regular intake of fresh fruits and vegetables stored under refrigeration systems, but also milks and green tea are reported to be associated with a decreased risk [7, 11, 15, 25, 40, 46, 58, 65, 67, 86, 88, 91-93]. Some constituents of these foods like ascorbic acid, ß-carotene, but certainly also flavonoids might behave as inhibitors of gastric cancer development and be responsible for the beneficial properties of these products [71, 93]. Furthermore, the regular use of aspirin can be related to the reduced incidence of gastric cancer [46], while the association of obesity with gastric carcinogenesis has remained unclear to date [46]. Therefore, dietary modification alone can represent some protective measures against tumorigenesis in stomach providing an important prevention strategy [67, 73, 91]. According to International Agency for Research on Cancer (IARC), cigarette smoking is an important risk factor and cause for malignant transformation and progression of gastric cancer [9, 58, 94]. Many ingredients found in tobacco smoke possess toxic properties contributing to its carcinogenic potential, including generating oxidative stress as well as promoting the invasion and migration of tumor cells [9, 65]. Similarly, alcohol is an etiological factor for gastric carcinogenesis being related to induction of mucosal lesions, including gastritis, peptic ulcer diseases and thereafter even gastric carcinoma [67, 95, 96]. Studies have shown that proinflammatory cytokines and oxidative stress play a crucial role in these processes [96]. In addition to these lifestyle factors, it has been demonstrated that some genetic alterations can also promote tumorigenesis in stomach by modifying the effects of environmental exposures at different biological events and impact thus the individual susceptibility to gastric cancer development. For instance, individuals with germ-line mutation in the adenomatous polyposis coli (APC), a tumor suppressor gene, have a tenfold increased risk for gastric cancer development compared to normal persons [84, 97].

A major established risk factor for gastric cancer is the infection with *Helicobacter pylori* (*H. pylori*). This microaerobic Gram-negative pathogen was first identified in the stomachs of patients by Warren and Marshall in 1983 [58, 98]. In 1994, the World Health Organization (WHO) classified *H. pylori* as a group I carcinogen for gastric cancer [71, 99]. Throughout the world, more than 50% of the population is infected by this bacterium [71, 98]. About 10% to 50% of infected individuals develop peptic ulcer disease, and 1% to 3% of them progress to gastric cancer or in other words, *H. pylori* infection is associated with a fivefold increased risk of gastric cancer [71, 94]. Infection with *H. pylori* results in inflammation of the gastric mucosa, and further progresses to chronic atrophic gastritis, intestinal metaplasia, dysplasia, and ultimately adenocarcinoma [71, 98, 100-103]. Various biological mechanisms are involved in the gastric mucosal damage including the direct activity of virulence factors generated by pathogen, augmentation of inflammation associated with the induction of different inflammatory cytokines, oxidative stress, and DNA damage [53, 69, 75, 99, 103, 104]. Several *in vitro* studies have shown that the growth of *H. pylori* can be suppressed by phytochemicals, including flavonoids, suggesting that *H. pylori*-induced gastric cancer might be prevented by a flavonoid-rich diet [58, 105]. This is especially important considering that the triple therapy usually recommended for *H. pylori* eradication (consisting of a proton pump inhibitor and two antimicrobials) can fail for some reason or other, such as the occurrence of drug resistant bacterial strains or persistence of inflammation even after the eradication [99, 105]. Besides *H. pylori*, infection with latent Epstein-Barr virus (EBV), a human gamma-1 herpesvirus, can also lead to gastric carcinoma being one of the most common EBV-associated malignancies [106, 107]. In fact, about 10% of tumors in stomach are EBV-associated gastric cancers, whereas the virus infects almost all gastric tumor cells [106, 107]. To date, the knowledge about possible association between infection with human immunodeficiency virus and development of gastric cancer is still rather poor [46].

Although tumorigenesis in the stomach has remained highly prevalent, its treatment options are still limited. Current therapeutic modalities of gastric cancer include surgery, chemotherapy, and/or radiotherapy [8, 16, 31, 77, 82, 83, 85, 108, 109]. Surgery is the mainstay of the management of patients with localized early-stage gastric cancer, whereas gastrectomy (or partial gastric resection) remains the only strategy for curative treatment of this highly lethal malignancy [4, 6, 8, 11, 17, 32, 42, 76, 78, 80, 82, 110, 111]. However, overall five-year survival after curative surgical resection is still only about 30% due to the frequent postoperative recurrences at regional and/or distant sites [8, 23, 32, 42, 56, 111, 112]. Moreover, the majority of all patients is already at the advanced stage at initial diagnosis and presents with inoperable tumors contradicting the use of surgical approach [17, 21, 31, 49, 56, 80, 108, 110]. To improve the surgical outcomes of resectable gastric tumors but especially for patients with advanced stage gastric cancer, chemotherapy is an important therapeutic approach and the standard treatment choice [1, 2, 4, 27, 28, 30, 34, 42, 48, 56, 57, 76, 82]. Due to only modest activity of single agents like 5-fluorouracil (5-FU), doxorubicin, mitomycin, cisplatin, oxaliplatin, paclitaxel, irinotecan or adriamycin against gastric cancer, interest in the use of combination therapies is increased [7, 21, 22, 31, 80]. 5-FU, a first line antineoplastic treatment in clinical practice, exhibits its anticancer effects through suppression of thymidylate synthase and incorporation of its metabolites into DNA or RNA [21, 33, 48]. The current conventional chemotherapeutic regimens include 5-FU-based combined chemotherapies containing mitomycin C, anthracycline antibiotics, cisplatin, epirubicin or docetaxel [81, 82, 112]. However, many chemotherapeutic agents have numerous severe side effects and complications, including bone marrow suppression, neutropenia or anemia, which deteriorate the quality of life of patients and make these treatments often intolerable [1, 2, 4, 7, 17, 23, 49, 54, 55, 57, 62, 79, 80, 82, 110, 112]. This is especially devastating considering that gastric cancer patients reveal very frequent weight loss (incidence of cachexia is about 83%) that has practically no effective treatment [43]. Another serious concern and main barrier to the chemotherapeutic success is intrinsic and acquired resistance of gastric cancer cells to multiple classical drugs [10, 21, 27, 29, 31, 33, 34, 38, 42, 48, 50, 83, 85, 112]. A crucial factor in chemotherapeutic failure is also heterogeneity of tumor itself [42]. Thus, the chemosensitivity of advanced gastric cancer is still unsatisfactory implicating limitations in its clinical management and a lack of an effective standard therapeutic regimen [2, 7, 11, 24, 34, 39, 79, 113, 114]. Also, radiotherapy alone or combined with chemotherapy can only improve the symptoms and possibly prolong survival of patients with unresectable gastric cancer without curing the disease [5, 6, 51, 78, 115-117]. Thus, progress in the treatment of gastric cancer has been limited and to date, there is no effective therapy for patients with advanced stages of this highly malignant disease [2, 3, 13, 61, 81, 118].

Therefore, it is of utmost clinical importance to develop novel efficient therapeutic strategies for treatment of gastric cancer, especially advanced unresectable tumors [1, 4, 5, 8, 11, 14, 17, 20-22, 24, 27, 32, 34, 38, 39, 41, 42, 50, 54, 55, 60, 61, 82, 108, 110, 112, 119, 120]. New agents with better activity on gastric cancer cells and less toxicity on normal healthy cells are urgently needed to reduce the mortality caused by this highly lethal disease, and improve the prognosis and clinical outcome. In recent years, much attention has been focused on the potential therapeutic use of natural products as it is expected that compounds of natural origin, such as phytochemicals could lead to the development of innovative chemotherapeutic agents [8, 27, 31, 48, 57, 78, 80, 121]. It is strongly believed that such novel agents can not only exert effective anticancer effects but might also potentiate the cytotoxic action of conventional chemotherapeutic drugs with attenuating the development of resistance. For this goal, it is important to identify and understand the molecular cellular mechanisms underlying the pathogenesis of gastric cancer, including its invasion and metastasis, and to find potent therapeutic targets for novel candidate compounds [10, 11, 14, 29, 36, 38, 42, 62, 74]. The effective intervention with these molecular processes presumes an intensive exploration and unraveling of key factors related to stomach tumorigenesis. At that, increasing evidence has suggested the involvement of gastric cancer stem cells (GCSCs) in the initiation of this malignancy [36, 122]. Indeed, in 2009, GCSCs were isolated and identified using a cell surface marker CD44 [45, 122, 123]. This rare cell population displays stem cell properties, including high tumorigenicity and self-renewal, and possesses chemo- and radioresistant abilities, indicating new paradigms for gastric cancer treatment [36, 45].

Further efforts are also needed to improve the diagnosis of gastric cancer and detect this highly lethal malignancy at the earliest possible stage. In this context, the screening programs are required to be evolved. However, the most effective and promising means in cancer control around the world is still the prevention. This could be achieved through the avoidance of exposure to different risk factors and known carcinogens, but also by regular intake of an appropriate diet rich in different potential anticancer phytochemicals, including flavonoids, or development of natural herbal medicines [13, 35, 49, 51, 53, 105, 124-126].

6.2. Model Systems to Study the Effect of Flavonoids on Gastric Carcinogenesis

To study the potential protective effects of various flavonoids against tumorigenesis in stomach different *in vitro* (Table 6.1) and *in vivo* model systems (Table 6.2) have been used.

Both poorly differentiated (AGS, HGC27, KATO III, MKN45, MKN74, NUGC2, St-4, TMK-1), moderately differentiated (SGC7901) as well as well-differentiated human gastric cancer cell lines (MGC803, MKN7, MKN28) derived from primary tumor or metastatic tissues of either male or female patients have been studied. These cultured model cell lines differed also by the status of infection with EBV (EBV-negative AGS cells *vs* latently EBV-infected SNU719 cells) or resistance to diverse chemotherapeutics (5-FU, cisplatin, daunorubicin, multiple drugs). Moreover, lines with either wild type or mutated p53 tumor suppressor gene have been tested and also stem cell like cells were isolated and investigated (Table 6.1). As reference, the human normal gastric epithelial cell lines GES-1 and HFE-145 were used.

To explore the potential chemopreventive and/or chemotherapeutic action of a wide array of flavonoids on stomach tumorigenesis in *in vivo* conditions, different human gastric cancer cell lines were inoculated into experimental mice. In addition to human gastric cancer cell lines, patient-derived tumor specimens were also transplanted to mice to generate more appropriate preclinical model and potent tool for evaluation of anticancer properties of flavonoids [12]. Besides these engineered xenografts, tumors in the stomach of mice were also produced by intragastric intubation of benzo[α]pyrene (B[a]P). As this carcinogen requires metabolic activation, this and subsequent events within the carcinogenesis process can be sensitive to influences by dietary components [127]. In the experimental models of rats and gerbils, gastric tumors were induced by oral administration of N-methyl-N`-nitro-N-nitrosoguanidine (MNNG). It has been previously reported that sodium chloride (NaCl) could remarkably augment the carcinogenic effects of MNNG in the stomach by decreasing the amount and viscosity of gastric mucin and disturbing the mucosal integrity [69, 125, 128]. Therefore, initiation and promotion of experimental MNNG-induced gastric carcinogenesis in Wistar rats were often enhanced by NaCl (Table 6.2).

Table 6.1. Human gastric cancer cell lines used in *in vitro* studies of anticancer effects of flavonoids

Cell line name	Characterization of cell line	Additional information[1]	Ref.
AGS	Adenocarcinoma, poorly (or moderately) differentiated, highly metastatic; derived from non-metastatic tissue of a 54-year-old Caucasian female patient	EBV-negative, TRAIL-resistant, overexpressing survivin and STAT3, high expression of FOXM1, constitutive activity of ß-catenin/Tcf and Stat3, wt p53	[1, 3, 5, 6, 13, 15, 16, 22-24, 27-30, 33-35, 37, 41, 48, 49, 54, 62, 70, 73, 75, 79, 80, 82, 84, 85, 88, 97-99, 102, 105, 107, 113-116, 118, 132, 151-154, 158, 169, 170, 175, 190-192]
AGS/luteolin		Luteolin-resistant AGS	[27]
AGS-R		EGCG-resistant AGS	[132]
AGS-EBV-GFP		Recombinant EBV-infected AGS	[107]
AGS-wt EBV		wt EBV-infected AGS	[107]
AZ521	Adenocarcinoma	Multidrug-resistant	[113, 131]
BGC7901			[14]
BGC803	Adenocarcinoma		[52]
BGC823	Adenocarcinoma	CDDP-resistant, higher activation of STAT3	[2, 21, 27, 42, 44, 50, 52, 76, 124, 126, 138, 145, 161, 163, 193]
BGC823/luteolin		Luteolin-resistant BGC823	[27]
EPG85-257, EPG85-257P	Adenocarcinoma		[140, 156]
EPG85-257NOV		Atypical MDR phenotype, no enhanced expression of MDR1/P-gp	[140]
EPG85-257RDB		Classical MDR phenotype, overexpressing MDR1/P-gp and ABCB1, daunorubicin-resistant	[140, 156]
HGC27	Adenocarcinoma, un- (or poorly) differentiated	CDDP-resistant, higher activation of STAT3	[17, 31, 52, 55, 78, 143, 155, 173]
Hs 740.T	Carcinoma		[194]
Hs 746 T	Carcinoma		[10, 11, 194]
HSC-39K6			[166]
HSC-40A1			[166]
HSC-41E6			[166]
HSC-42H			[166]
HSC-43C1			[166]
HSC-45M2			[166]

Table 6.1. (Continued)

Cell line name	Characterization of cell line	Additional information[1]	Ref.
KATO III	Adenocarcinoma, poorly differentiated	p53-negative	[129, 133, 134, 139, 141, 148]
MGC803	Carcinoma, well-differentiated, derived from non-metastatic tissue		[4, 11, 19, 20, 31, 39, 52, 55, 60, 74, 91, 112, 116, 122, 161, 171, 172, 188]
GCSCs from MGC803		Spheroid body-forming cells, high expression of SC markers Oct4, Sox2, Nanog and CD44	[36]
MK1	Adenocarcinoma		[142, 195]
MKN1	Carcinoma	p53-mutated	[75, 129, 135, 160]
MKN7	Adenocarcinoma, well-differentiated	p53-mutated	[139, 155, 160]
MKN28	Adenocarcinoma, well-differentiated	p53-mutated	[5, 10, 22, 80, 110, 114, 118, 135, 155, 160, 174]
MKN45	Adenocarcinoma, poorly differentiated, derived from metastatic tissue of a 62-year-old Japanese female patient with liver metastasis	Moderately sensitive to 5-FU, high cMet expression, wt p53	[7, 12, 32, 48, 80, 81, 85, 113, 116, 129, 135-137, 148, 150, 160]
MKN45cl85		Cachexia-inducing MKN45	[43]
85As2mLuc		Highly metastatic cachexia-inducing MKN45	[43]
H. pylori-infected MKN45		Infected with *H. pylori* strain ATCC 70824, cagA/vacA positive	[103]
MGLVA1		Ascitic variant of MKN45G (gastrin overproducing MKN45)	[196]
GSCLCs from MKN45		Stem-like cells, CD44(+)	[123]
MKN74	Adenocarcinoma, poorly differentiated	EBV-negative	[106, 117, 129, 148, 160, 182-185]
N87, NCI-N87			[10, 42, 82, 130]
NUGC2	Adenocarcinoma, poorly differentiated		[155]
NUGC3		Multidrug-resistant, p53-mutated	[113, 135]
SCM1			[162]

Cell line name	Characterization of cell line	Additional information[1]	Ref.
SGC7901	Adenocarcinoma, (poorly or) moderately differentiated, derived from metastatic tissue	Relatively resistant to 5-FU, CDDP-susceptive, high expression of cMet and FOXM1, high levels of STAT3 phosphorylation	[8, 12, 17, 18, 24, 27, 31, 32, 38-40, 42, 50-52, 55-57, 61, 74, 77, 78, 80, 81, 83, 110, 116, 120-122, 138, 144, 146, 147, 157, 161, 163-165, 167, 186-189]
SGC7901/5-FU		5-FU-resistant SGC7901	[38]
SGC7901/cDDP, SGC7901/DDP		CDDP-resistant SGC7901	[52, 146]
GCSLCs from SGC7901		Sphere-forming cells, increased migratory and invasive capabilities, high expression of N-cadherin, low of E cadherin	[45]
SH101-P4			[166]
SK-GT-5	Adenocarcinoma	p53-mutated	[184]
SNU1	Carcinoma		[15, 48]
SNU5			[85, 113]
SNU16		p53-mutated	[48, 113, 149]
SNU216			[101]
SNU484	Carcinoma		[22, 75, 159]
SNU601	Carcinoma		[75]
SNU638	Carcinoma		[75, 105]
SNU719	Carcinoma	EBV-positive, latently infected with EBV	[106, 107]
St-4	Adenocarcinoma, poorly differentiated, derived from a 45-year-old female patient		[150, 160]
TMK-1	Adenocarcinoma, poorly differentiated, derived from a 21-year-old male patient	p53-mutated	[5, 135, 148, 150]

[1]5-FU, 5-fluorouracil; ABCB1, ATP-binding cassette subfamily B member 1; CDDP, cisplatin; EBV, Epstein-Barr Virus; FOXM1, forkhead box protein M1; GCSC, gastric cancer stem cell; GCSLC, gastric cancer stem-like cell; *H. pylori*, *Helicobacter pylori*; MDR, multidrug resistance; P-gp, P-glycoprotein; STAT3, signal transducer and activator of transcription 3; Tcf, T-cell factor; TRAIL, tumor necrosis factor-related apoptosis-inducing ligand; wt, wild type.

Table 6.2. Animal models used for *in vivo* studies of flavonoids against gastric carcinogenesis

Species	Strain	Gender	Age, weeks	Weight, g	Induction of tumorigenesis[1]		Application route and site	Ref.
					Chemical	Implanted cell line[2]		
Gerbils	Mongolian	Male, female	5…8	50…55	MNNG (with *H. pylori*)		MNNG in drinking water	[71]
Mice	BALB/c, NCr, NOD/SCID, SFF KM, Swiss, 615	Male, female	4…10	13…22	B[a]P	AGS, BGC823, BGC7901, HGC27, HSC-41E6, HSC-45M2, MFC, MGC803, MKN28, MKN45, MKN45cl85, 85As2mLuc, MKN74, SGC7901, SGC7901/5-FU, SGC7901/cDDP, SGC7901/DDP, SH101-P4, SNU5, SNU719, TMK-1, tumor specimens from patients	Intragastric intubation of B[a]P. Subcutaneous inoculation of cells into the back, right flank, right or left axillary fossa, right oxter, right thigh or dorsal area, or intraperitoneal injection of cells	[2, 5, 7, 8, 10, 12, 14, 17, 21, 27, 30-32, 38, 40, 43, 51, 52, 55, 85, 106, 109, 110, 112-114, 116, 119, 122, 127, 146, 147, 150, 166, 178, 180, 182]
Rats	Wistar, Wistar albino, Wistar/KOB	Male	6…8	~100	MNNG (NaCl-promoted)		MNNG in drinking water, by oral gavage	[69, 86, 87, 125, 128, 176]

[1]B[a]P, benzo[α]pyrene; *H. pylori*, *Helicobacter pylori*; MNNG, N-methyl-N`-nitro-N-nitrosoguanidine; NaCl, sodium chloride.
[2]see Table 6.1; MFC, mouse gastric carcinoma cell line.

6.3. Effect of Different Flavonoids on Gastric Cancer Development

In this chapter all the hitherto published data on the activity of various flavonoids in human gastric cancer model systems are compiled and analyzed. At that, the activity of only single compounds and not mixtures of different flavonoids (like silymarin) is explored, and only monomers and not their polymers (like thearubigins or proanthocyanidins) are regarded.

6.3.1. *In Vitro* Results

To focus more on the tumorigenesis process in human stomach, only the cultured cell lines derived from human gastric cancer patients and not murine gastric cancer cell line (MFC) were included in the following analysis. Better understanding of the mechanisms of flavonoids in gastric carcinogenesis will probably contribute to the more efficient development of specific preventive and therapeutic strategies. The data published to date about the effects of flavonoids in human gastric cancer cell lines are summarized in Table 6.3.

6.3.1.1. Flavanols

Flavanols are bioactive polyphenols in green tea. The beneficial health effects of green tea appear to be predominantly ascribed to epigallocatechin gallate (EGCG), the most abundant and potent polyphenolic component in green tea accounting for more than 40% of the total polyphenols [5, 6, 8, 40, 129]. The molecular mechanisms through which EGCG affects tumorigenesis in stomach and inhibits the growth of malignant neoplasms are still rather poorly understood [130]; however, it is clear that effects of this natural compound in human gastric cancer cells are multifactorial acting on different molecular targets and influencing diverse cell signaling pathways. The diversity of these molecular action mechanisms is compiled in Table 6.3.

EGCG was reported to be able to reduce the level of DEAD-box RNA helicase p68 in AZ521 cells leading to prevention of ß-catenin oncogenic signaling. The p68 protein as target for action of EGCG is implicated in cancer development as its expression was demonstrated to be correlated with malignant transformation and tumor progression [131]. Exposure of SGC7901 cells to EGCG also inhibited canonical Wnt/ß-catenin signaling. Deregulation of ß-catenin signaling is a crucial clinicopathological feature of

Table 6.3. ***In vitro*** **antiproliferative effects and respective molecular mechanisms of natural flavonoids in human gastric cancer cell lines**

Compound	Cell line[1]	Cell growth inhibition IC_{50}, μM[2]	Assay method[3]	Assay time, h	Action mechanism[4]	Effect in human normal oral cells[5]	Ref.
ANTHOCYANIDINS							
Cyanidin	AGS	NA up to 696 μM	MTT	48			[169]
Cyanidin 3-galactoside	AGS	NA at 445 μM	MTT	48			[169]
Cyanidin 3-glycoside	AGS	NA at 445 μM	MTT	48			[169]
Delphinidin	AGS	198	TB	48			[1]
	AGS	NA up to 660 μM	MTT	48			[169]
Delphinidin 3-galactoside	AGS	NA at 399 μM	MTT	48			[169]
Delphinidin 3-glycoside	AGS	161.2	MTT	48			[170]
Delphinidin 3-rutinoside	AGS	154.6	MTT	48			[170]
Malvidin	AGS	128	TB	48	G0/G1 phase cell cycle arrest, mitochondrial apoptosis, increase in p38 kinase expression, decrease in ERK activity		[1]
	AGS	255.7	MTT	48			[169]
Malvidin 3-glycoside (oenin)	AGS	149	TB	48			[1]
Pelargonidin	AGS	651.1	MTT	48			[169]
Pelargonidin 3-galactoside	AGS	NA at 427 μM	MTT	48			[169]
Petunidin	AGS	>630	MTT	48			[169]
CHALCONES							
2`,4`-Dihydroxychalcone	MGC803	24.89	MTT	48	Apoptosis		[60]
	MGC803				G2/M phase cell cycle arrest, apoptosis, downregulation of survivin expression		[172]
3`-Methyl-3-hydroxychalcone	HGC27	2.5	Counting	72	G0/G1 phase cell cycle arrest, inhibition of [^{3}H]-E2 binding to cytosol and nuclear type-II EBS		[173]

Compound	Cell line[1]	Cell growth inhibition IC_{50}, μM[2]	Assay method[3]	Assay time, h	Action mechanism[4]	Effect in human normal oral cells[5]	Ref.
Chalcone	AGS	~24…29	MTT	24	Decrease in cell invasion and migration, down regulation of MMP-2/-9 through inactivation of FAK/JNK signaling and reducing NF-κB DNA binding activity		[3]
		~19…24	MTT	48			
Isoliquiritigenin	AGS				No induction of JNK phosphorylation, no preventive effect on EBV infection		[107]
	AGS-wt EBV				Induction of JNK phosphorylation		[107]
	AGS-EBV-GFP				Induction of JNK phosphorylation		[107]
	MGC803				Apoptosis through increase in $[Ca^{2+}]i$ and decrease in ΔΨm		[171]
	SNU719	57	CCK-8	24	Non-significant apoptosis, no effect on cell cycle. No effect on DNMT1 and DNMT3A expression, maintaining EBV latent replication		[107]
		45	CCK-8	48			
	AGS	147.3	MTT	48		IC_{50} 251.1 μM in GES-1	[80]
	MKN28	167.8	MTT	48			
	MKN45	128.4	MTT	48			
	SGC7901	157.9	MTT	48			
Licochalcone A	BGC823	~70…80	MTT	24	Apoptosis, increase in ROS, activation of ERK, JNK and p38, suppression of PI3K/AKT	IC_{50}>100 μM in GES-1 at 48 h	[2]
		~90	MTT	48			

Table 6.3. (Continued)

Compound	Cell line[1]	Cell growth inhibition IC_{50}, μM[2]	Assay method[3]	Assay time, h	Action mechanism[4]	Effect in human normal oral cells[5]	Ref.
	MKN45		CCK-8	24 … 144	G2/M phase cell cycle arrest, apoptosis, increase in p21 and p53 expression		[81]
	SGC7901		CCK-8	24 … 144	G2/M phase cell cycle arrest, apoptosis, increase in p21 and p53 expression		[81]
	AGS	41.1	MTT	48	Mitochondrial apoptosis	IC_{50} 92.7 μM in GES-1	[80]
	MKN28	42.0	MTT	48	G2/M phase cell cycle arrest, decrease in cyclin A/B and MDM2 levels, increase in Rb; mitochondrial apoptosis	IC_{50} 92.7 μM in GES-1	[80]
	MKN45	40.7	MTT	48	G2/M phase cell cycle arrest, decrease in cyclin A/B and MDM2 levels, increase in Rb; mitochondrial apoptosis		
	SGC7901	40.8	MTT	48			
FLAVANOLS							
Catechin (C)	BGC823	NA	MTT	72			[138]
	KATO III	>>150	Counting	48	No apoptosis		[129]
	MKN45	>>150	Counting	48	No apoptosis		[129]
	SGC7901	NA	MTT	72			[138]
Catechin gallate (CG)	MGLVA1	~30	MTT	48			[196]
Epicatechin (EC)	AGS				No effect on PMA-induced MMP-9 expression and cell invasiveness. No effect on PMA-induced RON expression		[6]
	AGS						[5]

Compound	Cell line[1]	Cell growth inhibition IC_{50}, μM[2]	Assay method[3]	Assay time, h	Action mechanism[4]	Effect in human normal oral cells[5]	Ref.
	KATO III	>200	TB	24, 72	No apoptosis, no inhibition of okadaic acid-induced TNF-α release		[133]
	KATO III	>>150	Counting	48	No apoptosis. Synergistic increase in EGCG-induced apoptosis		[129]
	MKN45	>>150	Counting	48	No apoptosis. Synergistic increase in EGCG-induced apoptosis mediated by ROS production		[129]
Epicatechin gallate (ECG)	AGS				No effect on PMA-induced MMP-9 expression and cell invasiveness		[6]
	AGS				No effect on PMA-induced RON expression		[5]
	KATO III	140	TB	24, 72	Apoptosis, inhibition of okadaic acid-induced TNF-α release		[133]
	KATO III	~80…150	Counting	48	No apoptosis		[129]
	MKN45	~40	Counting	48	No apoptosis		[129]
Epigallocatechin (EGC)	AGS				No effect on PMA-induced RON expression		[5]
	KATO III	210	TB	24, 72	No apoptosis. Inhibition of okadaic acid-induced TNF-α release		[133]
	KATO III	~80…150	Counting	48	Weak apoptosis		[129]
	MKN45	~40	Counting	48	Weak apoptosis. Synergistic increase in EGCG-induced apoptosis involving ROS production		[129]

Table 6.3. (Continued)

Compound	Cell line[1]	Cell growth inhibition IC_{50}, μM[2]	Assay method[3]	Assay time, h	Action mechanism[4]	Effect in human normal oral cells[5]	Ref.
Epigallocatechin gallate (EGCG)	AGS		MTT	24	Cytoprotective effect against *H. pylori*-induced cytotoxicity via interference in *H. pylori*-induced TLR-4 signaling		[99]
	AGS		MTT	48	Inhibition of IL-1ß-induced IL-8 production and secretion partly mediated via reduction of NF-κB activity. No decrease in IL-1ß-induced ERK or JNK activity		[73]
	AGS	~175	CCK-8	24	Decrease in Id1 expression, apoptosis, S phase cell cycle arrest		[62]
		~153…175	CCK-8	72			
	AGS	19	UV	72			[132]
	AGS	NA up to 50 μM			Inhibition of PMA-induced MMP-9 expression, cell invasiveness, level of active AP-1, Erk and JNK phosphorylation		[6]
	AGS				Inhibition of IL-6-induced VEGF expression and production and angiogenesis by suppression of Stat3 activation, nuclear translocation and		[102]

Compound	Cell line[1]	Cell growth inhibition IC_{50}, μM[2]	Assay method[3]	Assay time, h	Action mechanism[4]	Effect in human normal oral cells[5]	Ref.
					Stat3-DNA binding activity		
	AGS				Decrease in PMA-induced RON expression and reduction of its promoter activity, partial abrogation of cell invasion		[5]
	AGS-R	30	UV	72	Development of cross-resistance to N-nitroso insecticides and ROS-generating substances		[132]
	AZ521	>10	WST-8	24	Preventing ß-catenin oncogenic signaling through proteasomal degradation of p68		[131]
	KATO III	140	TB	24, 72	Apoptosis, inhibition of okadaic acid-induced TNF-α expression and release mediated through inhibition of AP-1 and NF-κB activation		[133]
	KATO III	~80…150	Counting	48	Apoptosis		[129]
	KATO III				Apoptosis		[134]
	MKN1	>100	Cell Titer-Glo™	24	No suppression of survivin expression		[135]
	MKN1	~60…80	Counting	48			[129]
	MKN28	~50…75	Cell Titer-Glo™	24	Suppression of survivin expression		[135]
	MKN28				Decrease in PMA-induced RON expression and reduction of its promoter activity		[5]

Table 6.3. (Continued)

Compound	Cell line[1]	Cell growth inhibition IC_{50}, μM[2]	Assay method[3]	Assay time, h	Action mechanism[4]	Effect in human normal oral cells[5]	Ref.
	MKN45	>87.3	MTT	2			[7]
	MKN45	~75	Cell Titer-Glo™	24	Suppression of survivin expression, activation of p53 and p73		[135]
	MKN45	~40	Counting	48	Apoptosis, activation of casp-3, -8 and -9		[129]
	MKN45				Apoptosis		[137]
	MKN45				Mitochondrial-dependent apoptosis		[136]
	MKN74	~60	Counting	48			[129]
	NCI-N87	~50	MTT	48	Increase in KLF4 expression through activation of MEF2A, apoptosis, G0/G1 phase cell cycle arrest		[130]
	NUGC3	~50…75	Cell Titer-Glo™	24	Apoptosis, no effect on cell cycle, downregulation of survivin expression through p73 activation		[135]
	SGC7901	28.87±1.46	MTT	24	Apoptosis, G0/G1 phase cell cycle arrest, inhibition of Wnt/ß-catenin signaling		[8]
		17.35±1.24	MTT	48			
	SGC7901				Downregulation of VEGF expression and secretion, decrease in Stat3 activation		[40]
	TMK-1	~≥100	Cell Titer-Glo™	24	Suppression of survivin expression		[135]
	TMK-1				Decrease in PMA-induced RON expression and reduction of its promoter activity		[5]

Compound	**Cell line**[1]	**Cell growth inhibition IC_{50}, µM**[2]	**Assay method**[3]	**Assay time, h**	**Action mechanism**[4]	**Effect in human normal oral cells**[5]	**Ref.**
FLAVANOLIGNANS							
Silibinin	MGC803	>200	CCK-8	24	Apoptosis, G2/M phase cell cycle arrest; decrease in p-STAT3, Mcl-1, Bcl-xL, survivin expression, upregulation of casp-3/-9, decrease in CDK1 and cyclin B1 levels		[4]
		~200	CCK-8	48			
		~100…200	CCK-8	72			
	MKN28	104	MTT	24	Apoptosis, no protection against oxidative stress		[174]
	SGC7901	≥200	MTT	48	G2 phase cell cycle arrest, decrease in p34cdc2 expression, increase in p53 and p21 expression, casp-3-, 8-, -9-independent apoptosis		[61]
	SNU216				Suppression of TNF-α-induced MMP-9 expression through inhibition of MER/ERK pathway		[101]
FLAVANONES							
Alpinetin	AGS	120	MTT	24, 48, 72	Apoptosis through mitochondrial pathways, G2/M phase cell cycle arrest		[82]
	N87	120	MTT	24, 48, 72	G2/M phase cell cycle arrest		[82]
Flavanone	AGS				Inhibition of ß-catenin/Tcf signaling, downregulation of its target gene cyclin D1		[97]
Hesperetin	AGS	>328	MTT	24			[105]
	SGC7901		CCK-8	12, 24, 36, 48, 60, 72	Production of ROS, mitochondrial apoptosis		[31]
	SNU16	>200	MTT	48			[48]
Hesperidin	SNU638	>328	MTT	24			[105]

Table 6.3. (Continued)

Compound	Cell line[1]	Cell growth inhibition IC_{50}, μM[2]	Assay method[3]	Assay time, h	Action mechanism[4]	Effect in human normal oral cells[5]	Ref.
Liquiritigenin	AGS	339.3	MTT	48		IC_{50} 384.9 in GES-1	[80]
	MKN28	314.7	MTT	48			
	MKN45	376.7	MTT	48			
	SGC7901	358.0	MTT	48			
Naringenin	AGS	>184	MTT	72			[191]
	AGS				Inhibition of ß-catenin/Tcf signaling with unknown mechanisms		[84]
	EPG85-257	95.0±8.6	SRB	120			[140]
	EPG85-257RDB	27.0±2.4	SRB	120	Apparently not extruded by ABC-transporters		[140]
	EPG85-257NOV	6.2±0.8	SRB	120	Apparently not extruded by ABC-transporters		[140]
	KATO III	407±58	MTT	48			[139]
	MKN7	320±40	MTT	48			[139]
	SGC7901	~80…160	CCK-8	48	Inhibition of colony formation, PARP cleavage, decrease in Akt activation, increase in p53		[83]
	SGC7901		MTT	48	Downregulation of PCNA, decrease in migration and invasion, inhibition of MMP-2/-9 expression, apoptosis, suppression of AKT phosphorylation		[77]
Naringin	AGS	>345	MTT	24			[105]
	AGS	2000	MTT	24	Non-apoptotic cell death, autophagy, downregulation of PI3K/Akt/mTOR cascade via activation of MAPKs		[29]
			MTT	48			

Compound	Cell line[1]	Cell growth inhibition IC_{50}, μM[2]	Assay method[3]	Assay time, h	Action mechanism[4]	Effect in human normal oral cells[5]	Ref.
	SNU638	>345	MTT	24			[105]
Neohesperidin	AGS	>328	MTT	24			[105]
	SNU638	>328	MTT	24			[105]
Ponciretin	KATO III	NA up to 10 μM	MTT	48	Inhibition of ethanol- and LPS-induced IL-8, TNF, IFN-γ and COX-2 expression and NF-κB activation		[141]
Poncirin	AGS	130	MTT	24	Extrinsic (mitochondrial-independent) apoptosis, upregulation of FasL		[54]
	AGS	>336	MTT	24			[105]
	KATO III	NA up to 10 μM	MTT	48	Inhibition of ethanol- and LPS-induced IL-8, TNF, IFN-γ and COX-2 expression and NF-κB activation		[141]
	SGC7901	≤42	MTT	48			[120]
	SNU638	>336	MTT	24			[105]
FLAVONES							
5,4`-Dihydroxy-3,6,3`-trimethoxyflavone 7-glycoside	SGC7901	NA up to 63 μM	MTT	24, 48, 72			[57]
5,4`-Dihydroxy-6,7,3`,5`-tetramethoxyflavone	MK1	>267	MTT	48			[142]
5,6-Dihydroxy-7,3`,4`-trimethoxyflavone	MK1	33	MTT	48			[142]
3,5,4`-Trihydroxy-6,7,3`-trimethoxyflavone (TTF)	SGC7901	394.57	MTT	24	Apoptosis, G1 phase cell cycle arrest		[57]
		81.12	MTT	48			
		23.12	MTT	72			

Table 6.3. (Continued)

Compound	Cell line[1]	Cell growth inhibition IC_{50}, µM[2]	Assay method[3]	Assay time, h	Action mechanism[4]	Effect in human normal oral cells[5]	Ref.
5,6,3`-Trihydroxy-7,4`-dimethoxyflavone	MK1	55	MTT	48			[142]
5,3`,4`-Trihydroxy-6,7,5`-trimethoxyflavone	MK1	22	MTT	48			[142]
5,6,4`-Trihydroxy-7,3`,5`-trimethoxyflavone	MK1	73	MTT	48			[142]
5-Hydroxy-3,6,7,8,3`,4`-hexamethoxyflavone (5HHMF)	AGS	2.4…12	MTT	48	Production of ROS, apoptosis, increase in Fas, FasL, DR4 levels, cleavage of casp-3, -8, -9 and PARP		[154]
	BGC7901	5~10	MTT	24	Production of ROS, intrinsic apoptosis, decrease in PI3K and p-Akt expression		[14]
6-Hydroxyluteolin	MK1	26	MTT	48			[142]
	MK1	26	MTT	48			[195]
6-Methoxyflavone	MK1	398	MTT	48			[142]
7-Methoxyflavone	MK1	119	MTT	48			[142]
Acacetin	AGS		MTT	24	Increase in ROS, apoptosis through mitochondria- and Fas-mediated pathways		[88]
Apigenin	AGS	18.9±3.0	MTT	72			[191]
	HGC27	~18.5…37	CellTiter96	24	Apoptosis, increase in Bax, decrease in Bcl-2, decline of ΔΨm; no effect on cell cycle	IC_{50} >74 µM in GES-1 at 24, 48 and 72 h, no apoptosis	[78]
		~9.3…18.5	CellTiter96	48			
		~9.3	CellTiter96	72			
	MK1	22	MTT	48			[142]
	H. pylori-infected MKN45				Antiinflammatory activity. Increase in *H. pylori*-decreased IκBα expression and *H. pylori*-induced MUC-2 expression;		[103]

Compound	Cell line[1]	Cell growth inhibition IC_{50}, µM[2]	Assay method[3]	Assay time, h	Action mechanism[4]	Effect in human normal oral cells[5]	Ref.
					decrease in *H. pylori*-induced COX-2 and ICAM-1 expression, ROS, IL-8 and IL-6 levels		
	SGC7901	~37	CellTiter96	24	Apoptosis, increase in Bax, decrease in Bcl-2, decline of ΔΨm; no effect on cell cycle	IC_{50} >74 µM in GES-1 at 24, 48 and 72 h, no apoptosis	[78]
		~37	CellTiter96	48			
		~9.3…18.5	CellTiter96	72			
	SGC7901		MTT	Various	Suppression of clone formation, apoptosis		[144]
Baicalein	AGS	>80	MTT, under hypoxia	24	Inhibition of glucose uptake and lactate production, retarding the accelerated glycolytic rate, attenuation of expression of hypoxia-induced glycolysis-related enzymes (HK2, LDHA, PDK1), inhibition of hypoxia-dependent HIF-1α expression via PTEN/Akt pathway		[33]
	AGS	~60…80	MTT, under hypoxia	48			
	AGS	59.5	XTT	72			[192]
	AGS				Inhibition of migration and invasion by suppressing TGF-ß/Smad4 signaling, reduction of expression of metastasis-associated N-cadherin, vimentin, ZEB1/2		[34]
	MGC803				Decrease in invasion, suppression of MMP-2/-9 expression		[74]
	MK1	26	MTT	48			[142]

Table 6.3. (Continued)

Compound	Cell line[1]	Cell growth inhibition IC_{50}, μM[2]	Assay method[3]	Assay time, h	Action mechanism[4]	Effect in human normal oral cells[5]	Ref.
	MKN45	~60	MTT	24			[32]
		~15…30	MTT	48			
		~15	MTT	72			
	SGC7901	~160…200	MTT	24	Decrease in invasion and migration, suppression of MMP-2/-9 expression and activity through inhibition of p38 pathway		[74]
	SGC7901	>125	MTT	24	Suppression of colony formation, S phase cell cycle arrest, mitochondrial apoptosis		[32]
		~30	MTT	48			
		~15…30	MTT	72			
	SNU16	~50…100	MTT	48			[48]
Baicalin	AGS	64.3	XTT	72			[192]
Calycopterin	AGS	>134	MTT	72			[191]
Chrysin	AGS	NA up to 50 μM	MTT	24	Decrease in invasiveness via inhibiting MMP-9 expression through suppression of JNK/c-Jun and ERK/c-Fos pathways		[35]
	AGS		MTT	24	Inhibition of endogenous and PMA-induced RON expression by suppressing the activation of Egr-1 and NF-κB, decrease in invasion		[115]
	AGS	68.24	MTT	24	Decrease in miR-18a, miR-21, miR-221 expression		[79]
		56.16	MTT	48			
		42.32	MTT	72			

Compound	Cell line[1]	Cell growth inhibition IC_{50}, µM[2]	Assay method[3]	Assay time, h	Action mechanism[4]	Effect in human normal oral cells[5]	Ref.
	AGS	68.24	MTT	24	Increase in miR-34a expression		[28]
		56.16	MTT	48			
		42.32	MTT	72			
	SGC7901	40.56	MTT	48	Apoptosis		[187]
	SGC7901	5.8	MTT	144			[186]
	MK1	63	MTT	48			[142]
Cirsilineol	MK1	17	MTT	48			[142]
Cirsiliol	MK1	18	MTT	48			[142]
Cirsimaritin	AGS	45.8±1.9	MTT	72			[191]
Desmethoxycentaureidin	MK1	24	MTT	48			[195]
	MK1	24	MTT	48			[142]
Eucalyptin	AGS	174.94±9.07	MTT	48	G2/M phase cell cycle arrest		[23]
Eupafolin	MK1	29	MTT	48			[142]
	MK1	29	MTT	48			[195]
Eupatilin	AGS	~1000	MTT	24	Suppression of H_2O_2-induced apoptosis through inhibition of ERK, JNK, NF-κB (at lower doses)		[153]
	AGS	50…100	CCK-8	72			[75]
	AGS		CCK-8	24, 48, 72	G1 phase cell cycle arrest, differentiation mediated by activation of ERK and induction of cellular differentiation regulating proteins (TFF1, ZO-1, occludin), followed by apoptosis		[152]

Table 6.3. (Continued)

Compound	Cell line[1]	Cell growth inhibition IC_{50}, μM[2]	Assay method[3]	Assay time, h	Action mechanism[4]	Effect in human normal oral cells[5]	Ref.
	MKN1	50…100	CCK-8	72	Apoptosis, decrease in release of proinflammatory cytokines (IL-1ß, TNF-α, IL-6, IL-8), inhibition of NF-κB activity, suppression of invasion, downregulation of MMP-2/-9 expression		[75]
	SNU484	50…100	CCK-8	72			[75]
	SNU601	50…100	CCK-8	72			[75]
	SNU638	50…100	CCK-8	72			[75]
	MK1	55	MTT	48			[142]
Eupatorin	MK1	58	MTT	48			[142]
Flavone	HGC27				G1 phase cell cycle arrest		[143]
Hispidulin	MK1	83	MTT	48			[142]
Jaceosidin	MK1	27	MTT	48			[142]
Luteolin	AGS		CCK-8	24, 48	Upregulation of miR-34a expression, increase in p53 and p21 expression, decrease in MEK and ERK phosphorylation		[27]
	AGS	50	MTT	24	G2/M phase cell cycle arrest, decrease in Cdc2, Cyclin B1, Cdc25C, increase in p21/cip1 level; intrinsic apoptosis		[49]
		29.6±3.8	MTT	48			
		23.5±2.4	MTT	72			
	AGS	24.7	XTT	72			[192]
	AGS	32.1±2.1	MTT	72			[191]
	AGS/luteolin	16.15	CCK-8		Decrease in miR-34a could contribute to luteolin resistance		[27]

Compound	Cell line[1]	Cell growth inhibition IC_{50}, μM[2]	Assay method[3]	Assay time, h	Action mechanism[4]	Effect in human normal oral cells[5]	Ref.
	BGC803	21.57	CellTiter96	24			[52]
	BGC823	6.742	CellTiter96	24			[52]
	BGC823		CCK-8	24, 48	Upregulation of miR-34a expression, increase in p53 and p21 expression, decrease in MEK and ERK phosphorylation		[27]
	BGC823		MTT	24, 48	Downregulation of Bcl-2 via miR-34a-mediated pathway, apoptosis		[50]
	BGC823/ luteolin	10.34	CCK-8		Decrease in miR-34a could contribute to luteolin resistance		[27]
	HGC27	5.778	CellTiter96	24	Apoptosis, inhibition of STAT3 activation through disrupting the binding of HSP-90 to STAT3, which promoted its interaction to SHP-1, resulting in the dephosphorylation of STAT3		[52]
	HGC27				G1 phase cell cycle arrest		[143]
	Hs 746 T		CCK-8	96, 120	Apoptosis by inhibiting Akt signaling, inhibition of motility, migration and invasion, reversing EMT, increase in E-cadherin, decrease in N-cadherin, vimentin and Snail, inhibition of Notch1 signaling		[10]

Table 6.3. (Continued)

Compound	Cell line[1]	Cell growth inhibition IC_{50}, μM[2]	Assay method[3]	Assay time, h	Action mechanism[4]	Effect in human normal oral cells[5]	Ref.
	Hs 746 T				Inhibition of angiogenesis and VM tube formation through suppressing VEGF secretion dependent on Notch1 expression		[11]
	MGC803	17.36	CellTiter96	24			[52]
	MGC803				Inhibition of VM tube formation, decrease in the number of tubes and intersections of VM tubes		[11]
	MK1	31	MTT	48			[142]
	MKN28		CCK-8	96, 120	Apoptosis by inhibiting Akt signaling, inhibition of motility, migration and invasion, reversing EMT, increase in E-cadherin, decrease in N-cadherin, vimentin and Snail, inhibition of Notch1 signaling		[10]
	MKN45		MTT	24, 48, 72	Apoptosis, inhibition of migration and invasion, decrease in MMP-9 expression, downregulation of cMet/Akt/ERK signaling		[12]
	NCI-N87		CCK-8		Suppression of motility		[10]
	SGC7901	41.4	CellTiter96	24			[52]

Compound	**Cell line[1]**	**Cell growth inhibition IC_{50}, μM[2]**	**Assay method[3]**	**Assay time, h**	**Action mechanism[4]**	**Effect in human normal oral cells[5]**	**Ref.**
	SGC7901		CCK-8	24, 48	Upregulation of miR-34a expression, increase in p53 and p21 expression, decrease in MEK and ERK phosphorylation		[27]
	SGC7901		MTT	24, 48	Downregulation of Bcl-2 expression via miR-34a-mediated pathway, apoptosis		[50]
	SGC7901		MTT	24, 48, 72	Apoptosis, inhibition of migration and invasion, decrease in MMP-9 expression, downregulation of cMet/Akt/ERK signaling		[12]
	SGC7901				Mitochondrial apoptosis, reduction of PGE2 synthesis, decrease in the expression of VEGF and HIF-1α		[51]
	SGC7901/ DDP	12.8	CellTiter96	24	Apoptosis, inhibition of STAT3 activation through disrupting the binding of HSP-90 to STAT3, which promoted its interaction to SHP-1, resulting in the dephosphorylation of STAT3		[52]
Nobiletin	AGS	~3.5…4	MTT	24	Suppression of invasion and migration, downregulation of FAK/PI3K/Akt signaling and reduction of MMP-2/-9 expression		[13]
		~3.5	MTT	48			

Table 6.3. (Continued)

Compound	Cell line[1]	Cell growth inhibition IC_{50}, μM[2]	Assay method[3]	Assay time, h	Action mechanism[4]	Effect in human normal oral cells[5]	Ref.
	AGS	>200	MTT	48			[48]
	KATO III	146.3	MTT	48			[148]
	MKN45	~100…200	MTT	48			[48]
	MKN45	226.3	MTT	48			[148]
	MKN45	494	MTT	48			[150]
	MKN74	189.7	MTT	48			[148]
	SNU1	~200	MTT	48			[48]
	SNU16		MTT	24, 48	ER stress-mediated apoptosis and protective autophagy via downregulation of Akt/mTOR signaling		[149]
	SNU16	~25…50	MTT	48	Mitochondrial apoptosis, G1 phase cell cycle arrest via $p21^{WAF1/CIP1}$ and cyclin D1-mediated pathway in p53-independent manner		[48]
	St-4	655	MTT	48			[150]
	TMK-1	134.8	MTT	48	Apoptosis, G0/G1 phase cell cycle arrest		[148]
	TMK-1	156	MTT	48	Inhibition of MMP-9 activity in medium obtained from co-culture of TMK-1 and mouse fibroblasts		[150]
Oroxylin A	BGC823	321±36	MTT	24	G2/M phase cell cycle arrest, decrease in Cdk7, cdc2/p34 and cyclin A/B1 expression		[145]
		105±15	MTT	48			
		51±10	MTT	72			
Pectolinaringenin	MK1	115	MTT	48			[142]
Penduletin	AGS	28.2±2.0	MTT	72			[191]

Compound	Cell line[1]	Cell growth inhibition IC_{50}, μM[2]	Assay method[3]	Assay time, h	Action mechanism[4]	Effect in human normal oral cells[5]	Ref.
Tangeretin	AGS	>240	MTT	24	Apoptosis through p53-dependent mitochondrial pathway and Fas/FasL-mediated extrinsic pathway		[151]
		~120…240	MTT	48			
		~30…60	MTT	72			
	SGC7901	>100	MTT	24			[116]
Wogonin	MGC803	109.05±9.44	MTT	48	Apoptosis, decrease in NF-κB expression		[112]
	SGC7901				Apoptosis; downregulation of the expression of immune-regulatory molecules (B7H1, MICA, VEGF), release of dangerous signals (CRT, HMGB1) alarming immune system to eradicate tumors; regulation of B7H1, MICA and CRT expression by STAT3 signaling pathway		[147]
	SGC7901	71.20±4.25	MTT	72	Apoptosis, S phase cell cycle arrest, increase in ROS, reduction of GSH level		[146]
	SGC7901/ cDDP	89.49±7.37	MTT	72	Apoptosis, S phase cell cycle arrest, increase in ROS, reduction of GSH level, inhibition of COX-1/-2, suppression of CK2 activity		[146]
Xanthomicrol	AGS	13.1±2.9	MTT	72			[191]

Table 6.3. (Continued)

Compound	Cell line[1]	Cell growth inhibition IC_{50}, μM[2]	Assay method[3]	Assay time, h	Action mechanism[4]	Effect in human normal oral cells[5]	Ref.
FLAVONOLS							
Astragalin	AGS	67.89±9.16	MTT	48			[23]
	SCM1	>1000	MTT	24	Cell death prevention by inhibiting Cu(II)-induced apoptosis and necrosis		[162]
Casticin	BGC823	~10	MTT	24	Potentiation of TRAIL-induced apoptosis by upregulating DR5 through ROS-ER stress-CHOP pathway and downregulating expression of Bcl-2, XIAP, cFLIP, survivin	IC_{50} >>10 μM in GES-1, no effect on DR5 expression, no effect with TRAIL	[161]
	MGC803	~10	MTT	24	Increase in DR5 expression. Sensitization to TRAIL-induced apoptosis with activation of casp-3, -8, -9		
	SGC7901	~10	MTT	24			
Chrysosplenetin	MKN1	0.41	SRB	48			[160]
	MKN7	3.02	SRB	48			[160]
	MKN28	0.49	SRB	48			[160]
	MKN45	2.19	SRB	48			[160]
	MKN74	0.55	SRB	48			[160]
	St-4	58.9	SRB	48			[160]
Dihydromyricetin	AGS		MTT	48, 72	Mitochondrial apoptosis, upregulation of p53 expression		[16]

Compound	Cell line[1]	Cell growth inhibition IC_{50}, μM[2]	Assay method[3]	Assay time, h	Action mechanism[4]	Effect in human normal oral cells[5]	Ref.
Fisetin	AGS		MTT	24, 48, 72	ROS production through MRC complex I, DNA damage, G1 phase cell cycle arrest, mitochondrial p53-independent apoptosis		[15]
	SNU1		MTT	24, 48, 72	ROS production through MRC complex I, DNA damage, G1 phase cell cycle arrest, mitochondrial p53-independent apoptosis		[15]
Galangin	SGC7901	160	MTT	24	Extrinsic and intrinsic apoptosis integrated by Bid		[18]
		100	MTT	48			
		75	MTT	72			
	SNU484	100	MTT	24	Mitochondrial apoptosis, increase in Uch-L1, decrease in GSTP expression, regulation of ERK1/2 and JNK		[159]
		~50	MTT	48			
Hyperoside	AGS	>500	MTT	24			[158]
Isokaempferide	AGS	35.0±3.3	MTT	72			[191]
Isorhamnetin	AGS		MTT	24, 48, 72	Apoptosis (downregulation of Bcl-2/-xL), suppression of migration and invasion, anticancer effects mediated through activation of PPAR-γ signaling	IC_{50} >50 μM in HFE-145	[85]
	AGS				Inhibition of constitutive NF-κB activation		[113]

Table 6.3. (Continued)

Compound	Cell line[1]	Cell growth inhibition IC_{50}, μM[2]	Assay method[3]	Assay time, h	Action mechanism[4]	Effect in human normal oral cells[5]	Ref.
	AZ521	~25	MTT	24			[113]
		~25	MTT	48			
		~50	MTT	72			
	NUGC3	>50	MTT	24			[113]
		~50	MTT	48			
		~50	MTT	72			
	SNU5				Inhibition of expression of diverse NF-κB-regulated gene products (COX-2, CXCR4), PARP cleavage		[113]
	SNU16				Inhibition of expression of diverse NF-κB-regulated gene products (CXCR4, VEGF, Cyclin D1), PARP cleavage		[113]
Kaempferin	AGS	82.17±11.63	MTT	48			[23]
Kaempferol	AGS	>500	MTT	24			[158]
	BGC823	NA	MTT	72			[138]
	MKN28	124.86	CCK-8	24	G2/M phase cell cycle arrest, mitochondrial apoptosis, reduction of COX-2, p-AKT and p-ERK levels	NA up to 200 μM in GSE-1 cells at 24, 48, 72 h	[110]
		80.52	CCK-8	48			
		55.52	CCK-8	72			
	SGC7901	128.62	CCK-8	24	G2/M phase cell cycle arrest, mitochondrial apoptosis, reduction of COX-2, p-AKT and p-ERK levels		
		79.87	CCK-8	48			
		63.42	CCK-8	72			
	SGC7901	≥350	MTT	48			[157]
	SGC7901	NA	MTT	72			[138]

Compound	Cell line[1]	Cell growth inhibition IC_{50}, µM[2]	Assay method[3]	Assay time, h	Action mechanism[4]	Effect in human normal oral cells[5]	Ref.
Myricetin	HGC27		CCK-8	24, 48, 72	Apoptosis, G2/M phase cell cycle arrest, binding to RSK2 and leading to increased expression of Mad1		[17]
	SGC7901		CCK-8	24, 48, 72	Apoptosis, G2/M phase cell cycle arrest, binding to RSK2 and leading to in-creased expression of Mad1		[17]
Quercetin	AGS	37.5	MTT	24	Inhibition of survivin and pSTAT3 at higher doses		[37]
	AGS	40	MTT	48	Mitochondrial apoptosis, HIF-1α accumulation-mediated autophagy by inhibiting Akt-mTOR signaling pathway. Initiated autophagy antagonized apoptotic cell death		[114]
	AGS	50	AB	48	Modulation of GSK-3ß/ß-catenin signaling		[30]
	AGS	~400…500	MTT	24	Intrinsic apoptosis through MAPK signaling and inhibition of TRPM7 channels		[41]
		~400	MTT	48			
		267	MTT	72			
	AGS	39.12±3.23	MTT		Apoptosis, increase in autophagy. Synergistic anticancer efficacy with miR-143 through autophagy inhibition		[118]
	AGS				No effect on JNK phosphorylation or PARP expression. Suppression of EBV transfer to cells		[107]

Table 6.3. (Continued)

Compound	Cell line[1]	Cell growth inhibition IC_{50}, μM[2]	Assay method[3]	Assay time, h	Action mechanism[4]	Effect in human normal oral cells[5]	Ref.
	AGS-EBV-GFP				Phosphorylation of JNK, inhibition of PARP expression		[107]
	AGS-wt EBV				Phosphorylation of JNK, inhibition of PARP expression		[107]
	BGC823				Mitochondrial apoptosis, S phase cell cycle arrest		[124]
	EPG85-257P	12	SRB	72	Apoptosis		[156]
	EPG85-257RDB	12	SRB	72	Apoptosis, decrease in P-gp expression and its drug transport abilities, reduction of ABCB1 expression		[156]
	HGC27	52	TB	96	Suppression of DNA synthesis, G1 phase cell cycle arrest		[155]
	MGC803	>40	MTT	24			[39]
		>40	MTT	48			
	MGC803	80.6	MTT	48			[188]
	MGC803	23.35±2.14	MTT	72	No apoptosis, no effect on AKT and ERK phosphorylation. In combination with curcumin stronger cytotoxic effect, inhibition of AKT and ERK phosphorylation, mitochondrial apoptosis		[19]
	MGC803-derived GCSCs	53	MTT	24	Mitochondrial apoptosis via blockade of PI3K-Akt signaling		[36]

Compound	Cell line[1]	Cell growth inhibition IC_{50}, μM[2]	Assay method[3]	Assay time, h	Action mechanism[4]	Effect in human normal oral cells[5]	Ref.
	MKN7	55	TB	96			[155]
	MKN28	160	MTT	48	Mitochondrial apoptosis, HIF-1α accumulation-mediated autophagy by inhibiting Akt-mTOR signaling pathway. Initiated autophagy antagonized apoptotic cell death		[114]
	MKN28	35	TB	96			[155]
	MKN28	148.77±9.42	MTT		Apoptosis, increase in autophagy. Synergistic anticancer efficacy with miR-143 through autophagy inhibition		[118]
	NUGC2	32	TB	96			[155]
	SGC7901	>40	MTT	24			[39]
		>40	MTT	48			
	SGC7901	75.7	MTT	48			[188]
	SGC7901	200	MTT	24	Extrinsic and intrinsic apoptosis, integrated by Bid		[18]
		135	MTT	48			
		112	MTT	72			
	SNU16	~100	MTT	48			[48]
	SNU719	125	CCK-8	24	Apoptosis mediated by JNK/ERK pathway, S/G2 phase cell cycle arrest, elimination of DNMT1 and DNMT3A expression, induction of EBV lytic reactivation		[107]
		62	CCK-8	48			
Quercitrin	AGS	>500	MTT	24			[158]
Rutin	AGS	>500	MTT	24			[158]

Table 6.3. (Continued)

Compound	Cell line[1]	Cell growth inhibition IC_{50}, μM[2]	Assay method[3]	Assay time, h	Action mechanism[4]	Effect in human normal oral cells[5]	Ref.
Troxerutin	SGC7901	>80	MTT	24	Suppression of cell migration, decrease in clonogenic growth, apoptosis. Reduction of p65 expression, high expression of IκBα	NA up to 80 μM in GES-1	[38]
	SGC7901/ 5-FU		MTT	24	No change in cell migration, no effect on clonogenic growth, no apoptosis. No effect on p65 level or IκBα expression		
FLAVANONOLS							
Aromadendrin	EPG85-257	>100	SRB	120			[140]
	EPG85-257RDB	64.0±6.6	SRB	120			[140]
	EPG85-257NOV	29.0±2.5	SRB	120			[140]
ISOFLAVONES							
2``,6``-O-Diacetyloninin	Hs 740.T	14.79±0.19	MTT	144			[194]
	Hs 746 T	17.28±0.35	MTT	144			[194]
4`-Demethyltoxicarol isoflavone	AGS				Apoptosis in combination with TRAIL caused by increase in DR5 expression		[190]
Biochanin A	HSC-39K6	54.5	CV	96			[166]
	HSC-40A1	49.2	CV	96			[166]
	HSC-41E6	26.7	CV	96	Apoptosis		[166]
	HSC-42H	38.7	CV	96			[166]
	HSC-43C1	62.3	CV	96			[166]
	HSC-45M2	34.5	CV	96	Apoptosis		[166]
	SH101-P4	35.9	CV	96	Apoptosis		[166]

Compound	Cell line[1]	Cell growth inhibition IC_{50}, μM[2]	Assay method[3]	Assay time, h	Action mechanism[4]	Effect in human normal oral cells[5]	Ref.
Daidzein	BGC823		MTT	24, 48, 72	Mitochondrial apoptosis		[44]
	HGC27				G1 phase cell cycle arrest		[143]
	HSC-39K6	>157	CV	96			[166]
	HSC-40A1	>157	CV	96			[166]
	HSC-41E6	>157	CV	96			[166]
	HSC-42H	>157	CV	96			[166]
	HSC-43C1	>157	CV	96			[166]
	HSC-45M2	130	CV	96			[166]
	MKN45cl85	>100	CV	72			[43]
	85As2mLuc	>100	CV	72			[43]
	SH101-P4	143.6	CV	96			[166]
Dihydrodaidzin	Hs 740.T	41.52±1.03	MTT	144			[194]
	Hs 746 T	35.67±0.72	MTT	144			[194]
Dihydrogenistein	Hs 740.T	34.81±0.78	MTT	144			[194]
	Hs 746 T	28.18±0.64	MTT	144			[194]
Equol	MGC803	180.10	MTS	24	G0/G1 phase cell cycle arrest and apoptosis by blockade of Akt signaling		[91]
		19.98	MTS	36			
		13.93	MTS	48			
	MGC803		MTS	48	Mitochondrial apoptosis via sustained activation of ERK1/2 pathway		[20]
Genistein	AGS	NA at 50 μM	MTT	24	Sensitization of cells to TRAIL-induced apoptosis mediated by upregulation of DR5 pathway		[70]
	AGS	45.9	MTT	24	G2/M phase cell cycle arrest, decrease in expression of FOXM1 and its target genes CDK1, Cdc25B, cyclin B, increase in $p27^{KIP1}$		[24]

Table 6.3. (Continued)

Compound	Cell line[1]	Cell growth inhibition IC_{50}, μM[2]	Assay method[3]	Assay time, h	Action mechanism[4]	Effect in human normal oral cells[5]	Ref.
	BGC823	~80	MTT	24	G2/M phase cell cycle arrest, inhibition of Akt activation by upregulating PTEN		[163]
	BGC823	>200	MTT	24	Apoptosis, inactivation of NF-κB signaling resulting in reduced levels of COX-2 protein		[126]
		~200	MTT	48			
		~100	MTT	72			
	HGC27	20	Counting	48	G2/M phase cell cycle arrest		[143]
	Hs 740.T	16.21±0.26	MTT	144			[194]
	Hs 746 T	21.54±0.33	MTT	144			[194]
	HSC-39K6	40.7	CV	96			[166]
	HSC-40A1	38.9	CV	96			[166]
	HSC-41E6	27.8	CV	96	Apoptosis		[166]
	HSC-42H	68.5	CV	96			[166]
	HSC-43C1	57.4	CV	96			[166]
	HSC-45M2	26.3	CV	96	Apoptosis		[166]
	MGC803				Inhibition of self-renewal, stem cell-like properties (sphere formation) and markers (OCT4, Sox2, Nanog, CD44, CD90)		[122]
	MKN45-derived GCSLCs				Decrease in stem cell marker CD44 and stem cell-related genes (Oct4, Bmi, Nestin, ABCG2), reduction of sphere formation ability and migration. Affecting HH signaling through suppressing Gli1 expression		[123]

Compound	Cell line[1]	Cell growth inhibition IC_{50}, μM[2]	Assay method[3]	Assay time, h	Action mechanism[4]	Effect in human normal oral cells[5]	Ref.
	MKN45cl85	~50	CV	72			[43]
	85As2mLuc	~50	CV	72			[43]
	Primary GC cells		MTT	24, 48, 72	Apoptosis, decrease in Bcl-2/Bax ratio		[168]
	SGC7901	46.3	MTT	24	G2/M phase cell cycle arrest, decreasing in expression of FOXM1 and its target genes CDK1, Cdc25B, cyclin B, increase in $p27^{KIP1}$		[24]
	SGC7901	~80	MTT	24	G2/M phase cell cycle arrest, inhibition of Akt activation by upregulating PTEN		[163]
	SGC7901	~100	MTT	24			[189]
		~80	MTT	48			
		~40…80	MTT	72			
	SGC7901	~37.0	MTT	168	G2/M phase cell cycle arrest, increase in cyclin B1 and $p21^{waf/cip1}$, decrease in cyclin D1		[164]
	SGC7901	~37.0	MTT	168	Apoptosis, decrease in NF-κB		[167]
	SGC7901	~40	MTT		G2/M phase cell cycle arrest mediated partly by downregulation of KIF20A kinesin protein		[165]
	SGC7901				Inhibition of self-renewal capacity, decrease in sphere formation		[122]
	SH101-P4	25.2	CV	96	Apoptosis		[166]

Table 6.3. (Continued)

Compound	Cell line[1]	Cell growth inhibition IC_{50}, μM[2]	Assay method[3]	Assay time, h	Action mechanism[4]	Effect in human normal oral cells[5]	Ref.
Genistin	HSC-39K6	>93	CV	96			[166]
	HSC-40A1	>93	CV	96			[166]
	HSC-41E6	>93	CV	96			[166]
	HSC-42H	>93	CV	96			[166]
	HSC-43C1	>93	CV	96			[166]
	HSC-45M2	41.6	CV	96			[166]
	SH101P4	>93	CV	96			[166]
Glabridin	AGS	69.6	MTT	48		IC_{50} 123.8 μM in GES-1	[80]
	MKN28	61.0	MTT	48			
	MKN45	83.7	MTT	48			
	SGC7901	84.2	MTT	48			
Prunectin	HSC-39K6	>34	CV	96			[166]
	HSC-40A1	>34	CV	96			[166]
	HSC-41E6	23.5	CV	96			[166]
	HSC-42H	>34	CV	96			[166]
	HSC-43C1	>34	CV	96			[166]
	HSC-45M2	>34	CV	96			[166]
	SH101-P4	>34	CV	96			[166]
Pseudobaptigen	HSC-39K6	>35	CV	96			[166]
	HSC-41E6	>35	CV	96			[166]
	HSC-42H	>35	CV	96			[166]
	HSC-43C1	>35	CV	96			[166]
	SH101-P4	30.1	CV	96			[166]
Puerarin	BGC823	~600…1600	CCK-8	48	Apoptosis		[21]
	HSC-39K6	>192	CV	96			[166]
	HSC-40A1	>192	CV	96			[166]
	HSC41E6	>192	CV	96			[166]
	HSC-42H	>192	CV	96			[166]
	HSC-43C1	>192	CV	96			[166]
	HSC-45M2	>192	CV	96			[166]

	SH101-P4	>192	CV	96			[166]
PRENYLATED FLAVONOIDS							
Cudraflavone C	BGC823	24.9±1.7	MTT	72			[138]
	SGC7901	35.5±3.7	MTT	72			[138]
Heterophyllin	AGS				Sensitization of cells to TRAIL, induction of CHOP-dependent upregulation of DR5 resulting in apoptosis		[175]
Icariin	BGC823	189	MTT	48	Suppression of invasion and migration via Rac1-dependent VASP signaling		[76]
Kurarinone	SGC7901	>10	MTT	24	Synergism with TRAIL enhancing TRAIL-induced apoptosis and G2/M phase cell cycle arrest. Inactivation of STAT3, downregulation of antiapoptotic proteins (Mcl-1, c-FLIP)		[56]
Morusin	BGC823	12.6±1.1	MTT	72			[138]
	SGC7901	75.8±4.4	MTT	72			[138]
Xanthohumol C	BGC823	380±80.0	MTT	72			[193]
Xanthohumol D	BGC823	1866±297.5	MTT	72			[193]
ROTENOIDS							
Deguelin	AGS	1.38	MTT	72			[22]
	MKN28	0.06	MTT	72			[22]
	SNU484	0.45	MTT	72	G2/M phase cell cycle arrest, apoptosis; decrease in Akt phosphorylation, VEGF expression and HIF-1α accumulation		[22]

[1]GC, gastric cancer; GCSC, gastric cancer stem cell; GCSLC, gastric cancer stem-like cell.
[2]NA, not active.

[3]AB, Alamar Blue; CCK-8, cell counting kit-8; CV, crystal violet; MTS, 3-(4,5-dimethylthiazol-2-yl)-5-(3-carboxymethoxtphenyl)-2-(4-sulfophenyl)-2H-tetrazolium; MTT, 3-(4,5-dimethylthiazol-2-yl)-2,5-diphenyltetrazolium bromide; SRB, sulforhodamine B; TB, trypan blue exclusion; UV, ultraviolet absorption using method; WST-1, water soluble tetrazolium-1; XTT, 2,3-bis(2-methoxy-4-nitro-5-sulfophenyl)-2H-tetrazolium-5-carboxanilide.

[4]ΔΨm, mitochondrial transmembrane potential; ABC, ATP-binding cassette; Akt, protein kinase B; AP-1, activator protein 1; Bcl-2, B-cell lymphoma-2; c-FLIP, cellular FLICE-inhibitory protein; casp, caspase; CDK, cyclin-dependent kinase; CHOP, CCAAT-enhancer-binding protein homologous protein; CK2, casein kinase 2; cMet, tyrosine-protein kinase Met; COX-2, cyclooxygenase-2; CRT, calreticulin; CXCR4, C-X-C chemokine receptor type 4; DNMT, DNA (cytosine-5-)-methyltransferase; DR, death receptor; E2, estradiol; EBS, estrogen binding site; EBV, Epstein-Barr virus; Egr-1, early growth response protein 1; EMT, epithelial-mesenchymal transition; ER, endoplasmic reticulum; ERK, extracellular signal-regulated kinase; FAK, focal adhesion kinase; FasL, factor associated suicide ligand; FOXM1, fork-head box M1; GSH, reduced glutathione; GSK-3ß, glycogen synthase kinase-3ß; GST, glutathione S-transferase; H_2O_2, hydrogen peroxide; *H. pylori*, *Helicobacter pylori*; HH, Hedgehog; HIF-1α, hypoxia-inducible factor 1-α; HK2, hexokinase 2; HMGB1, high-mobility group protein 1; Hsp-90, heat shock protein-90; ICAM-1, intercellular adhesion molecule-1; IFN-γ, interferon-γ; IL, interleukin; JNK, c-Jun amino-terminal kinase; KLF4, Kruppel-like factor 4; LDHA, lactate dehydrogenase A; LPS, lipopolysaccharide; MAPK, mitogen-activated protein kinase; Mcl-1, myeloid cell leukemia-1; MDM2, murine double minute 2; MEF2A, myocyte enhancer factor-2A; MEK, mitogen-activated protein kinase kinase; MICA, MHC class I chain-related protein A; miR, microRNA; MMP, matrix metalloproteinase; MRC, mitochondrial respiratory chain; mTOR, mammalian target of rapamycin; MUC-2, mucin-2; NF-κB, nuclear factor κB; PARP, poly(ADP-ribose) polymerase; PCNA, proliferating cell nuclear antigen; PDK1, pyruvate dehydrogenase kinase-1; P-gp, P-glycoprotein; PGE2, prostaglandin E2; PI3K, phosphatidylinositol 3-kinase; PMA, phorbol-12-myristate-13-acetate; PPAR-γ, peroxisome proliferator-activated receptor γ; PTEN, phosphatase and tensin homolog; Rb, retinoblastoma; RON, recepteur d`origine nantais; ROS, reactive oxygen species; RSK2, ribosomal S6 kinase 2; SHP-1, SH2-containing protein tyrosine phosphatase-1; STAT, signal transducer and activator of transcription; Tcf, T-cell factor; TFF-1, trefoil factor 1; TGF-ß, transforming growth factor-ß; TLR-4, Toll-like receptor 4; TNF-α, tumor necrosis factor-α; TRAIL, TNF-related apoptosis-inducing ligand; TRPM7, transient receptor potential melastatin 7; Uch-L1, ubiquitin carboxy-terminal hydrolase isozyme L1; VASP, vasodilator-stimulated phosphoprotein; VEGF, vascular endothelial growth factor; VM, vasculogenic mimicry; ZEB, zinc finger E-box-binding homeobox; ZO-1, zonula occludens-1.

[5]GES-1 (GSE-1), human normal gastric epithelial cell line; HFE-145, human normal gastric epithelial cell line.

gastric cancer as its expression has been associated with tumor recurrence and poor disease-free survival. In addition, EGCG caused apoptosis and G0/G1 phase cell cycle arrest in SGC7901 cells [8]. EGCG inhibited SGC7901 cell growth also by downregulating vascular endothelial growth factor (VEGF) expression and secretion and angiogenesis related to decrease in signal transducer and activator of transcription 3 (Stat3) activation [40]. Gastric malignancy has generally high levels of Stat3 and VEGF which are independent prognostic factors of poor survival [40, 102]. Similarly, EGCG treatment suppressed interleukin (IL)-6-induced VEGF expression and production and angiogenesis in AGS cells via inhibition of Stat3 activation, nuclear translocation and Stat3-DNA binding activity [102]. Thus, EGCG can be a potent antiangiogenic agent for gastric cancer treatment. Moreover, EGCG can play also a crucial antiinflammatory role as in AGS cells, exposure to this green tea polyphenol suppressed IL-1ß-initiated IL-8 production and secretion through at least in part by decrease in nuclear factor-κB (NF-κB) activity [73]. Gastric malignancies often overexpress both IL-8 and NF-κB which strongly correlate with a poor prognosis [73]. In the same adenocarcinoma cell line, treatment with EGCG inhibited Id1 mRNA and protein expression, that is usually high in neoplastic tissues of stomach and related to malignant potential and metastatic ability, promoted apoptosis and induced S phase cell cycle arrest [62]. One more target regulated by EGCG in human gastric tumorigenesis is recepteur d`origine nantais (RON). This receptor plays a central role in development, progression and metastasis of gastric tumors and its increased expression is highly correlated with neoplastic invasiveness. EGCG suppressed phorbol 12-myristate 13-acetate (PMA)-induced expression and promoter activity of RON in AGS, MKN28 and TMK-1 cells and partially abrogated AGS cell invasion [5]. EGCG can exert its antiinvasive action also by suppressing PMA-induced matrix metalloproteinase (MMP)-9 expression in AGS cells via the decrease in MAPK and AP-1 activation [6]. Interestingly, EGCG-resistant AGS subline AGS-R revealed a cross-resistance to adverse effects of gastric-targeting N-nitroso insecticides showing that chronic exposure to EGCG can protect stomach epithelial tissue against potential damage by environmental risk factors [132]. This flavanol could exert gastroprotective activities also by preventing *H. pylori*-induced cytotoxicity and mucosal damage avoiding thus the deleterious consequences of infection [99]. In another poorly differentiated human gastric adenocarcinoma cell line KATO III, EGCG inhibited okadaic acid-induced tumor necrosis factor-α (TNF-α) expression and release via suppression of AP-1 and NF-κB activation. TNF-α is a central mediator and promoter of carcinogenesis [133]. Moreover, exposure to this green tea flavanol brought KATO III cells on apoptosis [129, 134]. The antitumorigenic function of EGCG in NCI-N87 cells was exerted by upregulation of the transcription factor Kruppel-like factor 4 (KLF4), expression of which is significantly decreased in gastric cancer tissues, arresting the cell cycle at the G0/G1 phase and induction of apoptosis [130]. In several other gastric cancer cell lines (MKN28, MKN45, NUGC3, TMK-1, but not MKN1), EGCG was shown to suppress the level of survivin downstream of p73. At that, NUGC3 cells were most sensitive to EGCG treatment and underwent apoptotic death [135]. Exposure to EGCG resulted in apoptosis also in MKN45 cells by regulating mainly the intrinsic mitochondrial-mediated pathway via decrease in the expression of antiapoptotic proteins Bcl-2 and Bcl-xL, increase in the levels of proapoptotic members Bad, Bid and Bax, and activation of caspases-3 and -9 [129, 136, 137]. Altogether, data presented above convincingly demonstrate that EGCG is a versatile antitumor agent with diverse modes of action in human gastric cancer cells. The major molecular targets and cellular signaling pathways of this flavanol are summarized and depicted in Figure 6.1. It is still important to mention that no data have been published to date about the potential effect of EGCG in normal human gastric epithelial cell lines precluding thus the estimation of specificity of EGCG responses towards malignant cells (Table 6.3).

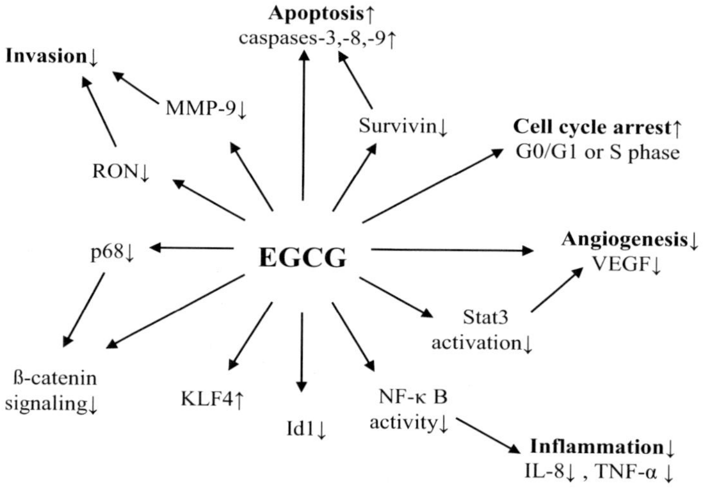

Figure 6.1. Major molecular targets and cellular signaling mechanisms of EGCG in human gastric cancer cells (IL-8, interleukin-8; KLF4, Kruppel-like factor 4; MMP-9, matrix metalloproteinase-9; NF-κB, nuclear factor-κB; RON, recepteur d`origine nantais; Stat3, signal transducer and activator of transcription 3; TNF-α, tumor necrosis factor-α; VEGF, vascular endothelial growth factor).

The current knowledge about anticancer role of other flavanols on gastric carcinogenesis is still much scarcer as compared to EGCG and reveals significantly lower or even no activities (Table 6.3). In KATO III cells epicatechin gallate (ECG) induced apoptotic cell death, and ECG and epigallocatechin (EGC) inhibited okadaic acid-induced TNF-α release from cells [133]. A weak apoptotic signal was described also by treating KATO III and MKN45 cells with EGC [129]. Although epicatechin (EC) itself was unable to trigger cell death in these two gastric cancer cell lines, it significantly enhanced the induction of apoptosis caused by EGCG through a mechanism involving production of reactive oxygen species (ROS) [129]. None of these three flavanols (EGC, ECG, EC) affected PMA-induced RON expression [5] or PMA-induced MMP-9 expression in AGS cells [6]. Furthermore, no effects have been described in treating human gastric cancer cells with catechin (C) [129, 138] (Table 6.3).

6.3.1.2. Flavanones

Flavanones are a subclass of flavonoids occurring mostly in citrus fruits. As can be seen in Table 6.3, these plant polyphenols have rather high half-maximal inhibitory concentrations (IC_{50} values) in suppressing the growth of different human gastric cancer cells. However, the activity depends considerably on the specific cell line and its molecular characteristics, as well as on the structure of certain flavanones.

Dysregulation of ß-catenin/T-cell factor (Tcf) signaling can play an important role in tumorigenesis and mutations in ß-catenin gene have been reported in both poorly as well as well-differentiated gastric cancers. In AGS line, missence mutation in this gene and constitutive transcriptional activity of ß-catenin/Tcf were detected [84, 97]. In these cells, ß-catenin/Tcf signaling was efficiently suppressed by both flavanone and naringenin without affecting nuclear levels of ß-catenin [84, 97]. In SGC7901 cells, naringenin suppressed cell proliferation and colony formation, migration and invasion, and induced apoptosis by increasing proapoptotic proteins Bax and cleaved caspase-3 and decreasing antiapoptotic proteins Bcl-2 and survivin [77, 83]. These anticancer effects might be related to inhibition of protein kinase B (Akt) activation [77, 83]. Although naringenin exerted some cytotoxicity also in KATO III and MKN7 lines these responses revealed at rather high concentration range [139]. Interestingly, this citrus flavonoid was significantly more potent in multidrug-resistant sublines of human gastric cancer line EPG85-257 than in parent cells, probably due to non-extrusion of naringenin from cells by ATP-binding cassette (ABC) transporters [140].

Apoptosis of human gastric cancer cells was triggered also by other flavanones. Alpinetin induced mitochondria-dependent apoptosis in AGS cells with translocation of Bax and Bcl-2 proteins, release of cytochrome (cyt) c into cytoplasm and activation of caspases-3 and -9. This flavanone also blocked the cell cycle at G2/M phase in both AGS and N87 cells [82]. Likewise, hesperetin caused mitochondrial-driven apoptosis in SGC7901 cells by increasing the levels of intracellular ROS [31]. Ponciretin, and to somewhat lesser extent also its 7-O-neohesperidoside, poncirin, strongly inhibited ethanol- and lipopolysaccharide (LPS)-stimulated expression of IL-8, TNF, interferon-γ (IFN-γ), and cyclooxygenase-2 (COX-2), and also suppressed the activation of NF-κB in KATO III cells, providing thus protection against ethanol-induced mucosal damage in the stomach [141]. As compared to other flavanones, poncirin revealed a rather potent inhibitory effect on the growth of SGC7901 cells being active already at low micromolar doses [120]. Somewhat surprisingly, poncirin triggered apoptotic death though extrinsic (and not mitochondrial intrinsic) signaling pathway that was led by upregulation of factor associated suicide ligand (FasL), sequential activation of caspases-8 and -3, and cleavage of poly(ADP-ribose) polymerase (PARP) in AGS cells [54]. Furthermore, exposure of AGS cells to another flavanone glycoside, naringin or 7-O-neohesperidoside of naringenin, did not initiate apoptotic cell death, but on the other hand, induced autophagy by downregulating phosphatidylinositol 3-kinase (PI3K)/Akt/mammalian target of rapamycin (mTOR) cascade via activation of MAPKs [29] (Table 6.3). These data show that flavanones may regulate important events in gastric tumorigenesis and can therefore be considered as potential lead compounds for development of chemopreventive and/or chemotherapeutic agents in the future.

6.3.1.3. Flavones

Various structurally different flavones have been investigated regarding to their potential *in vitro* activities against human gastric tumorigenesis (Table 6.3). These studies clearly demonstrate that substitutions in the flavone backbone considerably impact the anticancer potencies of these natural agents [57, 142].

Luteolin is the most studied flavone concerning its inhibitory effects on development and progression of gastric cancer. It is a highly pleiotropic compound that modulates diverse molecular targets and an overview of luteolin-regulated cellular signaling pathways in human gastric cancer cells is depicted in Figure 6.2. In AGS adenocarcinoma line, this flavone initiated mitochondrial apoptosis with shifting the Bax/Bcl ratio in favor of apoptosis, increasing in active caspases-3, -6, and -9 (but not -8) and releasing cyt c to the cytosol, and arrested cells at G2/M phase [49]. However, treatment of HGC27 cells with luteolin promoted the arrest of cell cycle at G1 phase [143]. Inhibition of cell proliferation and apoptotic death was induced also in several other human gastric cancer lines after exposure to luteolin, such as Hs 746 T, MKN28, MKN45, SGC7901 and BGC823 [10, 12, 50, 51]. In SGC7901 cells, apoptosis was executed by intrinsic pathway via downregulation of Bcl-2 and activation of caspases-3 and -9 [51]. In addition, luteolin was a more potent inhibitor of STAT3-overactivated cell growth. Tumor cells have been frequently characterized by constitutive activation of STAT3 and upregulation of this transcription factor is often related to a worse prognosis and drug resistance. Luteolin exerted its antitumor effects in cisplatin-resistant gastric cancer cells HGC27 and SGC7901/DDP by suppressing STAT3 activation [52]. In Hs 746 T and MKN28 cells, this natural polyphenol inhibited invasion and migration and reversed epithelial-mesenchymal transition (EMT) by increasing the epithelial biomarker E-cadherin and decreasing the mesenchymal biomarkers N-cadherin, vimentin and Snail, through suppressing Notch1 signaling [10]. Overexpression of Notch1 has been indeed correlated with poor overall survival of patients and tumor progression [10, 11]. In MKN45 and SGC7901 cells, invasiveness was suppressed by downregulating MMP-9 and repressing tyrosine-protein kinase Met (cMet)/Akt/extracellular signal-regulated kinase (ERK) signaling [12]. Treatment of Hs 746 T and MGC803 cells with luteolin resulted in inhibition of angiogenesis and formation of vasculogenic mimicry (VM) tubes [11]. At that, VEGF secretion was significantly decreased via suppression of Notch1 expression in Hs 746 T cells [11]. A decrease in the expression of VEGF and hypoxia-inducible factor 1-α (HIF-1α) was detected also in SGC7901 cells [51]. In several cell lines (AGS, BGC823, SGC7901), luteolin remarkably upregulated miR-34a, that could be a prognostic marker for gastric cancer [27, 50]. Indeed, this microRNA has been shown to be significantly downregulated in primary gastric tumor tissues as compared to adjacent non-tumor tissues and its low levels were related to shorter overall and disease-free survival rates. Moreover, decreased miR-34a was detected in luteolin-resistant sublines of AGS and BGC cells (AGS/luteolin and BGC/luteolin) and level of miR-34a could mediate the sensitivity of cells to luteolin [27]. In BGC823 and SGC7901 cells, luteolin-upregulated miR-34a expression led to downregulation of Bcl-2 playing an important role in promotion of apoptosis [50]. However, to date the data about possible action of luteolin in human normal gastric epithelial cell lines are missing (Table 6.3). Nevertheless, the above presented results strongly suggest the potential clinical application of luteolin in treatment of gastric cancer in the future. Moreover, as luteolin was also shown to inhibit the growth of *H. pylori*, this flavone can substantially contribute to chemoprevention of stomach tumorigenesis [100].

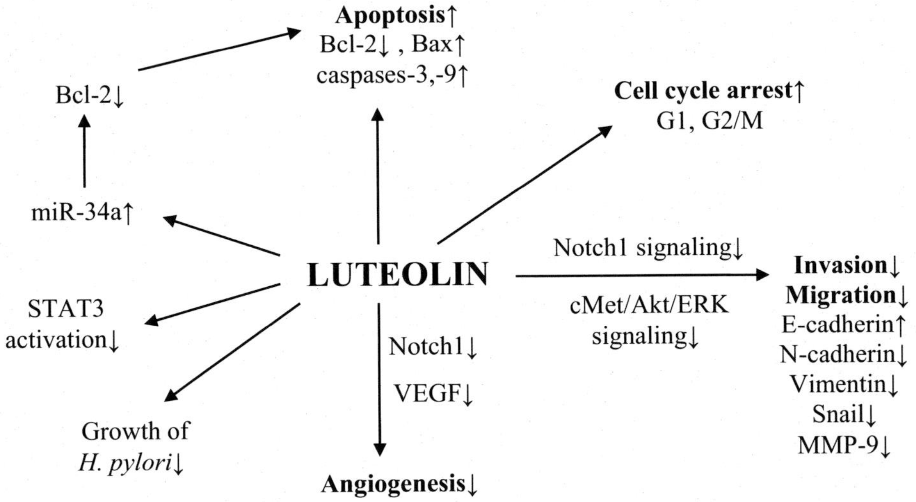

Figure 6.2. Major cellular signaling pathways regulated by luteolin in human gastric cancer cells (Akt, protein kinase B; Bcl-2, B-cell lymphoma-2; cMet, tyrosine-protein kinase Met; ERK, extracellular signal-regulated kinase; *H. pylori*, *Helicobacter pylori*; miR, microRNA; MMP-9, matrix metalloproteinase-9; STAT3, signal transducer and activator of transcription 3; VEGF, vascular endothelial growth factor).

Another commonly occurring flavone, apigenin, inhibited proliferation and induced apoptosis in human gastric cancer cells HGC27 and SGC7901 with increase in Bax/Bcl-2 ratio, decrease in mitochondrial membrane potential and activation of caspase-3, but without affecting the cell cycle progression [78, 144]. At that, the growth inhibition was more effective in undifferentiated HGC27 cells than in moderately differentiated SGC7901 cells, whereas normal human gastric cells GES-1 were almost not influenced by the exposure to apigenin [78]. Thus, apigenin holds therapeutic potential for development of a novel anticancer drug, especially for undifferentiated gastric tumors with poor prognosis. Moreover, this flavone could also effectively exhibit antiinflammatory effects in gastric epithelial cells by inhibition of NF-κB activation and expression of relevant inflammatory factors (ROS, IL-6, IL-8, COX-2, ICAM-1), and stimulation of mucin-2 (MUC-2) secretion in *H. pylori*-infected MKN45 adenocarcinoma cells [103] (Table 6.3).

Baicalein exerted its anticancer effects in SGC7901 cells by triggering apoptosis and S phase cell cycle arrest. This natural polyphenol broke the balance of Bcl-2 and Bax proteins in favor of apoptosis and disrupted mitochondrial membrane potential [32]. Moreover, baicalein could also suppress the invasion and migration in SGC7901 and MGC803 cells by reducing the expression and activity of MMP-2 and -9 via direct inhibition of p38 signaling [74]. Elevated activation of p38 pathway has been detected in gastric tumors and is correlated with metastatic ability [74]. In AGS cells, baicalein suppressed motility, migration and invasion by inactivating the transforming growth factor-ß (TGF-ß)/Smad4 signaling and reducing the expression of metastasis-associated downstream target genes, including N-cadherin, vimentin, ZEB1 and ZEB2 [34]. Increased expression of multifunctional cytokine TGF-ß has been observed in advanced stages of different malignancies and in gastric cancer, it is considered as a marker for poor prognosis [34]. Furthermore, under hypoxic conditions where the glycolytic rate of malignant cells is accelerated, baicalein suppressed and returned glucose uptake and lactate production to the levels detected under normoxia in AGS cells. This process was regulated by downregulation of the hypoxia-triggered glycolysis-related enzymes (hexokinase 2 (HK2), lactate dehydrogenase a (LDHA), pyruvate dehydrogenase kinase-1 (PDK1)), and inhibition of HIF-1α through regulation of the PTEN/Akt signaling [33] (Table 6.3). The current knowledge suggests that baicalein might be able to modulate several crucial signaling pathways in gastric tumorigenesis making this natural compound an attractive potential therapeutic lead for further treatment of this highly lethal malignancy.

Likewise, also chrysin appears to be an important flavone in suppression of metastatic ability of human gastric cancer cells. In AGS line, chrysin inhibited PMA-induced MMP-9 expression via AP-1 suppression resulting from blocking of the JNK1/2 and ERK1/2 pathways and leading to a significant decrease in invasiveness [35].

Antiinvasion action of chrysin in AGS cells was attained also through inhibition of RON expression by attenuating the activities of NF-κB and early growth response protein-1 (Egr-1). The expression of RON is usually several-fold higher in tumor tissues than in normal cells being associated with the metastatic phenotype [115]. In addition, chrysin was shown to upregulate the expression of miR-34a [28] and downregulate miR-18a, miR-21 and miR-221 in AGS gastric cancer cells [79] (Table 6.3).

It is generally believed that methylation of hydroxyl groups in flavone backbone can modulate the anticancer properties. The action of some monomethylated flavones (acacetin, oroxylin A, wogonin) has been studied in human gastric cancer cells (Table 6.3). Acacetin (4`-methoxyapigenin) inhibited the proliferation of AGS cells, whereas its cytotoxic activity was due to induction of apoptosis via both mitochondrial signaling (increase in intracellular ROS, loss of mitochondrial membrane integrity, release of cyt c into cytosol, triggering of caspase-9) as well as Fas-mediated pathway (upregulation of Fas and FasL and activation of caspase-8) [88]. Treatment of BGC823 cells with oroxylin A (6-methoxybaicalein) led to the arrest of cell cycle in G2/M phase with downregulation of cyclin-dependent kinase 7 (Cdk7), Cdc2/p34, cyclin A and cyclin B1 [145]. Wogonin (8-methoxynorwogonin) induced apoptosis in MGC803 cells with inhibition of NF-κB level [112]. Apoptosis, intracellular ROS production and S phase cell cycle arrest were initiated by wogonin also in SGC7901 cells and its cisplatin-resistant subline [146]. Moreover, in SGC7901 cells, this methoxyflavone exhibited dual antitumor effects via direct cytotoxicity by inducing apoptosis and indirect immune enhancement by displaying of signals alarming immune system to eradicate tumors, such as increase in membrane expression of annexin A1 and calreticulin (CRT) and release of ATP and high-mobility group protein 1 (HMGB1). At that, downregulation of immunoregulatory substances (B7H1 and MHC class I chain-related protein A (MICA)), and translocation of CRT to cell membrane were regulated through inhibition of STAT3 activation [147] (Table 6.3). Immunomodulative effect of wogonin was described also in murine MFC cells representing this methoxyflavone as an effective candidate for cancer immunotherapy [109].

Furthermore, potential anticancer action of several polymethoxyflavones has also been investigated in different human gastric cancer cell lines. Nobiletin (5,6,7,8,3`,4`-hexamethoxyflavone) exerted dose-dependent cytotoxicity in several gastric cancer lines, such as KATO III, MKN45, MKN74 and TMK-1; and induced apoptotic death and arrested cell cycle at G0/G1 phase in TMK-1 cells [148]. In SNU16 cells, nobiletin triggered apoptosis via intrinsic pathway as evidenced by elevation of Bax/Bcl-2 ratio, activation of caspases-9, -7 and -3 and cleavage of PARP protein, and blocked the cell cycle progression at G1 phase [48]. In addition, nobiletin-induced apoptosis in SNU16 cells was mediated by endoplasmic stress, whereas protective autophagy was followed by apoptosis, suggesting that treatment of gastric cancer patients with combination of nobiletin and an autophagy inhibitor might improve the clinical outcome [149]. Exposure of highly metastatic AGS cells to nobiletin led to suppression of cell adhesion, invasion and migration via inhibition of focal adhesion kinase (FAK)/PI3K/Akt pathway and expression and activity of MMP-2 and -9 [13]. The enzymatic activity of MMP-9 was completely inhibited by nobiletin also in culture medium derived from a co-culture of TMK-1 cells and murine fibroblasts [150]. Tangeretin (5,6,7,8,4`-pentamethoxyflavone) induced both p53-dependent mitochondrial as well as Fas/FasL-mediated extrinsic apoptosis with activation of caspases-3, -8 and -9 in AGS cells [151]. Eupatilin (5,7-dihydroxy-6,3`,4`-trimethoxyflavone) initiated apoptosis in MKN1 cells regulating PI3K/AKT and MAPK pathways, suppressed the release of proinflammatory cytokines (IL-1ß, IL-6, IL-8, TNF-α), inhibited NF-κB activation, and reduced invasive potential via downregulation of MMP-2 and -9 [75]. It is well known that proinflammatory cytokines are upregulated by chronic infection with *H. pylori* playing a critical role in inflammation and tumorigenesis of gastric mucosa via NF-κB-mediated pathway [75]. In AGS cells, eupatilin blocked the cell cycle at G1 phase with decrease in cdk2 and cdk6 expression and exerted antiproliferative activity. Moreover, eupatilin also triggered differentiation of AGS cells influenced by cell-matrix and cell-cell communications and mediated by ERK signaling, followed by ultimate apoptotic cell death. Indeed, eupatilin increased cell size and granularity, upregulated trefoil factor 1 (TFF1), a gene that participates in gastrointestinal cell differentiation, and increased and redistributed occludin and zonula occludens-1 (ZO-1) to the areas of cell-cell contacts [152]. In addition, in the same cells, eupatilin was also shown to suppress H_2O_2-induced apoptosis via inhibition of ERK, JNK and NF-κB, providing thus protective effects against oxidant-initiated injuries on gastric mucosa [153]. Another polymethoxyflavone, 5-hydroxy-3,6,7,8,3`,4`-hexamethoxyflavone (5HHMF) exhibited somewhat different effects in two human gastric cancer cell lines. In AGS cells, it induced apoptosis through elevation of death receptor-related mediators, including Fas, FasL and death receptor 4 (DR4), and cleavage of caspases-3, -8, -9 and PARP [154]. In BGC7901 cells, treatment with 5HHMF initiated apoptosis through targeting PI3K/Akt signaling cascade that is often overactive in malignant tumors suppressing apoptosis and promoting proliferation. 5HHMF inhibited this pathway, and induced higher Bax/Bcl-2 ratio, release of cyt c into cytosolic fraction and activation of caspases-3 and -9 (but not -8) [14]. Other tested polymethoxyflavones similarly revealed the dependence of anticancer potential on both molecular structure as well as investigated gastric cancer cell line. 3,5,4`-Trihydroxy-6,7,3`-trimethoxyflavone (TTF) inhibited the growth of SGC7901 cells, and induced apoptosis and G1 phase cell cycle arrest [57]. However, exposure of AGS cells to eucalyptin (5-hydroxy-7,4`-dimethoxy-6,8-dimethylflavone) blocked the cell cycle in G2/M phase [23] (Table 6.3).

6.3.1.4. Flavonols

Quercetin as an abundantly occurring natural compound is a main dietary flavonoid. This flavonol was described to suppress the growth of different human gastric cancer cell lines (Table 6.3) and one of its action mechanisms is through inducing serious defects in progression of normal cell cycle. Indeed, quercetin blocked cell cycle at G1 phase in poorly differentiated HGC27 cells [155], S phase in BGC823 cells [124] and S/G2 phase in SNU719 cells [107], showing that there must be some different mechanisms in action of this flavonol in different cell lines. In addition, quercetin initiated apoptotic cell death in stomach tumor cells, including parental EPG85-257P cells and its daunorubicin-resistant subline, EPG85-257RDB [156]. In BGC823 cells, quercetin induced apoptotic death through mitochondrial signaling with a drop in Bcl-2/Bax ratio and increase in caspase-3 [124]. However, in SGC7901 cells, this flavonol activated both extrinsic as well as intrinsic apoptotic pathways by activating caspase-8 that led to cleavage of Bid, translocation of Bax into mitochondria, decrease in mitochondrial membrane potential, release of cyt c into cytosol, and upregulation of caspases-9 and -3 [18]. The molecular mechanism of quercetin-triggered apoptosis has been further studied in different gastric cancer cell lines. In AGS cells quercetin downregulated survivin and pSTAT3 levels. It is generally accepted that overexpression of survivin plays a crucial role in appearance of resistance to apoptotic events being also a major factor in development of chemoresistance in gastric cancer, whereas STAT3 is a transcription factor required for survivin expression [37]. In the same cell line, quercetin induced mitochondrial apoptosis by activation of JNK and ERK signaling, and inhibition of transient receptor potential melastatin 7 (TRPM7) currents [41]. Interestingly, in MGC803 cells, quercetin triggered mitochondrial apoptosis only in combination with curcumin [19]. Furthermore, in both poorly differentiated AGS as well as well-differentiated MKN28 human gastric cancer cells, quercetin induced mitochondria-mediated apoptosis, but initiated also autophagy progression with formation of autophagic vacuoles via modulation of Akt-mTOR and HIF-1α signaling [114]. At that, autophagy displayed a protective mechanism against apoptosis and inhibition of autophagy could increase flavonol-induced apoptotic death [114]. Indeed, the cytotoxic efficacy of quercetin was synergistically enhanced when combined with an autophagy inhibitor, miR-143. The levels of miR-143 have been reported to be substantially downregulated or even lost in several malignant tissues, including stomach tumors as compared to the normal gastric mucosa [118]. In addition, quercetin stimulated apoptosis via JNK/ERK signaling and reduced EBV latency in virus-associated SNU719 cells, behaving thus also as an antiviral agent [107]. Moreover, this flavonol induced apoptosis only in EBV-transfected AGS cells and not in parent AGS line indicating a crucial role of EBV infection in quercetin-mediated apoptosis. Also, quercetin severely prevented the transfer of EBV into virus-negative AGS cells [107]. Last but not least, quercetin revealed the potential to effectively suppress the survival of GCSCs derived from MGC803 cells and initiated mitochondrial-dependent apoptosis with decrease in Bcl-2/Bax ratio and activation of caspases -3 and -9, via the blockade of PI3K-Akt pathway [36] (Table 6.3). All these signaling processes triggered by quercetin in gastric cancer cells are summarized in Figure 6.3.

So far, the potential effects of other flavonols against gastric tumorigenesis have been much less studied (Table 6.3). Fisetin induced ROS production mostly via mitochondrial respiratory chain (MRC) complex I and caused DNA damage in both AGS and SNU1 cells. In response to harm to DNA, the cells were blocked in G1 phase of the cell cycle with decreased expression of CDK2, CDK4 and cyclin D1 proteins, and destined to intrinsic apoptotic death as evidenced by a significant increase in Bax/Bcl-2 ratio and activation of caspase-3 [15]. Exposure of MKN28 and SGC7901 cells to kaempferol arrested the cell cycle at G2/M phase accompanied by decreased levels of cell cycle regulatory factors cyclin B1, Cdk1 and Cdc25C, and induced apoptosis in mitochondrial-dependent manner with drop in Bcl-2/Bax ratio, activation of caspases-3 and -9, and cleavage of PARP. At that, the growth of human normal gastric epithelial cells GSE-1 was not influenced by kaempferol [110]. Somewhat contrary, other authors did not confirm the cytotoxic activity of kaempferol in SGC7901 cells [138, 157], whereas very low or undetectable effects at physiologically relevant doses were reported also in AGS [158] and BGC823 cells [138]. Treatment of HGC27 and SGC7901 cells with myricetin led to G2/M phase cell cycle arrest with reduction of cyclin B1, cyclin D1, Cdk1 and Cdc25C, and triggered apoptosis as evidenced by decrease in antiapoptotic Bcl-2 and increase in proapoptotic Bax and cleaved caspase-3. Binding to ribosomal S6 kinase 2 (RSK2) could be involved in antiproliferative activities of myricetin [17]. Dihydromyricetin or ampelopsin promoted intrinsic mitochondrial p53-related apoptotic pathway by reducing Bcl-2 in AGS cells [16]. Galangin initiated mitochondrial apoptosis (drop in Bcl-2/Bax ratio, stimulation of caspases-3 and -9, cleavage of PARP) via increase in ubiquitin carboxyl-terminal hydrolase isozyme L1 (Uch-L1), decrease in glutahione S-transferase P (GSTP) and subsequent regulation of MAPKs in SNU484 cells [159]. However, in SGC7901 cells, this flavonol promoted both extrinsic and intrinsic apoptotic pathways regulated by activation of caspases-8, -9 and -3, and integrated by Bid protein. At that, galangin revealed more potent antiproliferative and proapoptotic efficiencies than quercetin [18] (Table 6.3).

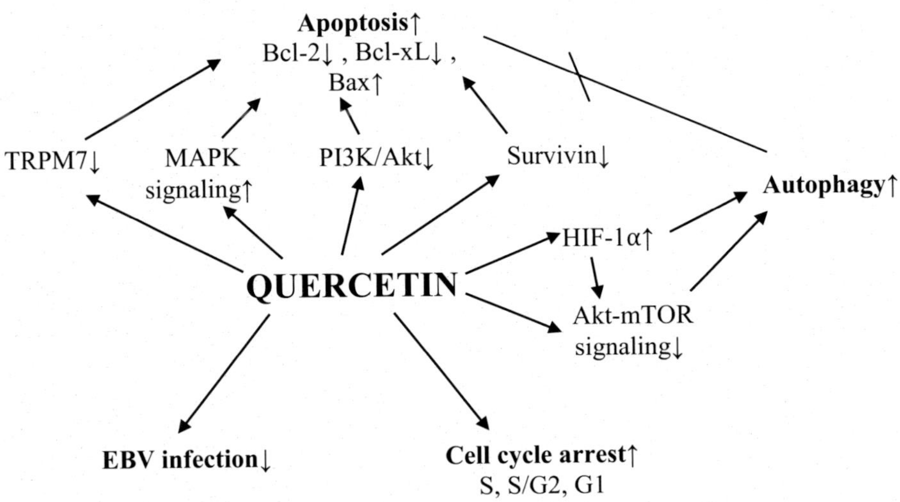

Figure 6.3. Major quercetin-mediated molecular pathways regulating antitumor events in human gastric cancer cells (Akt, protein kinase B; Bcl-2, B-cell lymphoma-2; EBV, Epstein-Barr virus; HIF-1α, hypoxia-inducible factor 1-α; MAPK, mitogen-activated protein kinase; mTOR, mechanistic target of rapamycin; PI3K, phosphatidylinositol 3-kinase; TRPM7, transient receptor potential melastatin 7).

In addition, the potential anticancer effects of several methylated flavonols (isorhamnetin, isokaempferide, casticin and chrysosplenetin) have also been studied (Table 6.3) Isorhamnetin (3`-methoxyquercetin) exerted antiproliferative and proapoptotic effects (with downregulation of Cyclin D1, Bcl-2 and Bcl-xL) through stimulation of the peroxisome proliferator-activated receptor-γ (PPAR-γ) signaling and suppressed migratory and invasive abilities of AGS cells. At that, this methylated derivative revealed only minimal influence on the proliferation of human normal gastric epithelial cells HFE-145 [85]. In different gastric cancer cell lines (AGS, SNU5 and SNU16) isorhamnetin abrogated NF-κB activation. Activation of NF-κB can be caused by *H. pylori* infection in gastric epithelial cells and this transcription factor plays a crucial role in initiation and progression of stomach tumors being related to several oncogenic features, such as abnormal growth and resistance to apoptosis. Moreover, NF-κB could also behave as a prognostic marker for grade IV stomach tumors [113]. Two polymethoxyflavonols (casticin or 3,6,7,4`-tetramethoxyquercetagetin and chrysosplenetin or 3,6,7,3`-tetramethoxyquercetagetin) displayed high potency in suppression of growth of different human gastric cancer cell lines. These activities were especially strong for chrysosplenetin with half-maximal inhibitory concentrations in low or submicromolar range both in poorly differentiated cells (MKN45, MKN74) as well as well-differentiated cells (MKN7, MKN28) [160]. Subtoxic doses of casticin potentiated TNF-related apoptosis-inducing ligand (TRAIL)-initiated apoptosis in several gastric cancer lines (BGC823, MGC803, SGC7901) by upregulating death receptor 5 (DR5) level on the cell surface and suppressing various cell survival proteins, whereas no effects were detected by treatment of normal gastric epithelial cells GES-1 with combination of casticin and TRAIL. In BGC823 cells, casticin promoted ROS production and triggered endoplasmic reticulum (ER) stress which led to increase in the level of CHOP protein that was involved in DR5 upregulation [161] (Table 6.3).

It is well accepted that sugar moiety in the flavonoid skeleton can markedly change its anticancer properties. Different quercetin sugar conjugates (hyperoside or quercetin 3-glycoside, quercitrin or quercetin 3-rhamnoside and rutin or quercetin 3-rutinoside) revealed no detectable cytotoxic activity in AGS cells [158]. A structurally modified rutin, troxerutin or 7,3`,4`-trihydroxyethylrutin, decreased the viability of 5-FU-resistant SGC7901 cells (SGC7901/5-FU) somewhat more efficiently than parental SGC7901 cells [38]. Kaempferin (kaempferol 3-rhamnoside) and astragalin (kaempferol 3-glycoside) exhibited antiproliferative effects in AGS cells [23]. However, in SCM1 cells, astragalin decreased cell viability only at very high doses. At that, this kaempferol sugar conjugate inhibited Cu(II)-induced apoptotic and necrotic cell death [162] (Table 6.3).

Interestingly, the antimetastatic efficacy of flavonols has still been explored very rarely and the current knowledge about antiinvasive and antimigratory activities of flavonols is essentially limited with the data of isorhamnetin in AGS cells [85]. Nevertheless, the cytotoxic and cytostatic potential described above clearly indicate that certain flavonols can be attractive therapeutic candidates for further treatment of gastric cancer.

6.3.1.5. Isoflavones

Isoflavones are principally found in soy products, whereas genistein is the most well known natural isoflavone. Several reports have been shown that genistein inhibited cell growth and proliferation and blocked the cell cycle progression at G2/M phase in different human gastric cancer lines, including AGS [24], HGC27 [143], BGC823 [163], and SGC7901 cells [24, 163-165]. In SGC7901 cells, the possible mechanism involved genistein-induced decrease in the overexpression of cyclin D1, and promotion of cyclin B1 and $p21^{waf/cip1}$ levels [164]. Additionally, the proteins of kinesin family, especially KIF20, behaved as important targets in triggering the cell cycle arrest in SGC7901 cells [165]. Treatment of both SGC7901 and BGC823 cells with genistein resulted in inactivation of Akt via upregulation of PTEN, decrease in cdc2, and increase in cyclin B1 and Wee1 proteins [163]. In SGC7901 and AGS cells, G2/M phase cell cycle arrest was mediated also by downregulation of forkhead box protein M1 (FOXM1) and its downstream genes, such as CDK1, Cdc25B and cyclin B [24]. It has been indeed reported that gastric cancer tissues often possess upregulated FOXM1 signaling that plays a crucial role in tumorigenesis, and FOXM1 level has been considered a marker of poor overall survival [24]. Furthermore, genistein induced human gastric cancer cells also to undergo apoptosis. Activation of apoptotic signal transduction pathway was described in HSC-41E6, HSC-45M2 and SH101-P4 cells after exposure to genistein [166]. In SGC7901 cells, this soy isoflavone triggered cell death through downregulation of NF-κB and upregulation of caspase-3 [167]. In BGC823 cells, inactivation of NF-κB signaling pathway led to reduced levels of COX-2 protein contributing to antiproliferative and proapoptotic activities [126]. Genistein was able to trigger apoptotic cell death with typical apoptotic characteristics, downregulation of Bcl-2 and upregulation of Bax expression also in primary gastric cancer cells [168]. Moreover, subtoxic doses of this common isoflavone sensitized AGS cells (that are normally resistant to TRAIL) to TRAIL-mediated apoptosis by increase in DR5 expression, activation of caspases-8 and -3, and cleavage of PARP [70]. In addition, genistein displayed the ability to suppress different stemness-related markers (expression of CD44, Oct4, Bmi, Nestin, ABCG2) and attenuated cancer stem-like properties (sphere formation ability and invasive capacity) via reduction of Gli1-activated Hedgehog signaling in $CD44^+$-sorted MKN45 cells [123]. Similarly, genistein inhibited gastric cancer stem cell-like characteristics also in MGC803 and SGC7901 cells by attenuating tumor cell self-renewal and sphere formation abilities [122] (Table 6.3). Thus, genistein could be applied as a potential effective chemopreventive and chemotherapeutic agent in gastric cancer treatment.

To date, other soy isoflavones have been studied much more scarce concerning their potential action against gastric tumorigenesis. Differently from genistein, daidzein inhibited the growth of HGC27 cells by arresting the cell cycle progression at G1 phase [143]. Treatment of BGC823 cells with daidzein led to inhibition of cell growth and induction of apoptotic death by disruption of mitochondrial membrane potential, increase in Bax/Bcl-2 ratio, upregulation of caspases-9 and -3, and cleavage of PARP [44]. Both daidzein and genistein were able to inhibit the growth of cachexia-inducing gastric cancer lines (MKN45cl85 and 85As2mLuc); however, genistein was considerably more efficient [43]. Daidzein was less potent than genistein also in inhibiting the growth of a set of different established human gastric cancer lines (HSC-39K6, HSC-40A1, HSC-41E6, HSC-42H, HSC-43C1, HSC-45M2 and SH101-P4) [166]. On the contrary, biochanin A suppressed the growth and induced apoptosis in HSC-41E6, HSC-45M2 and SH101-P4 cells [166]. Equol as a gut bacteria-produced daidzein metabolite suppressed the growth of MGC803 cells at low micromolar doses and blocked the cell cycle progression at G0/G1 phase, probably via restrain of the Akt signaling pathway. This process was accompanied by decrease in CDK2, CDK4, Cyclin D1 and Cyclin E1 levels and increase in $P21^{WAF1}$ level [91]. Moreover, exposure of MGC803 cells to equol resulted also in mitochondrial-dependent apoptosis with decrease in Bcl-xL and cleavage of caspases-9 and -3 and PARP, mediated by the sustained activation of ERK1/2 cascade [20, 91]. Both hitherto investigated isoflavone sugar conjugates (genistin or genistein 7-glycoside and puerarin or daidzein 8-C-glycoside) were almost ineffective at physiologically relevant doses to affect tumorigenesis in different human gastric cancer cell lines [21, 166] (Table 6.3). Thus, studies have provided some support for further use of certain isoflavones, such as genistein or equol, as promising candidates for chemoprevention and therapy of stomach tumors.

6.3.1.6. Other Flavonoid-Related Compounds

Numerous structurally related compounds to flavonoid backbone have been explored regarding to their inhibitory potential on gastric cancer initiation and progression (Table 6.3).

Some anthocyanidins, such as malvidin, delphinidin and pelargonidin, revealed growth inhibitory effects on AGS cells at high micromolar doses (with IC_{50} values more than 100 μM) [1, 169]. At that, malvidin arrested the cells at G0/G1 phase and promoted apoptosis with elevation of Bax/Bcl-2 ratio, activation of caspase-3 and cleavage of PARP, via upregulation of p38 enzyme and inactivation of ERK [1]. Although some glycosides like malvidin 3-glycoside, delphinidin 3-glycoside and delphinidin 3-rutinoside also exerted some antiproliferative activity in AGS cells [1, 170], other tested anthocyanidins (cyanidin and petunidin) and anthocyanins were practically inactive [169]. Thus, the biological activities of anthocyanidins largely depended on both the aglycone structure as well as the nature of conjugated glycoside moiety.

Also chalcones displayed some important anti-gastric cancer properties. Chalcone suppressed the adhesion, invasion and migration of highly metastatic AGS cells by downregulation of activity and expression levels of MMP-2 and -9 via inactivation of FAK/JNK signaling pathway and attenuation of NF-κB DNA binding activity [3]. Licochalcone A arrested the cell cycle of well-differentiated MKN28 and poorly differentiated MKN45 cells at G2/M phase with a decline in cyclin A, cyclin B and MDM2 expression and increase in Rb level. This natural polyphenol induced also apoptosis in MKN28, MKN45 and AGS cells in the canonical mitochondrial fashion by elevating Bax/Bcl-2 ratio and activating caspase-3 and PARP. Moreover, licochalcone A was significantly less cytotoxic in human normal gastric epithelial cells GES-1, pointing to selectivity of this compound towards malignant cells [80, 81]. Licochalcone A suppressed proliferation, triggered G2/M phase cell cycle arrest and apoptosis also in SGC7901 cells [81]. In BGC823 cells, licochalcone A induced apoptosis through generation of ROS and triggering oxidative stress, activation of ERK, JNK and p38 MAPKs, and inhibition of PI3K/Akt signaling cascade. At that, this licorice flavonoid did not significantly influence the human normal gastric mucosal cells GES-1 [2]. The PI3K/Akt pathway is overactive in various malignant tissues, restraining apoptotic death, allowing survival and proliferation, and conferring also resistance of tumor cells to drugs [2]. Another licorice chalcone, isoliquiritigenin, induced apoptotic death in MGC803 cells [171]. Although this chalcone was cytotoxic to SNU719 cells, exposure of EBV-associated SNU719 cells to isoliquiritigenin led to an increase in EBV latency behaving thus quite contrary to quercetin in these cells. Moreover, isoliquiritigenin did not inhibit the transfer of EBV to the EBV-negative AGS cells allowing the infection to succeed [107]. A natural derivative of chalcone, 2`,4`-dihydroxychalcone, blocked the cell cycle progression of MGC803 cells at G2/M phase and triggered apoptosis with increase in caspase-3 activation and suppression of antiapoptotic survivin expression [60, 172]. 3`-Methyl-3-hydroxychalcone (or methyl hydroxychalcone) displayed strong antiproliferative effect and perturbed the cell cycle of HGC27 cells at G0/G1 phase, and suppressed the binding of estradiol to both nuclear as well as cytosol type-II binding sites [173].

Flavanolignan silibinin inhibited the growth of MGC803 cells, arrested the cell cycle progression at G2/M phase and induced apoptosis, partly via suppression of STAT3 signaling pathway resulting in decrease in Mcl-1, Bcl-xL and survivin and increase in caspases-3 and -9 levels. Constitutive activation of STAT3 has been associated with tumorigenesis while STAT3 may promote cancer cell proliferation and survival, suppress apoptosis, and boost invasion, metastasis and angiogenesis. This transcription factor could be used as a molecular biomarker for poor prognosis of gastric cancer and inhibition of its signaling may thus be a potential therapeutic target [4]. Apoptosis was induced by silibinin also in MKN28 cells [174]. Treatment of SGC7901 cells with this flavanolignan led to blockade of cells in G2 phase accompanied by decrease in p34cdc2 and an apoptotic death that was independent on activation of caspases [61]. In SNU216 cells, silibinin suppressed TNF-α-induced expression of MMP-9 via inhibition of the MEK/ERK pathway representing thus a potential antimetastatic lead for treatment of gastric cancer. Abnormally high levels of pro-inflammatory cytokine TNF-α have been described in preneoplastic injuries, such as *H. pylori*-infected lesions, and could augment metastatic processes within malignancies [101].

Prenylated flavone heterophyllin sensitized TRAIL-resistant AGS cells to TRAIL-induced apoptosis via upregulation of CHOP-regulated DR5 expression and increase in caspases-3 and -7 activities [175]. Prenylated flavanone kurarinone also potentiated the cytotoxic activity of TRAIL by improving apoptotic death via inactivation of STAT3 and downregulation of antiapoptotic proteins Mcl-1 and c-FLIP, and arresting the cell cycle progression at G2/M phase with a significant decrease in cyclin B1 and cyclin A levels [56]. Prenylated flavonol glycoside icariin impeded the invasion and migration of BGC823 cells by suppressing the expression of cell motility-associated genes Rac1 and VASP [76].

Rotenoid deguelin inhibited the growth of gastric cancer cells SNU484, AGS and MKN28 with very high efficiency being active already at high nanomolar doses. Exposure of SNU484 cells to deguelin blocked the cell cycle progression at G2/M phase and induced apoptosis via caspases-9 and -3-mediated pathways. At that, phosphorylation of Akt, accumulation of HIF-1α and expression of VEGF were suppressed [22] (Table 6.3).

Altogether, there is no doubt that some flavonoids are promising candidates for further drug development and may offer potential application not only in chemoprevention but also in clinical medicine in the fight against gastric cancer. At that, a big advantage of natural flavonoids over the synthetic chemotherapeutics is their pleiotropic properties, acting simultaneously on various molecular targets and affecting diverse cellular signaling pathways. Indeed, structurally different flavonoids have been shown to exert antiproliferative, proapoptotic, antiangiogenic, antiinflammatory, and antimetastatic activities in human gastric cancer cell lines in *in vitro* conditions. Moreover, flavonoids have been demonstrated to exert their cytotoxic effects in human stomach tumor cells with different differentiation states (both poorly, moderately as well as well-differentiated cells) providing thus further therapeutic use for management of malignancies at different clinical stages. However, compared to numerous experimental studies performed with diverse human gastric cancer cell lines, potential action of these natural polyphenols in normal gastric mucosal cells has been explored relatively little, letting the selectivity issue still somewhat open. It is clear that flavonoids are worth further in-depth investigations to successfully intervene in gastric tumorigenesis in the future.

6.3.2. *In Vivo* Results

In addition to *in vitro* results obtained with established human gastric cancer cell lines, flavonoids were able to suppress the stomach tumorigenesis also in rodent models further confirming the anti-gastric cancer effectiveness of these natural polyphenols. At that, both the experimental mice with inoculated different human gastric cancer cells as well as chemically induced stomach tumors in experimental animals (gerbils, mice and rats) were applied (Table 6.2). Overview of these *in vivo* activities of different tested flavonoids is presented in Table 6.4.

Table 6.4. ***In vivo*** **effects of natural flavonoids in animal models of human gastric carcinogenesis**

Compound	Biological system[1]	Administration route and amount flavonoids	Assay time	Biological effects[2]	Ref.
CHALCONES					
Isoliquiritigenin	MKN74 cells injected into dorsum of mice	Oral gavage, 30 mg/kg for two weeks	21 days	Inhibition of tumor growth from day 13, increase in p53 expression, induction of p53-dependent apoptosis, cleavage of casp-3/-9 and PARP	[106]
	SNU719 cells injected into dorsum of mice	Oral gavage, 30 mg/kg for two weeks	21 days	Inhibition of tumor growth from day 5 through 11, no suppression of EBV proteins EBNA1 and LMP-2, induction of apoptosis, cleavage of casp-3/-9	[106]
Licochalcone A	BGC823 cells injected into flank of mice	Intratumoral injection, 200 or 400 µM, every two days	30 days	Inhibition of tumor growth	[2]
FLAVANOLS					
Catechin (C)	B[a]P-induced forestomach tumor in mice	Drinking water, 2.5 mg/day	18-20 weeks	Suppression of tumor burden and incidence. More effective inhibition in combination with turmeric. Increase in GST activity and GSH content	[127]
Epigallocatechin gallate (EGCG)	BGC823 cells injected into right flank of mice	Intraperitoneal injection, 1.5 mg/day	56 days	Retarding tumor growth, decrease in MVD. No signs of toxicity	[119]
	SGC7901 cells injected subcutaneously into mice	Injection through caudal vein; 0.5, 1 or 2 mg/kg, every two days	21 days	Inhibition of tumor growth by blocking ß-catenin signaling. No signs of toxicity	[8]
	SGC7901 cells injected into dorsal area of mice	Intraperitoneal injection, 1.5 mg/day	28 days	60.4% suppression of tumor growth, inhibition of tumor neovascularization, decrease in MVD, VEGF expression and Stat3 activation. No signs of toxicity	[40]
	TMK-1 cells injected into right flank of mice	Intraperitoneal injection, 1 mg twice in week	27 days	Inhibition of tumor growth in part by impairing the RON system. No signs of toxicity	[5]
	MNNG-induced gastric tumorigenesis in rats	Per-oral, 0.05%, for 15 weeks	44 weeks	Reduction of the number of adenocarcinomas, adenomas, hyperplasias and papillomas, decrease in BrdU labeling index and polyamine levels. No signs of toxicity	[86]
	MNNG-induced gastric tumorigenesis in rats	Per-oral, 0.05%, for 16 weeks	44 weeks	Decrease in the number of adenocarcinomas, adenomas and adenomatous hyperplasias	[176]

Table 6.4. (Continued)

Compound	Biological system[1]	Administration route and amount flavonoids	Assay time	Biological effects[2]	Ref.
FLAVANONES					
Hesperetin	SGC7901 cells injected into dorsal area of mice	Intraperitoneal injection: 10, 20 or 40 mg/kg thrice in week	36 days	Suppression of tumor growth, induction of apoptosis in tumor mass	[31]
Naringenin	MNNG-induced gastric tumorigenesis in rats	200 mg/kg	20 weeks	Gastroprotective effect, decrease in the level of plasma, liver, stomach and kidney glycoproteins	[125]
	MNNG-induced gastric tumorigenesis in rats	200 mg/kg	20 weeks	Reduction of tumor weight. Decrease in LPO and increase in enzymatic and nonenzymatic antioxidant defenses (SOD, CAT, GPx; GSH, vitamins C, E)	[69]
	MNNG-induced gastric tumorigenesis in rats	Per-oral, 200 mg/kg/day	20 weeks	Arresting MNNG-generated oxidative threats by restoring the activity of glutathione-metabolizing enzymes and phase I/II xenobiotic enzymes to near control levels	[87]
FLAVONES					
5-Hydroxy-3,6,7,8,3`,4`-hexamethoxyflavone (5HHMF)	BGC7901 cells injected into the back of mice	10 or 20 mg/kg/2 days	21 days	Suppression of tumor growth; 38.9% and 65.1% inhibitory rate at 10 and 20 mg/kg treatment. No signs of toxicity	[14]
Apigenin	MNNG- and *H. pylori*-induced gastric tumorigenesis in gerbils	Intragastric administration, 30 or 60 mg/kg/day	52 weeks	Decrease in dysplasia and tumor rates. Reduction in *H. pylori* colonization and *H. pylori*-induced histological changes of neutrophil and monocyte infiltrations and atrophic gastritis. No signs of toxicity	[71]
Baicalein	SGC7901 cells injected subcutaneously into mice	Intragastric administration, 15 or 50 mg/kg for 7 days	4 weeks	Inhibition of tumor growth, decrease in the size and weights of tumors	[32]
Luteolin	BGC823 cells injected into flank of mice	Intratumoral injection, 10 mg/kg for 2 weeks	6 weeks	Decrease in tumor size. Increase in miR-34a, p53 and p21 expression, decrease in phosphorylation level of MEK and ERK	[27]

Compound	Biological system[1]	Administration route and amount flavonoids	Assay time	Biological effects[2]	Ref.
	HGC27 cells injected into right flank of mice	Intraperitoneal injection, 20 mg/kg every two days	28 days	Decrease in tumor growth, inhibition of proliferation, induction of apoptosis. Downregulation of STAT3 phosphorylation, disruption of HSP-90 binding to STAT3 and enhancement of SHP-1 binding to STAT3, suppression of Ki-67. No signs of toxicity	[52]
	MKN28 cells injected subcutaneously into mice	Intraperitoneal injection, 10 mg/kg for 6 times	4 weeks	Suppression of tumor growth by inhibiting proliferation and inducing apoptosis. Decrease in expression of ß-catenin, Notch1 and Ki-67	[10]
	SGC7901 cells injected into right thigh of mice	Subcutaneous injection into tumors, 20 μM/kg	50 days	No significant change in survival of gastric cancer-implanted mice	[51]
	SGC7901 cells injected into right flank of mice	Intraperitoneal injection, 20 mg/kg every two days	28 days	No differences in tumor volumes between luteolin- or vehicle-treated mice	[52]
	SGC7901/DDP cells injected into right flank of mice	Intraperitoneal injection, 20 mg/kg every two days	28 days	Decrease in tumor growth, inhibition of proliferation, induction of apoptosis. Downregulation of STAT3 phosphorylation, disruption of HSP-90 binding to STAT3 and enhancement of SHP-1 binding to STAT3, suppression of Ki-67. No signs of toxicity	[52]
	Patients tumor specimens obtained from initial surgery implanted into flank of mice, serial transplantation	Intraperitoneal injection, 10 mg/kg/day	30 days	Inhibition of tumor growth, decrease in expression of metastatic (MMP-9) and proliferative markers (Ki-67). No signs of toxicity	[12]
Nobiletin	TMK-1 cells injected intraperitoneally in mice	Subcutaneously, 21 mg/kg/day	14 days	Inhibition of formation of peritoneal dissemination nodules. No signs of toxicity	[150]
Wogonin	MGC803 cells injected subcutaneously into mice	Intravenously; 15, 30 or 60 mg/kg/day	2 weeks	Suppression of tumor growth, 71.46% inhibition rate at 60 mg/kg. No signs of toxicity	[112]
	SGC7901 cells injected subcutaneously in mice, chopped into fragments and transplanted into right axillary fossa	Intravenously; 15, 30 or 60 mg/kg	12 days	Inhibition of tumor growth and angiogenesis, reduction of vascular density in tumors. No signs of toxicity	[178]
	SGC7901/cDDP cells injected into right oxter of mice	Intravenously, 5 mg/kg once every three days	21 days	Decrease in tumor volume	[146]

Table 6.4. (Continued)

Compound	Biological system[1]	Administration route and amount flavonoids	Assay time	Biological effects[2]	Ref.
FLAVONOLS					
Isorhamnetin	SNU5 cells injected into right flank of mice	Intraperitoneal injection, 1 mg/kg thrice in week	4 weeks	Increase in PPAR-γ expression, downregulation of Bcl-2 and CD31	[85]
	SNU5 cells injected into right flank of mice	Intraperitoneal injection, 1 mg/kg thrice in week	35 days	Decrease in tumor volume, downregulation of proliferative (Ki-67) and MVD biomarkers (CD31). No signs of toxicity	[113]
Kaempferol	SGC7901 cells injected into flanks of mice	Intraperitoneal administration, 20 mg/kg/day	3 weeks	Inhibition of tumor growth, reduction of Ki-67-positive tumor cells. Decrease in cyclin B1, Cdk1, Cdc25C, p-AKT, p-ERK expression and ratio of Bcl-2/ Bax; increase in casp-3/-9 levels, PARP cleavage. No signs of toxicity	[110]
Myricetin	HGC27-Luc cells injected into dorsal flank of mice	5 mg/kg/day	28 days	Suppression of tumor growth	[17]
Quercetin	AGS cells injected into right flank of mice	Intraperitoneal injection, 20 mg/kg thrice in week	4 weeks	Lowering VEGF-R2 level in tumor tissue, decrease in E-cadherin immunoreactive areas, reduction of Twist1 expression	[30]
	MKN28 cells injected subcutaneously into mice	Intraperitoneal injection, 50 mg/kg/day	24 days	~70% shrinking of average tumor size, induction of apoptosis and HIF-1α accumulation-mediated autophagy	[114]
	MKN74 cells injected into dorsum of mice	Oral gavage, 30 mg/kg for two weeks	21 days	Inhibition of tumor growth from day 13, increase in p53 expression, induction of p53-dependent apoptosis, cleavage of casp-3/-9 and PARP	[106]
	SNU719 cells injected into dorsum of mice	Oral gavage, 30 mg/kg for two weeks	21 days	Inhibition of tumor growth from day 5 to study endpoint, suppression of EBV proteins EBNA1 and LMP-2, increase in p53 and p21 expression, induction of apoptosis, cleavage of casp-3/-9 and PARP	[106]
Troxerutin	SGC7901/5-FU cells injected into right oxter of mice	Intraperitoneal injection, 75 mg/kg at two days interval	4 weeks	No reduction of tumor volume and weight. No signs of toxicity	[38]

Compound	Biological system[1]	Administration route and amount flavonoids	Assay time	Biological effects[2]	Ref.
ISOFLAVONES					
Biochanin A	HSC-41E6 cells injected into mice	Intraperitoneal injection, 400 μg/day for 10 times	6 weeks	No effect on tumor growth. No signs of toxicity	[166]
	HSC-45M2 cells injected into mice	Intraperitoneal injection, 400 μg/day for 10 times	6 weeks	Suppression of tumor growth, 46% inhibition of tumor weight. No signs of toxicity	[166]
	SH101-P4 cells injected into mice	Intraperitoneal injection, 400 μg/day for 10 times	6 weeks	Suppression of tumor growth, 61% inhibition of tumor weight. No signs of toxicity	[166]
Daidzein	MKN45cl85 cells injected subcutaneously into mice	Intraperitoneal injection, 400 μg/day every other day, 12 times in total	3 or 4 weeks	Suppression of proliferative growth, anticachectic effect	[43]
	85As2mLuc cells injected subcutaneously into mice	Intraperitoneal injection, 400 μg/day every other day, 12 times in total	3 or 4 weeks	Suppression of proliferative growth, anticachectic effect, no effect on remote metastasis, prolonged survival of animals	[43]
Genistein	HSC-41E6 cells injected into mice	Intraperitoneal injection, 400 μg/day for 10 times	6 weeks	No effect on tumor growth. No signs of toxicity	[166]
	HSC-45M2 cells injected into mice	Intraperitoneal injection, 400 μg/day for 10 times	6 weeks	No effect on tumor growth. No signs of toxicity	[166]
	MGC803 cells injected into right flank of mice	Intraperitoneal injection, 1.5 mg/kg/day	24 days	Reduction of tumor size and weight, attenuation of tumorigenicity	[122]
	MKN45cl85 cells injected subcutaneously into mice	Intraperitoneal injection, 400 μg/day every other day, 12 times in total	3 or 4 weeks	Suppression of proliferative growth, anticachectic effect	[43]
	85As2mLuc cells injected subcutaneously into mice	Intraperitoneal injection, 400 μg/day every other day, 12 times in total	3 or 4 weeks	Suppression of proliferative growth, no effect on remote metastasis, prolonged survival of animals	[43]
	SGC7901 cells injected subcutaneously into mice	Injection adjacent to tumor; 0.5, 1 or 1.5 mg/kg, six times at 2 days intervals	11 days	10.8%, 29.9% and 39.6% inhibition of tumor growth at 0.5, 1 and 1.5 mg/kg; induction of apoptosis, decrease in Bcl-2 and increase in Bax expression	[180]
	SH101-P4 cells injected into mice	Intraperitoneal injection, 400 μg/day for 10 times	6 weeks	No effect on tumor growth. No signs of toxicity	[166]

Table 6.4. (Continued)

Compound	Biological system[1]	Administration route and amount flavonoids	Assay time	Biological effects[2]	Ref.
	MNNG-induced and NaCl-enhanced gastric tumorigenesis in rats	Subcutaneous injection, 15 or 30 mg/kg every other day	52 weeks	Reduction of cancer incidence, induction of apoptosis, decrease in vascular counts of the antral mucosa and cancer (only at 30 mg/kg, not 15 mg/kg)	[128]
Puerarin	BGC823 cells injected in right dorsal area of mice	30 mg/kg/day thrice in week	3 weeks	68.37% inhibition of tumor volume, decrease in tumor weight, induction of apoptosis. No signs of toxicity	[21]

[1]see Tables 6.1 and 6.2; Luc, luciferase.

[2]Akt, protein kinase B; Bcl-2, B-cell lymphoma-2; BrdU, bromodeoxyuridine; casp, caspase; CAT, catalase, CDK, cyclin-dependent kinase; EBNA1, Epstein-Barr nuclear antigen 1; EBV, Epstein-Barr virus; ERK, extracellular signal-regulated kinase; GPx, glutathione peroxidase; GSH, reduced glutathione; GST, glutathione S-transferase; *H. pylori*, *Helicobacter pylori*; HIF-1α, hypoxia-inducible factor-1α; HSP-90, heat shock protein-90; LMP2, EBV latent membrane protein 2; LPO, lipid peroxidation; MEK, mitogen-activated protein kinase kinase; miR, microRNA; MMP-9, matrix metalloproteinase-9; MVD, microvessel density; PARP, poly(ADP-ribose) polymerase; PPAR-γ, peroxisome proliferator-activated receptor γ; RON, recepteur d`origine nantais; SHP-1, SH2-containing protein tyrosine phosphatase-1; SOD, superoxide dismutase; STAT3, signal transducer and activator of transcription 3; VEGF, vascular endothelial growth factor; VEGF-R2, VEGF receptor 2.

6.3.2.1. Flavanols

EGCG was able to retard the growth of xenografts engineered with different human gastric cancer cell lines in mice. This common green tea flavanol hindered tumor growth and lowered microvessel counts in BGC823 xenograft model [119]. Treatment of mice subcutaneously injected with SGC7901 cells with EGCG significantly suppressed the tumor growth and resulted in smaller tumors than those in controls by reducing the level of proliferating cell nuclear antigen (PCNA) and inactivating ß-catenin signaling [8]. In similar *in vivo* model (SGC7901-engineered xenografts in mice), exposure to EGCG led to a 60.4% inhibition of gastric cancer growth and an average of 38.2% decrease in microvessel density within tumor tissues. In addition, the level VEGF expression was suppressed being related to inactivation of Stat3 [40]. EGCG inhibited the tumor growth also in TMK-1 cells-based xenografts-bearing mice, whereas expression of RON was substantially downregulated within the tumors [5]. In all these models, administration of EGCG did not associate with any detectable signs of toxicity, such as weight loss, counts of white blood cells or unusual behavior [5, 8, 40, 119]. Moreover, EGCG suppressed the tumorigenesis process also in MNNG-induced gastric cancer in rats. Indeed, exposure of MNNG-treated rats to EGCG led to a significant decrease in incidence of gastric adenocarcinomas, adenomas, adenomatous hyperplasias and papillomas; reduced the bromodeoxyuridine (BrdU) labeling index and level of spermidine, without affecting the animal body weights [86, 176] (Table 6.4). In addition, this green tea flavanol (at doses of 100 mg/kg body weight/day) exerted also some protection against tobacco-induced carcinogenesis in experimental mice stomach by attenuating the tobacco smoke-induced epithelial-mesenchymal transition (EMT) via inactivation of ERK5 [9]. Thus, EGCG inhibited tumor growth in both human gastric cancer xenografts in mice as well as chemically induced stomach tumorigenesis in rats, representing thus a promising candidate for prevention and treatment of gastric cancer.

Similarly to the *in vitro* experiments the anti-gastric cancer potential of other flavanols has been studied very little also in *in vivo* conditions and to date, only one study with catechin is available. Although treatment of mice bearing B[a]P-induced forestomach tumors with catechin significantly suppressed tumor burden and incidence, this inhibition was much more efficient by cotreatment of animals with catechin and dietary turmeric [127] (Table 6.4).

6.3.2.2. Flavanones

Hesperetin was shown to be able to significantly and dose-dependently suppress the growth of SGC7901-engineered human gastric cancer xenografts in mice by inducing apoptotic cell death in the tumor mass [31]. Another common citrus flavanone, naringenin, revealed an important gastroprotective action by arresting the oxidative damage induced by carcinogenic chemicals in rats. In fact, naringenin decreased the levels of glycoproteins to near normalcy in MNNG-induced gastric cancer-bearing rats, maintaining thus the integrity of cell membranes and stomach mucosa against oxidative threat and suppressing malignant transformation [125]. Moreover, naringenin treatment arrested MNNG-caused oxidative damage in rats by reducing lipid peroxidation (LPO) and restoring the activities of glutathione enzyme and non-enzyme antioxidants as well as phase I and phase II xenobiotic enzymes [69, 87]. Also, treatment with this flavanone upregulated some other cellular enzymatic and nonenzymatic antioxidant systems, such as superoxide dismutase (SOD), catalase (CAT) and vitamins C and E, and reduced the tumor weight in MNNG-induced stomach cancer-bearing animals [69]. Thus, naringenin revealed mostly chemopreventive effects against MNNG-initiated stomach tumorigenesis in rats (Table 6.4). However, two flavanone glycosides, hesperidin or hesperetin 7-rutinoside and neohesperidin or hesperetin 7-neohesperidoside, exerted quite opposite effects in rats deteriorating the indomethacin-induced gastric mucosal lesions as evidenced by aggravated ulcer formation due to proinflammatory activities of these flavanone sugar conjugates. Therefore, their combination with indomethacin should be avoided [177].

6.3.2.3. Flavones

As *in vitro* gastric cancer cellular models, luteolin is the most studied flavone also in *in vivo* rodent systems. This flavone decreased the tumor growth in different human gastric cancer cells-engineered xenografts in experimental mice. In BGC823-injected mice, luteolin suppressed the tumor size and upregulated miR-34a level in malignant tissue [27]. In MKN28-based xenografts, this flavone suppressed tumor progression by inhibiting proliferation and initiating apoptosis. In addition, luteolin treatment also downregulated Notch1 and ß-catenin expression [10]. In HGC27 as well as SGC7901/DDP bearing mice, luteolin strongly inhibited tumor growth via a STAT3 inhibition mechanism and downregulation of its target genes (Mcl-1, Bcl-xL, survivin). At that, the body weight of animals was not influenced [52]. Moreover, luteolin was also able to inhibit the tumor growth in *in vivo* mice model implanted with cMet-overexpressing patient-obtained tumor specimens. This inhibition was accompanied by decrease in proliferative (Ki-47) and metastatic markers (MMP-9). As about 40% of all gastric tumor cases are characterized with c-Met overexpression correlating with aggressive phenotype, metastatic disease stage and poor prognosis, it is crucial to identify novel efficient therapies for management of this gastric cancer type [12]. Thus, luteolin was able to suppress the growth of tumors obtained by implanting mice with diverse human gastric cancer cells being at different differentiation stages, suggesting that this flavone could be therapeutically important for further treatment of both poorly as well as well-differentiated gastric cancers. On the other hand, exposure of SGC7901-engineered xenografts to luteolin did not affect the tumor volumes compared to controls

[52] or did also not substantially change the survival of these mice [51]. However, another common flavone baicalein dramatically suppressed the growth and decreased the dimensions of SGC7901-engineered xenografts in mice [32] (Table 6.4).

Apigenin was able to repress the tumor rates in MNNG- and *H. pylori*-induced gastric carcinogenesis in Mongolian gerbils by suppressing the bacterial colonization and damage, atrophic gastritis, dysplasia, and tumor initiation and progression [71]. Treatment with methylated flavone wogonin inhibited the growth of tumor xenografts obtained by transplanting MGC803 [112], SGC7901 [178] or drug-resistant SGC7901/cDDP cells to mice [146], without any detectable signs of toxicity. In SGC7901 bearing mice, wogonin decreased also the tumor angiogenesis and vascular density [178] (Table 6.4). Moreover, in the murine gastric cancer cells MFC-based xenografts in mice, wogonin could stimulate the infiltration of dendritic cells, T-cells and natural killers into malignant tissues to retard the tumor growth [147], whereas the pretreatment of MFC cells with wogonin led to a considerably lower rate of tumor-bearing mice [109]. In addition, this methylated flavone could also be applied as a chemoprotective agent as it was able to attenuate the ethanol-induced gastric injuries, such as hemorrhagic lesions and submucosal edema, in rat model [95].

Treatment of BGC7901-engineered tumors in mice with 5-hydroxy-3,6,7,8,3`,4`-hexamethoxyflavone (5HHMF) resulted in a strong, up to 65% inhibition of malignant growth, without affecting the average body weight of treated animals compared to control mice [14]. Another common citrus polymethoxyflavone nobiletin suppressed the number and weight of peritoneal dissemination nodules in TMK-1-inoculated mice behaving thus as an antimetastatic agent [150] (Table 6.4). As a whole, certain flavones can represent potential therapeutic lead compounds for development of future drugs against human gastric cancer. The only flavone glycoside studied so far, diosmin or diosmetin 7-O-rutinoside, displayed gastroprotective effects against ethanol-caused mucosal damage in rats as evidenced by attenuation of the severity of ulcer formation and reducing the area of gastric lesions. At that, diosmin behaved as an antiinflammatory agent, mainly via inhibition of NF-κB [96].

6.3.2.4. Flavonols

Studies with mice xenograft models provide additional evidence also about the anti-gastric cancer potential of flavonols. Treatment of MKN28 cells-injected mice with the most abundant natural flavonoid quercetin resulted in a substantial, about 70% decrease in the tumor size by inhibiting proliferation and inducing apoptotic death. In addition, upregulation of autophagy markers was also detected in tumor tissues, whereas accumulation of HIF-1α played a crucial role in these autophagic processes [114]. In AGS cells engineered xenografts, quercetin lowered the levels of VEGF-receptor 2 (VEGF-R2) in solid tumor mass related to its antiangiogenic properties [30]. Moreover, quercetin was able to suppress the expression of EBV viral proteins (EBNA1 and LMP2) in tumor tissues obtained by inoculation of EBV-positive SNU719 cells to mice, and induced p53-, p21- and caspases-dependent apoptosis. Apoptotic cell death was triggered also after exposure of EBV-negative MKN74 bearing mice to this flavonol [106] (Table 6.4). Thus, quercetin exerted its antitumor effects by regulating diverse cellular events in molecularly different gastric cancer cells, intervening both in angiogenic as well as viral processes. As it is generally accepted that EBV can cause and stimulate gastric carcinogenesis, quercetin might represent a promising novel drug candidate for treatment of EBV-associated tumors. Quercetin sugar conjugate, miquelianin or quercetin 3-glucuronopyranoside, exerted gastroprotective action in rats by mitigating indomethacin-induced mucosal damage, increasing mucus secretion and decreasing inflammatory reactions [179]; this effect was quite different from two hesperetin glycosides, hesperidin and neohesperidin [177].

Two other rather common natural flavonols, kaempferol and myricetin, inhibited tumor growth in SGC7901 and HGC27 cells implanted mice, respectively [17, 110]. At that, treatment with kaempferol led to significantly reduced levels of cyclin B1, Cdk1, Cdc25C, Bcl-2, p-Akt, p-ERK and COX-2, and elevated levels of Bax, cleaved caspases-3 and -9 and PARP in xenograft tissues [110]. Methylated flavonol isorhamnetin was also able to retard the growth of SNU5 cells engineered tumors in mice, whereas the potential mechanism involved upregulation of PPAR-γ and downregulation of the proliferative marker Ki-67, antiapoptotic factor Bcl-2 and microvessel density biomarker CD31 [85, 113]. On the other hand, troxerutin displayed almost no effects on the tumors obtained by inoculating 5-FU-resistant SGC7901 cells (SGC7901/5-FU) to mice [38] (Table 6.4).

6.3.2.5. Isoflavones

Soy isoflavone genistein suppressed the cancer incidence, increased apoptosis and decreased vascular counts in antral mucosa and tumors in rat model of MNNG-induced and NaCl-promoted carcinogenesis [128]. Treatment of MGC803 cells bearing mice with genistein inhibited the size and weight of tumors [122]. Exposure of SGC7901 cells injected mice to this isoflavone resulted in a significant suppression of tumor growth and induced apoptosis in transplanted malignant cells with an increase in Bax/Bcl-2 ratio [180]. Both genistein and another soy isoflavone daidzein inhibited malignant proliferation and attenuated cachectic weight loss in cachexia-inducing cells MKN45cl85 bearing mice; at that, daidzein was somewhat more potent than genistein [43]. Implantation of 85As2mLuc cells (derived from MKN45cl85 cells) led to not only cachectic mice but revealed also highly metastatic tumors, and treatment of these animals with genistein and daidzein prolonged the median survival rate of mice without suppression of remote metastasis; daidzein was more efficient than genistein also in this *in vivo* model [43]. While biochanin A was able to significantly inhibit the growth in xenografts of HSC-45M2 and SH101-P4 cells in mice, genistein did not affect the tumor growth in either of these transplanted rodents. The HSC-41E6 cells bearing mice were insensitive to the treatment of both genistein and biochanin A [166] (Table 6.4). These data show that anti-gastric cancer activity of isoflavones is substantially dependent on the structural peculiarities of compounds, but also on the characteristics of malignant cell types. It can not be excluded that genistein is rather a chemopreventive than a chemotherapeutic agent, as administration of this isoflavone significantly suppressed the stress (water immersion restraint)-induced gastric mucosal damage through its antioxidative and antiinflammatory properties and inhibition of acid secretion in rats [181]. The only tested isoflavone glycoside i.e., puerarin or daidzein 8-C-glycoside suppressed the volume of xenografts of BGC823 cells in mice and triggered apoptosis of tumor cells, whereas no adverse effects on the liver or renal functions were observed [21] (Table 6.4).

6.3.2.6. Flavonoid-Related Compounds

Among the different flavonoid-related compounds tested *in vitro* conditions with human gastric cancer cell lines, only chalcones have been explored also in xenograft mouse models. Licochalcone A significantly inhibited the progression of BGC823 cells engineered xenografts in mice [2]. Isoliquiritigenin suppressed the tumor growth and induced apoptosis in both EBV-positive SNU719 as well as EBV-negative MKN74 carcinoma bearing mice. However, differently from quercetin, isoliquiritigenin was not able to reduce the expression of EBV viral proteins (EBNA1 and LMP2) suggesting that this chalcone did not exhibit its antitumor action through the anti-EBV mechanism [106] (Table 6.4).

As a whole, *in vivo* studies performed with human gastric cancer xenografts in mice or chemically induced stomach tumorigenesis in rats, mice or gerbils provide important further evidence to the immense *in vitro* results about the inhibitory action of structurally different flavonoids on gastric carcinogenesis. Moreover, the *in vivo* results revealed no detectable toxic side effects of these natural polyphenols on normal rodent functioning. These data strongly encourage researchers to move on to the clinical trials to eventually identify effective treatment strategy for this highly fatal malignancy.

6.3.3. Combined Effects of Flavonoids and Conventional Cancer Treatments on Gastric Tumorigenesis

The two major clinical limitations in the successful application of current therapeutic regimens for treatment of malignant diseases are first the development of drug resistance and second the toxic side effects of current medications. Therefore, the anti-gastric cancer potential of flavonoids might be important also in the context of sensitizing tumor cells to conventional chemotherapeutics or irradiation. In the past, it has been indeed demonstrated that combinations of standard anticancer drugs with natural compounds can provide synergistic cytotoxic responses without enhancing adverse effects [77]. The data about combined activities of natural flavonoids and current classical treatment modalities for gastric cancer are presented in Table 6.5. However, these data show that beside important potentiation of the anticancer efficacies by certain flavonoids, some combinations remain inefficient or are even antagonistic.

Simultaneous administration of flavanol EGCG enhanced the tumor growth inhibitory and antiangiogenic effects of low-dose metronomic (LDM) docetaxel in BGC823 xenografts in mice. Such combinatorial treatment was effective and safe for animals in preclinical settings and can provide several advantages over the conventional application of docetaxel (maximum tolerated dose regimen). Continuous administration using a frequent schedule at low doses or the so-called metronomic dosing can target endothelial cells of newly forming blood vessels and induce antiangiogenic effects. LDM docetaxel chemotherapy was further improved by EGCG resulting in lower microvessel counts and delay of tumor growth [119].

Combined treatment of poorly differentiated adenocarcinoma AGS cells with flavone luteolin and cisplatin induced more effective inhibition of cell growth than either single agent alone. Cisplatin is one of the most frequently used drugs for the treatment of gastric cancer and potentiating its antitumor efficacy might be clinically important. However, luteolin did not chemosensitize AGS cells towards the action of 5-FU, gemcitabine, doxorubicin and paclitaxel [49]. Nevertheless, luteolin significantly enhanced the irradiation-caused clonogenic inhibition and increased irradiation-induced mitochondrial apoptosis in moderately differentiated adenocarcinoma SGC7901 cells and enhanced radioresponse, i.e., suppressed tumor volume and extended growth delay time in SGC7901 cells engineered xenografts in mice. At that, luteolin was able to intervene in several cellular processes implicated in radioresistance, as this flavone downregulated Bcl-2, activated caspases-3 and -9, released cyt c into cytosol, reduced prostaglandin E2 (PGE2) synthesis, and lowered VEGF and HIF-1α levels [51]. Another flavone, baicalein, enhanced the sensitivity of AGS cells toward treatment with 5-FU under hypoxia. In these conditions, baicalein decreased hypoxia-induced glycolytic rate and downregulated the glycolysis-related enzymes (HK2, LDHA, PDK1) via modulation of PTEN/Akt/HIF-1α signaling cascade. As hypoxia is a usual condition in most solid tumors and 5-FU is still a mainstay systemic drug for gastric cancer chemotherapy, the ability of baicalein to reverse hypoxia-induced chemotherapeutic resistance and improve

Table 6.5. Combined effects of natural flavonoids and standard cancer treatment modalities against gastric tumorigenesis

Compound	Drug, treatment	Biological system[1]	Assay conditions[2]	Co-effect[3]	Ref.
CHALCONES					
Licochalcone A	5-Fluorouracil (5-FU)	MKN45 cells	CCK-8, various times	Synergistic action with 120 μM 5-FU. Apoptosis, increase in p53 expression	[81]
	5-Fluorouracil (5-FU)	SGC7901 cells	CCK-8, various times	Synergistic action with 120 μM 5-FU. G0/G1 phase cell cycle arrest, apoptosis, increase in p53 expression	[81]
FLAVANOLS					
Epigallocatechin gallate (EGCG)	Docetaxel	BGC823 cells injected into right flank of mice	Intraperitoneally, 1.5 mg/day EGCG, 0.5 mg/kg LDM drug	Potentiation of antitumor and antiangiogenic activities, decrease in MVD. No signs of toxicity	[119]
FLAVONES					
Baicalein	5-Fluorouracil (5-FU)	AGS cells	MTT, under hypoxia, 48 h	Improvement of 5-FU efficacy, reversal of hypoxia-induced 5-FU resistance, suppression of glycolysis via regulation of PTEN/Akt/HIF-1α signaling	[33]
Luteolin	5-Fluorouracil (5-FU)	AGS cells	MTT, 72 h	No chemosensitization	[49]
	Cisplatin (CDDP)	AGS cells	MTT, 72 h	Synergistic effect on cell growth inhibition with combinatorial treatment at ratio of 1:1 for several doses	[49]
	Doxorubicin	AGS cells	MTT, 72 h	No chemosensitization	[49]
	Gemcitabine	AGS cells	MTT, 72 h	No chemosensitization	[49]
	Paclitaxel	AGS cells	MTT, 72 h	No chemosensitization	[49]
	Radiation	SGC7901 cells	Colony formation, 14 days	Enhancement of radiation (8 Gy)-induced clonogenic inhibition by 20 μM luteolin, inhibition of radiation-induced prosurvival (Bcl-2) and radioprotective factors (PGE2), enhancement of radiation-induced mitochondrial apoptosis	[51]
	Radiation	SGC7901 cells injected into right thigh of mice	Intratumoral injection 20 μM/kg luteolin, irradiation 8 Gy	Synergistic increase in radiation-induced reduction of tumor volume, delaying growth time	[51]
Nobiletin	5-Fluorouracil (5-FU)	SNU16 cells	MTT, 48 h	Synergistic decrease in cell viability. Modulation of different pathways by 5-FU (increase in p53 expression and phosphorylation) and nobiletin (increase in $p21^{WAF1/CIP1}$ expression)	[48]

Table 6.5. (Continued)

Compound	Drug, treatment	Biological system[1]	Assay conditions[2]	Co-effect[3]	Ref.
	Cisplatin (CDDP)	MKN45 cells	MTT, different schemes	Synergistic effect when treated with nobiletin 24 h prior to CDDP, additive effect when CDDP (24 h) was followed by nobiletin, antagonistic antitumor effect with simultaneous dosage (24 h)	[148]
	Cisplatin (CDDP)	TMK-1 cells	MTT, different schemes	Synergistic effect when treated with nobiletin 24 h prior to CDDP, additive effect when CDDP (24 h) was followed by nobiletin, antagonistic antitumor effect with simultaneous dosage (24 h)	[148]
Tangeretin	Radiation	SGC7901 cells	Colony formation, 24 h	Increase in radiation-induced cell clonogenic death, suppression of radiation-induced EMT and metastasis probably via inhibition of Notch1 signaling and switching on of miR-410	[116]
	Radiation	SGC7901 cells injected subcutaneously in mice	Intraperitoneally 30 mg/kg tangeretin, irradiation 2 Gy	Decrease in tumor size, almost complete inhibition of lung metastasis. Alleviation of radiation-induced weight loss and better physical condition of animals	[116]
Wogonin	5-Fluorouracil (5-FU)	MGC803 cells	MTT, 48 h	Synergistic inhibition of proliferation, sensitization of cells to 5-FU-induced apoptosis. Increase in 5-FU retention for prolonged catabolism by modulating drug metabolic enzymes, inhibition of NF-κB nuclear translocation	[112]
	5-Fluorouracil (5-FU)	MGC803 cells injected subcutaneously in mice	Intravenously 15, 30 or 60 mg/kg/day wogonin, intraperitoneally 10 mg/kg/day drug	Synergistic antitumor effect, 91.69% inhibition rate at combined treatment of 5-FU and wogonin (60 mg/kg). No signs of toxicity	[112]
FLAVONOLS					
Isorhamnetin	5-Fluorouracil (5-FU)	AGS cells	MTT, 24 h	1.63-fold enhancement of 5-FU (60 μM) cytotoxic activity by 10 μM isorhamnetin	[85]
	Capecitabine (CAPE)	AGS cells	MTT, 24 h	1.4-fold enhancement of CAPE (10 μM) cytotoxic activity by 10 μM isorhamnetin	[85]
	Capecitabine (CAPE)	MKN45 cells	MTT	Sensitization to 10 μM CAPE by 10 μM isorhamnetin, potentiation of apoptosis, attenuation of CAPE-induced NF-κB activation	[113]
	Capecitabine (CAPE)	SNU5 cells	MTT	Sensitization to 10 μM CAPE by 10 μM isorhamnetin, potentiation of apoptosis	[113]

Compound	Drug, treatment	Biological system[1]	Assay conditions[2]	Co-effect[3]	Ref.
	Capecitabine (CAPE)	SNU5 cells injected into right flank of mice	Intraperitoneally 1 mg/kg isorhamnetin, by gavage 60 mg/kg drug	Augmentation of antitumor efficacy, downregulation of proliferative (Ki-67) and MVD biomarkers (CD31), suppression of activation of NF-κB signaling and NF-κB-regulated proteins (p65, VEGF, COX2, MMP-9). No signs of toxicity	[113]
	Capecitabine (CAPE)	SNU16 cells	MTT	Sensitization to 10 μM CAPE by 10 μM isorhamnetin, potentiation of apoptosis	[113]
	Doxorubicin (DOX)	AGS cells	MTT, 24 h	1.89-fold enhancement of DOX (1 μM) cytotoxic activity by 10 μM isorhamnetin	[85]
Quercetin	5-Fluorouracil (5-FU)	AGS cells	MTT, 24 h	Only nominal changes in cell viability, levels of survivin and STAT3 with combined treatment as compared to either agent alone	[37]
	SN-38 (active metabolite of irinotecan)	AGS cells	AB, 48 h	Improvement of SN-38 efficacy, decrease in cell survival, increase in apoptosis, modulation of GSK-3ß/ß-catenin signaling	[30]
	Irinotecan	AGS cells injected into right flank of mice	Intraperitoneally, 20 mg/kg quercetin, 10 mg/kg drug	Enhancement of irinotecan efficacy, decrease in tumor size, downregulation of angiogenesis-associated (VEGF-A, VEGF-R2) and EMT-related factors (Twist1, ITGß6), inhibition of metastasis-related factors. No signs of toxicity	[30]
	Daunorubicin	EPG85-257P cells	SRB, 72 h	Synergistic cytotoxic action, enhancement of apoptosis	[156]
	Daunorubicin	EPG85-257RDB cells	SRB, 72 h	Reversal of resistance to daunorubicin, enhancement of apoptosis, decrease in P-gp expression and its drug transport abilities	[156]
Troxerutin	5-Fluorouracil (5-FU)	SGC7901 cells	MTT, 24 h	Promotion of 5-FU (20 μM) anticancer activity, suppression of migration, decrease in p65 level. No synergistic activity in normal gastric epithelial cell line GES-1	[38]
	5-Fluorouracil (5-FU)	SGC7901/5-FU cells	MTT, 24 h	Potentiation of 5-FU (20 μM)-induced antitumor effect, decrease in migration, inhibition of colony formation, induction of apoptosis through suppression of p-STAT3/NF-κB and Bcl-2	[38]
	5-Fluorouracil (5-FU)	SGC7901/5-FU cells injected into right oxter of mice	Intraperitoneally, 75 mg/kg troxerutin, 50 mg/kg drug	Potentiation of decrease in tumor volume and weight. Reduction of p-STAT3, p-NF-κB and Bcl-2 positive cells in tumor. No signs of toxicity	[38]

Table 6.5. (Continued)

Compound	Drug, treatment	Biological system[1]	Assay conditions[2]	Co-effect[3]	Ref.
ISOFLAVONES					
Calycosin	5-Fluorouracil (5-FU)	BGC823 cells	MTT, various times	Enhancement of 5-FU (100 μg/ml)-induced antiproliferation by 35 μM calycosin, promotion of 5-FU-induced migration inhibition	[42]
	5-Fluorouracil (5-FU)	NCI-N87 cells	MTT, various times	Enhancement of 5-FU (100 μg/ml)-induced antiproliferation by 35 μM calycosin, promotion of 5-FU-induced migration inhibition	[42]
	5-Fluorouracil (5-FU)	SGC7901 cells	MTT, various times	Enhancement of 5-FU (100 μg/ml)-induced antiproliferation by 35 μM calycosin, promotion of 5-FU-induced migration inhibition	[42]
	Adriamycin (ADM)	BGC823 cells	MTT, various times	Enhancement of ADM (50 μg/ml)-induced antiproliferation by 35 μM calycosin, promotion of ADM-induced migration inhibition	[42]
	Adriamycin (ADM)	NCI-N87 cells	MTT, various times	Enhancement of ADM (50 μg/ml)-induced antiproliferation by 35 μM calycosin, promotion of ADM-induced migration inhibition	[42]
	Adriamycin (ADM)	SGC7901 cells	MTT, various times	Enhancement of ADM (50 μg/ml)-induced antiproliferation by 35 μM calycosin, promotion of ADM-induced migration inhibition	[42]
	Cisplatin (CDDP)	BGC823 cells	MTT, various times	Enhancement of CDDP (20 μg/ml)-induced antiproliferation by 35 μM calycosin, G1 phase cell cycle arrest, enhancement of CDDP-induced apoptosis and inhibition of invasion by inhibiting Akt pathway	[42]
	Cisplatin (CDDP)	NCI-N87 cells	MTT, various times	Enhancement of CDDP (20 μg/ml)-induced antiproliferation by 35 μM calycosin, G1 phase cell cycle arrest, enhancement of CDDP-induced apoptosis and inhibition of invasion by inhibiting Akt pathway	[42]
	Cisplatin (CDDP)	SGC7901 cells	MTT, various times	Enhancement of CDDP (20 μg/ml)-induced antiproliferation by 35 μM calycosin, G1 phase cell cycle arrest, enhancement of CDDP-induced apoptosis and inhibition of invasion by inhibiting Akt pathway	[42]
Genistein	5-Fluorouracil (5-FU)	MGC803 cells	MTT, genistein (24 h) followed by drug (72 h)	Enhancement of chemosensitivity by 15 μM genistein, downregulation of ABCC1/5 and ABCG2 expression, inhibition of ERK1/2 activity	[122]
	Cisplatin (CDDP)	MGC803 cells	MTT, genistein (24 h) followed by drug (72 h)	Enhancement of chemosensitivity by 15 μM genistein, repression of ABCC1/5 and ABCG2 expression, inhibition of ERK1/2 activity	[122]

Compound	Drug, treatment	Biological system[1]	Assay conditions[2]	Co-effect[3]	Ref.
Puerarin	5-Fluorouracil (5-FU)	BGC823 cells	CCK-8, 48 h	Synergistic anticancer effect with 5-FU (80 μM), increase in apoptosis	[21]
	5-Fluorouracil (5-FU)	BGC823 cells injected into right dorsal area of mice	30 mg/kg/day puerarin, 12 mg/kg/day drug	Potentiation of 5-FU effect on suppression of tumor growth and apoptosis, 90.65% inhibition of tumor volume. No signs of toxicity	[21]

[1] see Tables 6.1 and 6.2.

[2] AB, Alamar Blue; CCK-8, cell counting kit-8; LDM, low dose metronomic; MTT, 3-(4,5-dimethylthiazol-2-yl)-2,5-diphenyltetrazolium bromide; SRB, sulforhodamine B.

[3] ABC, ATP-binding cassette; Akt, protein kinase B; Bcl-2, B-cell lymphoma-2; COX2, cyclooxygenase-2; ERK, extracellular signal-regulated kinase; GSK-3ß, glycogen synthase kinase-3ß; HIF-1α, hypoxia-inducible factor 1-α; ITGß6, integrin ß6; miR, microRNA; MMP-9, matrix metalloproteinase-9; MVD, microvessel density; NF-κB, nuclear factor κB; P-gp, P-glycoprotein; PGE2, prostaglandin E2; PTEN, phosphatase and tensin homolog; STAT3, signal transducer and activator of transcription 3; VEGF, vascular endothelial growth factor; VEGF-R2, VEGF receptor-2.

the cytotoxicity of 5-FU might be of clinical relevance [33]. In addition, the antitumor activity of 5-FU was synergistically potentiated also by methylated flavone wogonin in well-differentiated MGC803 cells. This effect was achieved through enhancement of apoptotic rate by blocking NF-κB action and downregulation of 5-FU metabolic enzyme dihydropyrimidine dehydrogenase (DPD) leading to drug retention for a prolonged period. Furthermore, combined treatment of wogonin and 5-FU led to a significant, more than 90% growth inhibition of tumors derived from implantation of MGC803 cells to mice, without any detectable toxic effects on animals [112]. Polymethoxyflavone nobiletin was also able to potentiate the anticancer effects of 5-FU by increasing the $p21^{WAF/CIP1}$ protein expression in p53-mutated SNU16 cells [48]. Development of chemoresistance is a crucial obstacle for the clinical use of 5-FU, whereas only 21% response rate has been reported for patients with advanced gastric cancer receiving of 5-FU single treatment [112]. Therefore, identification of natural agents with the ability to chemosensitize gastric tumor cells to this drug is very important and can advance current therapeutic regimens. Nobiletin could modulate also the chemosensitivity of MKN45 and TMK-1 cells toward cisplatin but this effect depended critically on the time schedule. In fact, exposure of gastric cancer cells to nobiletin 24 h prior to cisplatin revealed synergistic antitumor effects, while administration of nobiletin 24 h after the drug showed only additive action. Furthermore, simultaneous treatment of MKN45 and TMK-1 cells with nobiletin and cisplatin resulted in antagonistic antitumor responses [148]. Another citrus polymethoxyflavone tangeretin behaved as a radiosensitizer in SGC7901 cells and suppressed the irradiation-induced EMT processes, invasion and migration via blockade of Notch1 expression and elevation of miR-410 levels. Co-treatment of SGC7901 injected mice with tangeretin and irradiation led to considerably smaller tumors with almost no radiation-induced lung metastases. Moreover, tangeretin also alleviated the radiation-caused weight loss and improved general physical condition of mice [116]. Considering that radiotherapy has remained a standard treatment for patients suffering from stomach tumors, it is crucial to find radiosensitizing agents to minimize irradiation-induced serious adverse effects, including distant metastases. Two flavones, luteolin and tangeretin, could be potential attractive candidates for such combined regimens in further gastric cancer treatment schemes.

Flavonol quercetin did not significantly affect the anticancer action of 5-FU in AGS cells [37]. On the contrary, treatment of AGS cells with combination of quercetin and SN-38, a potent DNA topoisomerase I inhibitor and an active metabolite of irinotecan, substantially improved the drug efficacy to decrease the cell survival via regulation of GSK-3ß/ß-catenin signaling. Moreover, *in vivo* study, this combination (quercetin and irinotecan) revealed stronger suppressive effect on the growth of AGS cells engineered xenografts in mice than single agents alone, accompanied by downregulation of EMT- and angiogenesis-related markers in solid tumor tissues [30]. Synergistic cytotoxic action was observed also in EPG85-257P cells by cotreatment with quercetin and daunorubicin, a DNA topoisomerase II inhibitor. In daunorubicin-resistant subline of EPG85-257P cells (EPG85-257RDB), quercetin behaved as a chemosensitizer and reversed resistance to daunorubicin by reducing the expression of P-glycoprotein (P-gp) and inhibiting drug transport abilities of this cell membrane protein [156]. Treatment of AGS cells with methylated quercetin, isorhamnetin, enhanced the cytotoxic activity of three chemotherapeutic drugs, i.e., 5-FU, doxorubicin and capecitabine [85]. The antitumor proapoptotic efficacy of capecitabine was augmented by isorhamnetin also in MKN45, SNU5 and SNU16 cells. Moreover, combined treatment with isorhamnetin and capecitabine significantly suppressed the growth of SNU5 cells-based tumors in mice through the downregulation of NF-κB signaling and its oncogenic gene products [113]. Structurally modified flavonol glycoside, troxerutin, promoted the antitumor activity of 5-FU in SGC7901 cells and sensitized 5-FU-resistant subline (SGC7901/5-FU) to drug-induced apoptosis by inhibiting NF-κB activation and repressing Bcl-2. Additionally, combined treatment of troxerutin and 5-FU dramatically decreased the tumor size and weight in SGC7901/5-FU implanted mice without any adverse effects on body weight or liver weight of experimental rodents [38].

Pretreatment of well-differentiated adenocarcinoma MGC803 cells with isoflavone genistein resulted in a significant decrease in chemoresistance to both 5-FU and cisplatin. This phenomenon could be related to the downregulation of ABCG2 level and ERK1/2 activity [122]. A methylated isoflavone, calycosin or 3`-hydroxyformononetin, enhanced 5-FU- and adriamycin-induced antiproliferative action and antimigratory potential in different human gastric cancer cell lines, BGC823, NCI-N87 and SGC7901. Treatment of these three cell lines with calycosin augmented also the ability of cisplatin to block the cell cycle at G1 phase, trigger apoptosis and suppress invasion through inhibition of Akt signaling cascade. Hence, the therapeutic effects of these drugs could be better and achieved at lower doses with less side effects and improved tolerance [42]. The combined treatment of BGC823 cells with an isoflavone glycoside, puerarin, and 5-FU led to a synergistic anticancer action with increased proportion of apoptotic cells. Puerarin was also able to potentiate 5-FU-induced suppression of the growth of xenografts obtained by implanting BGC823 cells in mice, reducing thus the effective dose of drug. At that, liver and renal toxicities were not increased [21].

The only tested chalcone, licochalcone A, exerted synergistic anticancer action in combination with 5-FU in MKN45 cells that were moderately sensitive to 5-FU and SGC7901 cells with relative resistance to 5-FU [81] (Table 6.5).

As development of resistance to conventional cancer treatment modalities is a common phenomenon in clinical practice resulting in poor prognosis of patients, sensitizing the action of standard chemotherapeutic drugs has remained an important task in the field of oncotherapy. At that, potentiation of conventional treatments could lead to the reduction of drug dosages and minimization of accompanying toxic side effects. Several studies performed with human gastric cancer cell lines have suggested the ability of flavonoids to behave as chemo- or radiosensitizers. Moreover, preclinical works with xenograft mice have provided additional evidence about the augmentation of anticancer potential of standard treatments, encouraging thus the design of further clinical trials. As a result, the appropriate combinations of chemotherapeutics or irradiation and certain natural flavonoids administered according to the proven time schedules can provide more efficient and better tolerable clinical therapeutic strategies for treatment of gastric cancer in the future.

6.4. Current Shortcomings and Future Prospects

Despite the promising preclinical results and multitargeted action of flavonoids in gastric cancer experimental models, the clinical application of these natural polyphenols is still obstructed by several disadvantages, including poor water solubility and extensive metabolic conversion leading to low serum levels of parent flavonoids after peroral administration. One strategy to overcome these limitations is to synthesize structurally modified derivatives with improved physiological properties; the other solution could be to design the appropriate systems for drug delivery.

Therefore, natural flavonoids could be considered as potential lead compounds for the development and synthetic preparation of novel anticancer agents. Different flavonoid derivatives have been indeed designed and synthesized and their activities in human gastric cancer models are summarized in Table 6.6. Among these compounds, the synthetic flavone flavopiridol, a cyclin-dependent kinase inhibitor, is the most extensively studied agent, whereas several investigations have shown that flavopiridol can potentiate the anticancer action of conventional chemotherapeutics. In fact, treatment of poorly differentiated adenocarcinoma MKN74 cells with this synthetic flavone augmented manifold the apoptotic responses induced by docetaxel [182], paclitaxel [183], gemcitabine [184] or mitomycin C [185]. Flavopiridol potentiated gemcitabine-triggered cytotoxicity also in adenocarcinoma SK-GT-5 cells [184]. Moreover, exposure of docetaxel-treated MKN74 xenografts bearing mice to flavopiridol significantly decreased the tumor volume that was quite opposite to a 394% increase in tumor volume observed with docetaxel alone [182]. Furthermore, flavopiridol was able to potentiate also the rate of irradiation-induced apoptosis in MKN74 cells [117]. In all these cases, the sequence of treatments was absolutely critical, as flavopiridol had to follow the standard therapy to achieve the potentiation of cytotoxic effects. Indeed, no enhancement of mitomycin C-induced apoptosis was observed in MKN74 cells when treatment with flavopiridol preceded the drug [185]. Moreover, pretreatment with flavopiridol led even to antagonizing the effects of docetaxel [182] and paclitaxel [183]. Thus, when flavopiridol is combined with the conventional drugs or irradiation in a strictly sequence-specific manner, this synthetic flavone might significantly improve the treatment responses of gastric cancer therapy. However, phase II study performed with flavopiridol in patients suffering from metastatic stomach cancer revealed no objective responses on tumor regression but showed frequent toxic adverse effects, such as fatigue, diarrhea and development of vascular thromboses at the central catheter tip [111] (Table 6.6).

Other synthetic flavonoids exhibited enhanced activity levels or improved physicochemical properties as compared to their natural lead compounds. A hybrid obtained by conjugating wogonin with Pt(IV) complex revealed anticancer responses in SGC7901 cells which were superior to those of wogonin and cisplatin. In addition, this complex exerted potential to reverse the drug resistance via inhibiting NF-κB signaling in cisplatin-resistant subline of SGC7901 cells (SGC7901/cDDP) *in vitro* but also *in vivo* in mice xenograft model with no substantial toxicity in contrast to cisplatin [146]. To improve the cytotoxic action of chrysin, a series of synthetic derivatives of this flavone was prepared and tested in SGC7901 cells. As a result, the 8-substituted derivatives appeared to possess somewhat stronger activities than parent chrysin, as exemplified by 5,7-dimethoxy-8-iodochrysin and 5,7-dihydroxy-8-nitrochrysin [121, 186, 187]. 7-O-Geranylquercetin decreased the viability, induced mitochondrial-dependent apoptosis and blocked the cell cycle progression at G2/M phase in MGC803 and SGC7901 cells more potently than intact quercetin, whereas almost no apparent cytotoxicity was detected on human normal gastric epithelial cells GES-1. Moreover, this new synthetic alkylated derivative possessed also better lipid solubility than the parent quercetin, further suggesting its potential as an antitumor agent for gastric cancer treatment [39, 188]. A novel synthetic genistein analogue, 7-difluoromethoxyl-5,4`-di-n-octylgenistein (DFOG), revealed about tenfold higher inhibitory activity on the growth of AGS and SGC7901 cells as compared to genistein. This derivative induced G2/M phase cell cycle arrest, and downregulated FOXM1 and its downstream genes, including cell cycle-associated genes CDK1, Cdc25B and cyclin B. As FOXM1 can be a rational molecular target for anticancer agents, DFOG could possess an important therapeutic impact for further treatment of stomach malignancies [24]. Moreover, DFOG was also able to suppress the stem cell characteristics of SGC7901 cell line-derived gastric cancer stem-like cells (GCSLCs), including self-renewal capacity, and reversed the EMT phenotype, accompanied by downregulation of FOXM1 and Twist1 [45]. A brominated synthetic chalcone H72 displayed a rather high potency in restraining the growth and inducing intrinsic apoptosis in HGC27, MGC803 and SGC7901 cells with considerably less cytotoxicity in normal stomach epithelial cells GES-1. Administration of this synthetic chalcone analogue to MGC803 cell-based xenografts bearing mice led to about 60% tumor regression without any apparent toxicity [55]. Furthermore, another synthetic flavonoid, 7-carboxymethyloxy-5,3`,4`-trimethoxyflavone, exerted inhibitory action on *H. pylori*-induced proinflammatory responses [98] (Table 6.6). Therefore, structural analyses might provide valuable knowledge about certain moieties on flavonoid skeleton which are necessary for expression of anticancer activities. It can be expected that these studies can eventually lead to the discovering and development of a highly potent and selective pharmaceutical drug for further treatment of human gastric neoplasms.

However, despite the strong efforts done on synthetic chemistry in the past few decades, the bottlenecks of clinical application of flavonoids, i.e., poor water solubility and limited bioavailability, have still remained largely unresolved. An alternative strategy to overcome the current shortcomings and allow the biomedical application of flavonoids is to design the appropriate biocompatible systems for targeted and controlled drug delivery. At that, nanoscale materials (nanoparticles) have been recognized to hold a great potential for development of drug delivery carriers to tumor cells, by virtue of their many unique properties, such as large surface-to-volume ratios and easy of penetrating cell membrane. When drug is loaded to polymeric nanoparticles, it reveals improved retention and permeability with lower side effects [7, 28, 189]. However, to date, only a few nanoencapsulated flavonoids have been tested in human gastric cancer preclinical models and these results are compiled in Table 6.7. Controlled release of EGCG from nanoencapsulated system, based on fucose-conjugated chitosan and polyethylene glycol (PEG)-conjugated chitosan complex with gelatin, suppressed the growth and induced apoptosis in MKN45 cells more efficiently than free EGCG. Moreover, *in vivo* assay indicated that EGCG-loaded nanoparticles inhibited gastric cancer progression in luciferase-expressing MKN45 engineered xenografts in mice, prolonged survival rate of animals and decreased the inflammatory reactions in stomach and liver tissues [7]. Nanoencapsulated chrysin revealed significantly stronger antiproliferative efficacy in AGS cells as compared to free chrysin, and exhibited more extensive downregulation of miR-18a, miR-21 and miR-221 levels and upregulation of miR-34a level. At that, poly(lactic-co-glycolic acid) (PLGA)-PEG-PLGA nanoparticles improved the water solubility of chrysin and increased the entry of this flavone into the cells [28, 79]. Genistein nanoconjugate, designed by decorating Fe_3O_4 nanoparticles with carboxymethylated chitosan, exerted significantly higher inhibitory and proapoptotic effects in SGC7901 cells as compared to free genistein, accompanied by a slower release of genistein from the complex. This nanoconjugate revealed also rather good water solubility [189] (Table 6.7). As a whole, these results suggest that nanoparticles could represent efficient systems for delivery of anticancer flavonoids to malignant cells providing thus an attractive approach for further clinical management of human gastric neoplasms.

Table 6.6. Effect of synthetic flavonoids on human gastric cancer models and their combinations with standard anticancer treatment modalities

Synthetic flavonoid	Biological system[1]	Cell growth inhibition IC_{50}, μM	Assay conditions[2]	Biological effects[3]	Co-effects with drug/treatment[4]	Ref.
CHALCONES						
H72 brominated chalcone	GES-1 cells (non-malignant)	>>6	MTT, 24 h			[55]
	HGC27 cells	~3.57…5.61	MTT, 24 h	Increase in ROS, mitochondrial apoptosis		[55]
	MGC803 cells	3.57	MTT, 24 h	Increase in ROS, mitochondrial apoptosis		[55]
	MGC803 cells injected into right flank of mice		Intraperitoneally, 40 mg/kg/day, 21 days	Inhibition of tumor growth, decrease in tumor volume and weight. No signs of toxicity		[55]
	SGC7901 cells	~3.57…5.61	MTT, 24 h			[55]
FLAVANOLS						
8-Propylcatechin gallate	MGLVA1 cells	~30	MTT, 48 h			[196]
FLAVONES						
5,7-Dihydroxy-8-nitrochrysin	SGC7901 cells	2.8	MTT, 144 h			[186]
	SGC7901 cells	4.14	MTT, 48 h	Apoptosis, increase in PPAR-γ, decrease in Bcl-2/Bax ratio		[187]
5,7-Dimethoxy-8-iodochrysin	SGC7901 cells	2.2	MTT, 144 h			[186]
7-Carboxymethyloxy-5,3`,4`-tri-methoxyflavone (DA-6034)	AGS cells			Abrogation of *H. pylori*-induced proinflammatory responses, suppression of *H. pylori*-triggered iNOS induction and *H. pylori*-increased NF-κB DNA binding		[98]

Table 6.6. (Continued)

Synthetic flavonoid	Biological system[1]	Cell growth inhibition IC_{50}, μM	Assay conditions[2]	Biological effects[3]	Co-effects with drug/treatment[4]	Ref.
Flavopiridol	MKN74 cells		Docetaxel (100 nM) 18 h → flavopiridol (300 nM) 24 h		About 2-3-fold increase in docetaxel-induced apoptosis, accelerating mitotic exit. **Antagonistic effect when flavopiridol was added before docetaxel**	[182]
	MKN74 cells injected into flank of mice		Docetaxel (10 mg/kg) → 3-7 h later flavopiridol (2.5 mg/kg)		1-18% decrease in tumor volume, sequence dependent mechanism. No regression when flavopiridol was added before docetaxel	[182]
	MKN74 cells		Gemcitabine (100 nM) 24 h → flavopiridol (300 nM) 24 h		About 10-15-fold increase in gemcitabine-induced apoptosis, sequence dependent mechanism. Downregulation of RR-M2	[184]
	SK-GT-5 cells		Gemcitabine (100 nM) 24 h → flavopiridol (300 nM) 24 h		About 10-15-fold increase in gemcitabine-induced apoptosis, sequence dependent mechanism. Downregulation of RR-M2	[184]
	MKN74 cells		Mitomycin-C (1 μg/ml) 24 h → flavopiridol (300 nM) 24 h		Potentiation of mitomycin-C-induced apoptosis, sequence dependent mechanism. No increase in apoptosis when flavopiridol was added before mitomycin-C	[185]

Synthetic flavonoid	Biological system[1]	Cell growth inhibition IC_{50}, μM	Assay conditions[2]	Biological effects[3]	Co-effects with drug/treatment[4]	Ref.
	MKN74 cells		Paclitaxel (100 nM) 18 h → flavopiridol (300 nM) 24 h		About 2-3-fold increase in paclitaxel-induced apoptosis, accelerating mitotic exit by inhibiting cdc-2 kinase. Antagonistic effect when flavopiridol was added before paclitaxel	[183]
	MKN74 cells		Radiation (20 Gy) 24 h → flavopiridol (300 nM) 24 h		Enhancement of radiation-induced apoptosis, suppression of p21 expression	[117]
	Phase II study with 14 patients		Continuous infusion 50 mg/m^2/day for 72 h every 2 weeks	No objective effects, common toxicities included fatigue, diarrhea, venous thromboses at the central catheter tip		[111]
Wogonin Pt(IV) complex	SGC7901 cells	4.31±0.51	MTT, 72 h	Apoptosis, S phase cell cycle arrest, increase in ROS, reduction of GSH level, inhibition of NF-κB-dependent gene transcription		[146]
	SGC7901/ cDDP cells	3.29±0.42	MTT, 72 h	Apoptosis, S phase cell cycle arrest, increase in ROS, reduction of GSH level, decrease in COX-1/-2, inhibition of NF-κB-dependent gene transcription, suppression of CK2 activity		[146]

Table 6.6. (Continued)

Synthetic flavonoid	Biological system[1]	Cell growth inhibition IC_{50}, µM	Assay conditions[2]	Biological effects[3]	Co-effects with drug/treatment[4]	Ref.
	SGC7901/cDDP cells injected into right oxter of mice		Intravenously, 5 mg/kg once every three days, 21 days	Decrease in tumor volume. No signs of toxicity		[146]
FLAVONOLS						
7-O-Geranylquercetin	GES-1 cells (non-malignant)	>>20	MTT, 24 h			[39]
	MGC803 cells	~15 (24 h)	MTT; 12, 24, 48 h	Apoptosis via activation of ROS-MAPK-mediated mitochondrial pathway, G2/M phase cell cycle arrest		[39]
	MGC803 cells	25.4	MTT, 48 h			[188]
	SGC7901 cells	~15 (24 h)	MTT; 12, 24, 48 h	Apoptosis via activation of ROS-MAPK-mediated mitochondrial pathway, G2/M phase cell cycle arrest		[39]
	SGC7901 cells	18.5	MTT, 48 h			[188]
ISOFLAVONES						
7- Difluoromethoxyl-5,4`-di-n-octylgenistein (DFOG)	AGS cells	3.9	MTT, 24 h	Inhibition of colony formation, G2/M phase cell cycle arrest, decrease in FOXM1 expression		[24]
	SGC7901 cells	5.2	MTT, 24 h	Inhibition of colony formation, G2/M phase cell cycle arrest, decrease in FOXM1 expression		[24]

Synthetic flavonoid	Biological system[1]	Cell growth inhibition IC_{50}, μM	Assay conditions[2]	Biological effects[3]	Co-effects with drug/treatment[4]	Ref.
	SGC7901-derived GCSLCs			Inhibition of CSC characteristics, reversal of EMT phenotype by modulation of FoxM1 and decrease in Twist1 expression		[45]

[1]see Tables 6.1 and 6.2.
[2]MTT, 3-(4,5-dimethylthiazol-2-yl)-2,5-diphenyltetrazolium bromide.
[3]Bcl-2, B-cell lymphoma-2; CK2, casein kinase 2; COX, cyclooxygenase; CSC, cancer stem cell; EMT, epithelial-mesenchymal transition; FOXM1, fork-head box M1; *H. pylori*, *Helicobacter pylori*; GSH, reduced glutathione; iNOS, inducible nitric oxide synthase; MAPK, mitogen-activated protein kinase; NF-κB, nuclear factor κB; PPAR-γ, peroxisome proliferator-activated receptor γ; ROS, reactive oxygen species.
[4]RR-M2, ribonucleotide reductase M2 subunit.

Table 6.7. Effect of nanoencapsulated natural flavonoids on *in vitro* and *in vivo* human gastric cancer models

Parent compound	Flavonoid nanoparticles[1]	Biological system[2]	Assay conditions[3]	Comparative biological effects[4]	Ref.
FLAVANOLS					
Epigallocatechin gallate (EGCG)	Fucose-chitosan/PEG-chitosan/gelatin-encapsulated EGCG	MKN45 cells	MTT, 2h	Inhibition of growth, induction of apoptosis, decrease in VEGF level. IC_{50} ~87 μM for encapsulated EGCG, >87 μM for free EGCG	[7]
	Fucose-chitosan/PEG-chitosan/gelatin-encapsulated EGCG	MKN45-Luc cells injected into the gastric wall of mice	By intragastric gavage, 40 mg/kg EGCG	Inhibition of tumor growth, decrease in proliferation marker (Ki-67), increase in apoptotic marker (cleaved PARP), reduction of inflammatory reaction in gastric and liver tissues	[7]
FLAVONES					
Chrysin	PLGA-PEG-encapsulated chrysin	AGS cells	MTT; 24, 48, 72 h	Inhibition of growth, decrease in miR-18a, miR-21 and miR-221 expression, increase in miR-34a level. IC_{50} 58.24 μM for encapsulated chrysin, 68.24 μM for free chrysin (24 h); 44.21 μM for encapsulated chrysin, 56.16 μM for free chrysin (48 h); 36.8 μM for encapsulated chrysin, 42.32 μM for free chrysin (72 h)	[28, 79]

Table 6.7. (Continued)

Parent compound	Flavonoid nanoparticles[1]	Biological system[2]	Assay conditions[3]	Comparative biological effects[4]	Ref.
ISOFLAVONES					
Genistein	Fe_3O_4-CMCH-encapsulated genistein	SGC7901 cells	MTT; 24, 48, 72 h	Inhibition of proliferation, induction of apoptosis. IC_{50} >100 µM for encapsulated genistein, ~100 µM for free genistein (24 h); ~80…100 µM for encapsulated genistein, ~80 µM for free genistein (48 h); ~20…40 µM for encapsulated genistein, ~40…80 µM for free genistein (72 h)	[189]

[1] CMCH, carboxymethylated chitosan; PEG, polyethylene glycol; PLGA, poly(lactic-co-glycolic acid).
[2] see Tables 6.1 and 6.2; Luc, luciferase.
[3] MTT, 3-(4,5-dimethylthiazol-2-yl)-2,5-diphenyltetrazolium bromide.
[4] miR, microRNA; PARP, poly(ADP-ribose) polymerase; VEGF, vascular endothelial growth factor.

In parallel to advancing the strategies for more efficient application of flavonoids in the treatment of stomach tumors, either alone or in combination with current conventional therapeutic modalities, it is important to bear in mind that these natural polyphenols have also an remarkable chemopreventive potential alleviating the damages caused by different risk factors. At that, the ability of flavonoids to combat *H. pylori*-induced gastric mucosal injuries might be one of the crucial steps in the prevention of neoplasms in the stomach. Therefore, high dietary intake of flavonoids could possibly reduce the susceptibility to gastric tumorigenesis.

REFERENCES

[1] Shih PH, Yeh CT, Yen GC. Effects of anthocyanidin on the inhibition of proliferation and induction of apoptosis in human gastric adenocarcinoma cells. *Food Chem Toxicol* 2005; 43: 1557-66.

[2] Hao W, Yuan X, Yu L, Gao C, Sun X, Wang D, Zheng Q. Licochalcone A-induced human gastric cancer BGC-823 cells apoptosis by regulating ROS-mediated MAPKs and PI3K/AKT signaling pathways. *Sci Rep* 2015; 5: 10336.

[3] Lin SH, Shih YW. Antitumor effects of the flavone chalcone: inhibition of invasion and migration through the FAK/JNK signaling pathway in human gastric adenocarcinoma AGS cells. *Mol Cell Biochem* 2014; 391: 47-58.

[4] Wang YX, Cai H, Jiang G, Zhou TB, Wu H. Silibinin inhibits proliferation, induces apoptosis and causes cell cycle arrest in human gastric cancer MGC803 cells via STAT3 pathway inhibition. *Asian Pac J Cancer Prev* 2014; 15: 6791-8.

[5] Park JS, Khoi PN, Joo YE, Lee YH, Lang SA, Stoeltzing O, Jung YD. EGCG inhibits recepteur d`origine nantais expression by suppressing Egr-1 in gastric cancer cells. *Int J Oncol* 2013; 42: 1120-6.

[6] Kim HS, Kim MH, Jeong M, Hwang YS, Lim SH, Shin BA, Ahn BW, Jung YD. EGCG blocks tumor promoter-induced MMP-9 expression via suppression of MAPK and AP-1 activation in human gastric AGS cells. *Anticancer Res* 2004; 24: 747-53.

[7] Lin YH, Chen ZR, Lai CH, Hsieh CH, Feng CL. Active targeted nanoparticles for oral administration of gastric cancer therapy. *Biomacromolecules* 2015; 16: 3021-32.

[8] Yang C, Du W, Yang D. Inhibition of green tea polyphenol EGCG((-)-epigallocatechin-3-gallate) on the proliferation of gastric cancer cells by suppressing canonical wnt/ß-catenin signalling pathway. *Int J Food Sci Nutr* 2016; 67: 818-27.

[9] Lu L, Chen J, Tang H, Bai L, Lu C, Wang K, Li M, Yan Y, Tang L, Wu R, Ye Y, Jin L, Liang Z. EGCG suppresses ERK5 activation to reverse tobacco smoke-triggered gastric epithelial-mesenchymal transition in BALB/c mice. *Nutrients* 2016; 8: E380.

[10] Zang MD, Hu L, Fan ZY, Wang HX, Zhu ZL, Cao S, Wu XY, Li JF, Su LP, Li C, Zhu ZG, Yan M, Liu BY. Luteolin suppresses gastric cancer progression by reversing epithelial-mesenchymal transition via suppression of the Notch signaling pathway. *J Transl Med* 2017; 15: 52.

[11] Zang M, Hu L, Zhang B, Zhu Z, Li J, Zhu Z, Yan M, Liu B. Luteolin suppresses angiogenesis and vasculogenic mimicry formation through inhibiting Notch1-VEGF signaling in gastric cancer. *Biochem Biophys Res Commun* 2017; 490: 913-9.

[12] Lu J, Li G, He K, Jiang W, Xu C, Li Z, Wang H, Wang W, Wang H, Teng X, Teng L. Luteolin exerts a marked antitumor effect in cMet-overexpressing patient-derived tumor xenograft models of gastric cancer. *J Transl Med* 2015; 13: 42.

[13] Lee YC, Cheng TH, Lee JS, Chen JH, Liao YC, Fong Y, Wu CH, Shih YW. Nobiletin, a citrus flavonoid, suppresses invasion and migration involving FAK/PI3K/Akt and small GTPase signals in human gastric adenocarcinoma AGS cells. *Mol Cell Biochem* 2011; 347: 103-15.

[14] Wang X, Xia M. 5-Hydroxy-3,6,7,8,3`,4`-hexamethoxyflavone, a polymethoxyflavone, exerts antitumor effect on PI3K/Akt signaling pathway in human gastric cancer cell BGC-7901. *J Recept Signal Transduct Res* 2016; 36: 471-7.

[15] Sabarwal A, Agarwal R, Singh RP. Fisetin inhibits cellular proliferation and induces mitochondria-dependent apoptosis in human gastric cancer cells. *Mol Carcinog* 2017; 56: 499-514.

[16] Ji FJ, Tian XF, Liu XW, Fu LB, Wu YY, Fang XD, Jin HY. Dihydromyricetin induces cell apoptosis via a p53-related pathway in AGS human gastric cancer cells. *Genet Mol Res* 2015; 14: 15564-71.

[17] Feng J, Chen X, Wang Y, Du Y, Sun Q, Zang W, Zhao G. Myricetin inhibits proliferation and induces apoptosis and cell cycle arrest in gastric cancer cells. *Mol Cell Biochem* 2015; 408: 163-70.

[18] Xu YX, Wang B, Zhao XH. In vitro effects and the related molecular mechanism of galangin and quercetin on human gastric cancer cell line (SGC-7901). *Pak J Pharm Sci* 2017; 30: 1279-87.

[19] Zhang JY, Lin MT, Zhou MJ, Yi T, Tang YN, Tang SL, Yang ZJ, Zhao ZZ, Chen HB. Combinational treatment of curcumin and quercetin against gastric cancer MGC-803 cells in vitro. *Molecules* 2015; 20: 11524-34.

[20] Yang Z, Zhao Y, Yao Y, Li J, Wang W, Wu X. Equol induces mitochondria-dependent apoptosis in human gastric cancer cells via the sustained activation of ERK1/2 pathway. *Mol Cells* 2016; 39: 742-9.

[21] Guo XF, Yang ZR, Wang J, Lei XF, Lv XG, Dong WG. Synergistic antitumor effect of puerarin combined with 5-fluorouracil on gastric carcinoma. *Mol Med Rep* 2015; 11: 2562-8.

[22] Lee H, Lee JH, Jung KH, Hong SS. Deguelin promotes apoptosis and inhibits angiogenesis of gastric cancer. *Oncol Rep* 2010; 24: 957-63.

[23] Babu TM, Rammohan A, Baki VB, Devi S, Gunasekar D, Rajendra W. Development of novel HER2 inhibitors against gastric cancer derived from flavonoid source of Syzygium alternifolium through molecular dynamics and pharmacophore-based screening. *Drug Des Devel Ther* 2016; 10: 3611-32.

[24] Xiang HL, Liu F, Quan MF, Cao JG, Lv Y. 7-Difluoromethoxyl-5,4`-di-n-octylgenistein inhibits growth of gastric cancer cells through downregulating forkhead box M1. *World J Gastroenterol* 2012; 18: 4618-26.

[25] Golpour S, Rafie N, Safavi SM, Miraghajani M. Dietary isoflavones and gastric cancer: A brief review of current studies. *J Res Med Sci* 2015; 20: 893-900.

[26] Ferlay J, Soerjomataram I, Ervik M, Dikshit R, Eser S, Mathers C, Rebelo M, Parkin DM, Forman D, Bray F. *GLOBOCAN 2012 v1.1, Cancer Incidence and Mortality Worldwide: IARC CancerBase* No. 11[Internet]. Lyon, France: International Agency for Research on Cancer; 2014. Available from: http://globocan.iarc.fr.

[27] Zhou Y, Ding BZ, Lin YP, Wang HB. MiR-34a, as a suppressor, enhance the susceptibility of gastric cancer cell to luteolin by directly targeting HK1. *Gene* 2018; 644: 56-65.

[28] Mohammadian F, Abhari A, Dariushnejad H, Zarghami F, Nikanfar A, Pilehvar-Soltanahmadi Y, Zarghami N. Upregulation of miR-34a in AGS gastric cancer cells by a PLGA-PEG-PLGA chrysin nano formulation. *Asian Pac J Cancer Prev* 2015; 16: 8259-63.

[29] Raha S, Yumnam S, Hong GE, Lee HJ, Saralamma VV, Park HS, Heo JD, Lee SJ, Kim EH, Kim JA, Kim GS. Naringin induces autophagy-mediated growth inhibition by downregulating the PI3K/Akt/mTOR cascade via activation of MAPK pathways in AGS cancer cells. *Int J Oncol* 2015; 47: 1061-9.

[30] Lei CS, Hou YC, Pai MH, Lin MT, Yeh SL. Effects of quercetin combined with anticancer drugs on metastasis-associated factors of gastric cancer cells: in vitro and in vivo studies. *J Nutr Biochem* 2018; 51: 105-13.

[31] Zhang J, Wu D, Vikash, Song J, Wang J, Yi J, Dong W. Hesperetin induces the apoptosis of gastric cancer cells via activating mitochondrial pathway by increasing reactive oxygen species. *Dig Dis Sci* 2015; 60: 2985-95.

[32] Mu J, Liu T, Jiang L, Wu X, Cao Y, Li M, Dong Q, Liu Y, Xu H. The traditional Chinese medicine baicalein potently inhibits gastric cancer cells. *J Cancer* 2016; 7: 453-61.

[33] Chen F, Zhuang M, Zhong C, Peng J, Wang X, Li J, Chen Z, Huang Y. Baicalein reverses hypoxia-induced 5-FU resistance in gastric cancer AGS cells through suppression of glycolysis and the PTEN/Akt/HIF-1α signaling pathway. *Oncol Rep* 2015; 33: 457-63.

[34] Chen F, Zhuang M, Peng J, Wang X, Huang T, Li S, Lin M, Lin H, Xu Y, Li J, Chen Z, Huang Y. Baicalein inhibits migration and invasion of gastric cancer cells through suppression of the TGF-ß signaling pathway. *Mol Med Rep* 2014; 10: 1999-2003.

[35] Xia Y, Lian S, Khoi PN, Yoon HJ, Joo YE, Chay KO, Kim KK, Do Jung Y. Chrysin inhibits tumor promoter-induced MMP-9 expression by blocking AP-1 via suppression of ERK and JNK pathways in gastric cancer cells. *PLoS One* 2015; 10: e0124007.

[36] Shen X, Si Y, Wang Z, Wang J, Guo Y, Zhang X. Quercetin inhibits the growth of human gastric cancer stem cells by inducing mitochondrial-dependent apoptosis through the inhibition of PI3K/Akt signaling. *Int J Mol Med* 2016; 38: 619-26.

[37] Pandey A, Vishnoi K, Mahata S, Tripathi SC, Misra SP, Misra V, Mehrotra R, Dwivedi M, Bharti AC. Berberine and curcumin target survivin and STAT3 in gastric cancer cells and synergize actions of standard chemotherapeutic 5-fluorouracil. *Nutr Cancer* 2015; 67: 1293-304.

[38] Xu GY, Tang XJ. Troxerutin (TXN) potentiated 5-fluorouracil (5-Fu) treatment of human gastric cancer through suppressing STAT3/NF-κB and Bcl-2 signaling pathways. *Biomed Pharmacother* 2017; 92: 95-107.

[39] Zhu Y, Jiang Y, Shi L, Du L, Xu X, Wang E, Sun Y, Guo X, Zou B, Wang H, Wang C, Sun L, Zhen Y. 7-O-Geranylquercetin induces apoptosis in gastric cancer cells via ROS-MAPK mediated mitochondrial signaling pathway activation. *Biomed Pharmacother* 2017; 87: 527-38.

[40] Zhu BH, Zhan WH, Li ZR, Wang Z, He YL, Peng JS, Cai SR, Ma JP, Zhang CH. (-)-Epigallocatechin-3-gallate inhibits growth of gastric cancer by reducing VEGF production and angiogenesis. *World J Gastroenterol* 2007; 13: 1162-9.

[41] Kim MC, Lee HJ, Lim B, Ha KT, Kim SY, So I, Kim BJ. Quercetin induces apoptosis by inhibiting MAPKs and TRPM7 channels in AGS cells. *Int J Mol Med* 2014; 33: 1657-63.

[42] Zhou L, Wu Y, Guo Y, Li Y, Li N, Yang Y, Qin X. Calycosin enhances some chemotherapeutic drugs inhibition of Akt signaling pathway in gastric cells. *Cancer Invest* 2017; 35: 289-300.

[43] Yanagihara K, Takigahira M, Mihara K, Kubo T, Morimoto C, Morita Y, Terawaki K, Uezono Y, Seyama T. Inhibitory effects of isoflavones on tumor growth and cachexia in newly established cachectic mouse models carrying human stomach cancers. *Nutr Cancer* 2013; 65: 578-89.

[44] Tang S, Hu J, Meng Q, Dong X, Wang K, Qi Y, Chu C, Zhang X, Hou L. Daidzein induced apoptosis via down-regulation of Bcl-2/Bax and triggering of the mitochondrial pathway in BGC-823 cells. *Cell Biochem Biophys* 2013; 65: 197-202.

[45] Cao X, Ren K, Song Z, Li D, Quan M, Zheng Y, Cao J, Zeng W, Zou H. 7-Difluoromethoxyl-5,4`-di-n-octyl genistein inhibits the stem-like characteristics of gastric cancer stem-like cells and reverses the phenotype of epithelial-mesenchymal transition in gastric cancer cells. *Oncol Rep* 2016; 36: 1157-65.

[46] Gonzalez CA, Sala N, Rokkas T. Gastric cancer: epidemiologic aspects. *Helicobacter* 2013; 18 Suppl 1: 34-8.

[47] Sun CL, Yuan JM, Lee MJ, Yang CS, Gao YT, Ross RK, Yu MC. Urinary tea polyphenols in relation to gastric and esophageal cancers: a prospective study of men in Shanghai, China. *Carcinogenesis* 2002; 23: 1497-503.

[48] Moon JY, Cho M, Ahn KS, Cho SK. Nobiletin induces apoptosis and potentiates the effects of the anticancer drug 5-fluorouracil in p53-mutated SNU-16 human gastric cancer cells. *Nutr Cancer* 2013; 65: 286-95.

[49] Wu B, Zhang Q, Shen W, Zhu J. Anti-proliferative and chemosensitizing effects of luteolin on human gastric cancer AGS cell line. *Mol Cell Biochem* 2008; 313: 125-32.

[50] Wu H, Huang M, Liu Y, Shu Y, Liu P. Luteolin induces apoptosis by up-regulating miR-34a in human gastric cancer cells. *Technol Cancer Res Treat* 2015; 14: 747-55.

[51] Zhang Q, Wan L, Guo Y, Cheng N, Cheng W, Sun Q, Zhu J. Radiosensitization effect of luteolin on human gastric cancer SGC-7901 cells. *J Biol Regul Homeost Agents* 2009; 23: 71-8.

[52] Song S, Su Z, Xu H, Niu M, Chen X, Min H, Zhang B, Sun G, Xie S, Wang H, Gao Q. Luteolin selectively kills STAT3 highly activated gastric cancer cells through enhancing the binding of STAT3 to SHP-1. *Cell Death Dis* 2017; 8: e2612.

[53] Ekström AM, Serafini M, Nyren O, Wolk A, Bosetti C, Bellocco R. Dietary quercetin intake and risk of gastric cancer: results from a population-based study in Sweden. *Ann Oncol* 2011; 22: 438-43.

[54] Saralamma VV, Nagappan A, Hong GE, Lee HJ, Yumnam S, Raha S, Heo JD, Lee SJ, Lee WS, Kim EH, Kim GS. Poncirin induces apoptosis in AGS human gastric

cancer cells through extrinsic apoptotic pathway by up-regulation of Fas ligand. *Int J Mol Sci* 2015; 16: 22676-91.

[55] Zhang S, Li T, Zhang Y, Xu H, Li Y, Zi X, Yu H, Li J, Jin CY, Liu HM. A new brominated chalcone derivative suppresses the growth of gastric cancer cells in vitro and in vivo involving ROS mediated up-regulation of DR5 and 4 expression and apoptosis. *Toxicol Appl Pharmacol* 2016; 309: 77-86.

[56] Zhou W, Cao A, Wang L, Wu D. Kurarinone synergizes TRAIL-induced apoptosis in gastric cancer cells. *Cell Biochem Biophys* 2015; 72: 241-9.

[57] Luo Y, Yu H, Yang Y, Tian W, Dong K, Shan J, Ma X. A flavonoid compound from Chrysosplenium nudicaule inhibits growth and induces apoptosis of the human stomach cancer cell line SGC-7901. *Pharm Biol* 2016; 54: 1133-9.

[58] Woo HD, Lee J, Choi IJ, Kim CG, Lee JY, Kwon O, Kim J. Dietary flavonoids and gastric cancer risk in a Korean population. *Nutrients* 2014; 6: 4961-73.

[59] Wada K, Tsuji M, Tamura T, Konishi K, Kawachi T, Hori A, Tanabashi S, Matsushita S, Tokimitsu N, Nagata C. Soy isoflavone intake and stomach cancer risk in Japan: From the Takayama study. *Int J Cancer* 2015; 137: 885-92.

[60] Lou C, Wang M, Yang G, Cai H, Li Y, Zhao F, Yang H, Tong L, Cai B. Preliminary studies on anti-tumor activity of 2`,4`-dihydroxychalcone isolated from Herba Oxytropis in human gastric cancer MGC-803 cells. *Toxicol In Vitro* 2009; 23: 906-10.

[61] Zhang Y, Li Q, Ge Y, Chen Y, Chen J, Dong Y, Shi W. Silibinin triggers apoptosis and cell-cycle arrest of SGC7901 cells. *Phytother Res* 2013; 27: 397-403.

[62] Ma J, Shi M, Li G, Wang N, Wei J, Wang T, Ma J, Wang Y. Regulation of Id1 expression by epigallocatechin-3-gallate and its effect on the proliferation and apoptosis of poorly differentiated AGS gastric cancer cells. *Int J Oncol* 2013; 43: 1052-8.

[63] Hara A, Sasazuki S, Inoue M, Miura T, Iwasaki M, Sawada N, Shimazu T, Yamaji T, Tsugane S, Japan Public Health Center-Based Prospective Study Group. Plasma isoflavone concentrations are not associated with gastric cancer risk among Japanese men and women. *J Nutr* 2013; 143: 1293-8.

[64] Hara A, Sasazuki S, Inoue M, Iwasaki M, Shimazu T, Sawada N, Yamaji T, Tsugane S, Japan Public Health Center-Based Prospective Study Group. Isoflavone intake and risk of gastric cancer: a population-based prospective cohort study in Japan. *Am J Clin Nutr* 2012; 95: 147-54.

[65] Zamora-Ros R, Agudo A, Lujan-Barroso L, Romieu I, Ferrari P, Knaze V, Bueno-de-Mesquita HB, Leenders M, Travis RC, Navarro C, Sanchez-Cantalejo E, Slimani N, Scalbert A, Fedirko V, Hjartaker A, Engeset D, Skeie G, Boeing H, Förster J, Li K, Teucher B, Agnoli C, Tumino R, Mattiello A, Saieva C, Johansson I, Stenling R, Redondo ML, Wallström P, Ericson U, Khaw KT, Mulligan AA,

Trichopoulou A, Dilis V, Katsoulis M, Peeters PH, Igali L, Tjonneland A, Halkaer J, Touillaud M, Perquier F, Fagherazzi G, Amiano P, Ardanaz E, Bredsdorff L, Overvad K, Ricceri F, Riboli E, Gonzalez CA. Dietary flavonoid and lignan intake and gastric adenocarcinoma risk in the European Prospective Investigation into Cancer and Nutrition (EPIC) study. *Am J Clin Nutr* 2012; 96: 1398-408.

[66] Sun L, Subar AF, Bosire C, Dawsey SM, Kahle LL, Zimmerman TP, Abnet CC, Heller R, Graubard BI, Cook MB, Petrick JL. Dietary flavonoid intake reduces the risk of head and neck but not esophageal or gastric cancer in US men and women. *J Nutr* 2017; 147: 1729-38.

[67] Xie Y, Huang S, Su Y. Dietary flavonols intake and risk of esophageal and gastric cancer: A meta-analysis of epidemiological studies. *Nutrients* 2016; 8: 91.

[68] Petrick JL, Steck SE, Bradshaw PT, Trivers KF, Abrahamson PE, Engel LS, He K, Chow WH, Mayne ST, Risch HA, Vaughan TL, Gammon MD. Dietary intake of flavonoids and oesophageal and gastric cancer: incidence and survival in the United States of America (USA). *Br J Cancer* 2015; 112: 1291-300.

[69] Ekambaram G, Rajendran P, Magesh V, Sakthisekaran D. Naringenin reduces tumor size and weight lost in N-methyl-N`-nitro-N-nitrosoguanidine-induced gastric carcinogenesis in rats. *Nutr Res* 2008; 28: 106-12.

[70] Jin CY, Park C, Cheong J, Choi BT, Lee TH, Lee JD, Lee WH, Kim GY, Ryu CH, Choi YH. Genistein sensitizes TRAIL-resistant human gastric adenocarcinoma AGS cells through activation of caspase-3. *Cancer Lett* 2007; 257: 56-64.

[71] Kuo CH, Weng BC, Wu CC, Yang SF, Wu DC, Wang YC. Apigenin has anti-atrophic gastritis and anti-gastric cancer progression effects in Helicobacter pylori-infected Mongolian gerbils. *J Ethnopharmacol* 2014; 151: 1031-9.

[72] Ullah MF, Bhat SH, Husain E, Abu-Duhier F, Hadi SM, Sarkar FH, Ahmad A. Pharmacological intervention through dietary nutraceuticals in gastrointestinal neoplasia. *Crit Rev Food Sci Nutr* 2016; 56: 1501-18.

[73] Gutierrez-Orozco F, Stephens BR, Neilson AP, Green R, Ferruzzi MG, Bomser JA. Green and black tea inhibit cytokine-induced IL-8 production and secretion in AGS gastric cancer cells via inhibition of NF-κB activity. *Planta Med* 2010; 76: 1659-65.

[74] Yan X, Rui X, Zhang K. Baicalein inhibits the invasion of gastric cancer cells by suppressing the activity of the p38 signaling pathway. *Oncol Rep* 2015; 33: 737-43.

[75] Park BB, Yoon Js, Kim Es, Choi J, Won Yw, Choi Jh, Lee YY. Inhibitory effects of eupatilin on tumor invasion of human gastric cancer MKN-1 cells. *Tumour Biol* 2013; 34: 875-85.

[76] Wang Y, Dong H, Zhu M, Ou Y, Zhang J, Luo H, Luo R, Wu J, Mao M, Liu X, Zhang J, Wei L. Icariin exterts negative effects on human gastric cancer cell

invasion and migration by vasodilator-stimulated phosphoprotein via Rac1 pathway. *Eur J Pharmacol* 2010; 635: 40-8.

[77] Bao L, Liu F, Guo HB, Li Y, Tan BB, Zhang WX, Peng YH. Naringenin inhibits proliferation, migration, and invasion as well as induces apoptosis of gastric cancer SGC7901 cell line by downregulation of AKT pathway. *Tumour Biol* 2016; 37: 11365-74.

[78] Chen J, Chen J, Li Z, Liu C, Yin L. The apoptotic effect of apigenin on human gastric carcinoma cells through mitochondrial signal pathway. *Tumour Biol* 2014; 35: 7719-26.

[79] Mohammadian F, Pilehvar-Soltanahmadi Y, Mofarrah M, Dastani-Habashi M, Zarghami N. Down regulation of miR-18a, miR-21 and miR-221 genes in gastric cancer cell line by chrysin-loaded PLGA-PEG nanoparticles. *Artif Cells Nanomed Biotechnol* 2016; 44: 1972-8.

[80] Xiao XY, Hao M, Yang XY, Ba Q, Li M, Ni SJ, Wang LS, Du X. Licochalcone A inhibits growth of gastric cancer cells by arresting cell cycle progression and inducing apoptosis. *Cancer Lett* 2011; 302: 69-75.

[81] Lin X, Tian L, Wang L, Li W, Xu Q, Xiao X. Antitumor effects and the underlying mechanism of licochalcone A combined with 5-fluorouracil in gastric cancer cells. *Oncol Lett* 2017; 13: 1695-1701.

[82] Wang Z, Lu W, Li Y, Tang B. Alpinetin promotes Bax translocation, induces apoptosis through the mitochondrial pathway and arrests human gastric cancer cells at the G2/M phase. *Mol Med Rep* 2013; 7: 915-20.

[83] Zhang H, Zhong X, Zhang X, Shang D, Zhou YI, Zhang C. Enhanced anticancer effect of ABT-737 in combination with naringenin on gastric cancer cells. *Exp Ther Med* 2016; 11: 669-73.

[84] Lee JH, Park CH, Jung KC, Rhee HS, Yang CH. Negative regulation of beta-catenin/Tcf signaling by naringenin in AGS gastric cancer cell. *Biochem Biophys Res Commun* 2005; 335: 771-6.

[85] Ramachandran L, Manu KA, Shanmugam MK, Li F, Siveen KS, Vali S, Kapoor S, Abbasi T, Surana R, Smoot DT, Ashktorab H, Tan P, Ahn KS, Yap CW, Kumar AP, Sethi G. Isorhamnetin inhibits proliferation and invasion and induces apoptosis through the modulation of peroxisome proliferator-activated receptor γ activation pathway in gastric cancer. *J Biol Chem* 2012; 287: 38028-40.

[86] Yamane T, Takahashi T, Kuwata K, Oya K, Inagake M, Kitao Y, Suganuma M, Fujiki H. Inhibition of N-methyl-N`-nitro-N-nitrosoguanidine-induced carcinogenesis by (-)-epigallocatechin gallate in the rat glandular stomach. *Cancer Res* 1995; 55: 2081-4.

[87] Ganapathy E, Peramaiyan R, Rajasekaran D, Venkataraman M, Dhanapal S. Modulatory effect on naringenin on N-methyl-N`-nitro-N-nitrosoguanidine- and

saturated sodium chloride-induced gastric carcinogenesis in male Wistar rats. *Clin Exp Pharmacol Physiol* 2008; 35: 1190-6.

[88] Pan MH, Lai CS, Hsu PC, Wang YJ. Acacetin induces apoptosis in human gastric carcinoma cells accompanied by activation of caspase cascades and production of reactive oxygen species. *J Agric Food Chem* 2005; 53: 620-30.

[89] Ko KP, Park SK, Park B, Yang JJ, Cho LY, Kang C, Kim CS, Gwack J, Shin A, Kim Y, Kim J, Yang HK, Kang D, Chang SH, Shin HR, Yoo KY. Isoflavones from phytoestrogens and gastric cancer risk: a nested case-control study within the Korean Multicenter Cancer Cohort. *Cancer Epidemiol Biomarkers Prev* 2010; 19: 1292-300.

[90] Woo HD, Kim J. Dietary flavonoid intake and risk of stomach and colorectal cancer. *World J Gastroenterol* 2013; 19: 1011-9.

[91] Yang ZP, Zhao Y, Huang F, Chen J, Yao YH, Li J, Wu XN. Equol inhibits proliferation of human gastric carcinoma cells via modulating Akt pathway. *World J Gastroenterol* 2015; 21: 10385-99.

[92] Garcia-Closas R, Gonzalez CA, Agudo A, Riboli E. Intake of specific carotenoids and flavonoids and the risk of gastric cancer in Spain. *Cancer Causes Control* 1999; 10: 71-5.

[93] Rossi M, Rosato V, Bosetti C, Lagiou P, Parpinel M, Bertuccio P, Negri E, La Vecchia C. Flavonoids, proanthocyanidins, and the risk of stomach cancer. *Cancer Causes Control* 2010; 21: 1597-604.

[94] Sasazuki S, Inoue M, Miura T, Iwasaki M, Tsugane S, Japan Public Health Center-based Prospective Study Group. Plasma tea polyphenols and gastric cancer risk: a case-control study nested in a large population-based prospective study in Japan. *Cancer Epidemiol Biomarkers Prev* 2008; 17: 343-51.

[95] Park S, Hahm KB, Oh TY, Jin JH, Choue R. Preventive effect of the flavonoid, wogonin, against ethanol-induced gastric mucosal damage in rats. *Dig Dis Sci* 2004; 49: 384-94.

[96] Arab HH, Salama SA, Omar HA, Arafa el-SA, Maghrabi IA. Diosmin protects against ethanol-induced gastric injury in rats: novel anti-ulcer actions. *PLoS One* 2015; 30: e0122417.

[97] Park CH, Hahm ER, Lee JH, Jung KC, Yang CH. Inhibition of beta-catenin-mediated transactivation by flavanone in AGS gastric cancer cells. *Biochem Biophys Res Commun* 2005; 331: 1222-8.

[98] Lee JS, Kim HS, Hahm KB, Sohn MW, Yoo M, Johnson JA, Surh YJ. Inhibitory effects of 7-carboxymethyloxy-3`,4`,5-trimethoxyflavone (DA-6034) on Helicobacter pylori-induced NF-kappa B activation and iNOS expression in AGS cells. *Ann N Y Acad Sci* 2007; 1095: 527-35.

[99] Lee KM, Yeo M, Choue JS, Jin JH, Park SJ, Cheong JY, Lee KJ, Kim JH, Hahm KB. Protective mechanism of epigallocatechin-3-gallate against Helicobacter pylori-induced gastric epithelial cytotoxicity via the blockage of TLR-4 signaling. *Helicobacter* 2004; 9: 632-42.

[100] Chung JG, Hsia TC, Kuo HM, Li YC, Lee YM, Lin SS, Hung CF. Inhibitory actions of luteolin of the growth and arylamine N-acetyltransferase activity in strains of Helicobacter pylori from ulcer patients. *Toxicol In Vitro* 2001; 15: 191-8.

[101] Kim S, Choi MG, Lee HS, Lee SK, Kim SH, Kim WW, Hur SM, Kim JH, Choe JH, Nam SJ, Yang JH, Kim S, Lee JE, Kim JS. Silibinin suppresses TNF-alpha-induced MMP-9 expression in gastric cancer cells through inhibition of the MAPK pathway. *Molecules* 2009; 14: 4300-11.

[102] Zhu BH, Chen HY, Zhan WH, Wang CY, Cai SR, Wang Z, Zhang CH, He YL. (-)-Epigallocatechin-3-gallate inhibits VEGF expression induced by IL-6 via Stat3 in gastric cancer. *World J Gastroenterol* 2011; 17: 2315-25.

[103] Wang YC, Huang KM. In vitro anti-inflammatory effect of apigenin in the Helicobacter pylori-infected gastric adenocarcinoma cells. *Food Chem Toxicol* 2013; 53: 376-83.

[104] Lee SY, Shin YW, Hahm KB. Phytoceuticals: mighty but ignored weapons against Helicobacter pylori infection. *J Dig Dis* 2008; 9: 129-39.

[105] Lee JH, Lee SH, Kim YS, Jeong CS. Protective effects of neohesperidin and poncirin isolated from the fruits of Poncirus trifoliate on potential gastric disease. *Phytother Res* 2009; 23: 1748-53.

[106] Lee HH, Lee S, Shin YS, Cho M, Kang H, Cho H. Anti-cancer effect of quercetin in xenograft models with EBV-associated human gastric carcinoma. *Molecule*s 2016; 21: E1286.

[107] Lee M, Son M, Ryu E, Shin YS, Kim JG, Kang BW, Cho H, Kang H. Quercetin-induced apoptosis prevents EBV infection. *Oncotarget* 2015; 6: 12603-24.

[108] Lee DH, Park KI, Park HS, Kang SR, Nagappan A, Kim JA, Kim EH, Lee WS, Hah YS, Chung HJ, An SJ, Kim GS. Flavonoids isolated from Korea Citrus aurantium L. induce G2/M phase arrest and apoptosis in human gastric cancer AGS cells. *Evid Based Complement Alternat Med* 2012; 2012: 515901.

[109] Yang Y, Li XJ, Chen Z, Zhu XX, Wang J, Zhang LB, Qiang L, Ma YJ, Li ZY, Guo QL, You QD. Wogonin induced calreticulin/annexin A1 exposure dictates the immunogenicity of cancer cells in a PERK/AKT dependent manner. *PLoS One* 2012; 7: e50811.

[110] Song H, Bao J, Wei Y, Chen Y, Mao X, Li J, Yang Z, Xue Y. Kaempferol inhibits gastric cancer tumor growth: An in vitro and in vivo study. *Oncol Rep* 2015; 33: 868-74.

[111] Schwartz GK, Ilson D, Saltz L, O`Reilly E, Tong W, Maslak P, Werner J, Perkins P, Stoltz M, Kelsen D. Phase II study of the cyclin-dependent kinase inhibitor flavopiridol administered to patients with advanced gastric carcinoma. *J Clin Oncol* 2001; 19: 1985-92.

[112] Zhao Q, Wang J, Zou MJ, Hu R, Zhao L, Qiang L, Rong JJ, You QD, Guo QL. Wogonin potentiates the antitumor effects of low dose 5-fluorouracil against gastric cancer through induction of apoptosis by down-regulation of NF-kappaB and regulation of its metabolism. *Toxicol Lett* 2010; 197: 201-10.

[113] Manu KA, Shanmugam MK, Ramachandran L, Li F, Siveen KS, Chinnathambi A, Zayed ME, Alharbi SA, Arfuso F, Kumar AP, Ahn KS, Sethi G. Isorhamnetin augments the anti-tumor effect of capecitabine through the negative regulation of NF-κB signaling cascade in gastric cancer. *Cancer Lett* 2015; 363: 28-36.

[114] Wang K, Liu R, Li J, Mao J, Lei Y, Wu J, Zeng J, Zhang T, Wu H, Chen L, Huang C, Wei Y. Quercetin induces protective autophagy in gastric cancer cells: involvement of Akt-mTOR- and hypoxia-induced factor 1α-mediated signaling. *Autophagy* 2011; 7: 966-78.

[115] Xia Y, Lian S, Khoi PN, Yoon HJ, Han JY, Chay KO, Kim KK, Jung YD. Chrysin inhibits cell invasion by inhibition of Recepteur d`origine Nantais via suppressing early growth response-1 and NF-κB transcription factor activities in gastric cancer cells. *Int J Oncol* 2015; 46: 1835-43.

[116] Zhang X, Zheng L, Sun Y, Wang T, Wang B. Tangeretin enhances radiosensitivity and inhibits the radiation-induced epithelial-mesenchymal transition of gastric cancer cells. *Oncol Rep* 2015; 34: 302-10.

[117] Jung C, Motwani M, Kortmansky J, Sirotnak FM, She Y, Gonen M, Haimovitz-Friedman A, Schwartz GK. The cyclin-dependent kinase inhibitor flavopiridol potentiates gamma-irradiation-induced apoptosis in colon and gastric cancer cells. *Clin Cancer Res* 2003; 9: 6052-61.

[118] Du F, Feng Y, Fang J, Yang M. MicroRNA-143 enhances chemosensitivity of quercetin through autophagy inhibition via target GABARAPL1 in gastric cancer cells. *Biomed Pharmacother* 2015; 74: 169-77.

[119] Wu H, Xin Y, Xiao Y, Zhao J. Low-dose docetaxel combined with (-)-epigallocatechin-3-gallate inhibits angiogenesis and tumor growth in nude mice with gastric cancer xenografts. *Cancer Biother Radiopharm* 2012; 27: 204-9.

[120] Zhu X, Luo F, Zheng Y, Zhang J, Huang J, Sun C, Li X, Chen K. Characterization, purification of Poncirin from edible citrus Ougan (Citrus reticulate cv. Suavissima) and its growth inhibitory effect on human gastric cancer cells SGC-7901. *Int J Mol Sci* 2013; 14: 8684-97.

[121] Zheng X, Cao JG, Meng WD, Qing FL. Synthesis and anticancer effect of B-ring trifluoromethylated flavonoids. *Bioorg Med Chem Lett* 2003; 13: 3423-7.

[122] Huang W, Wan C, Luo Q, Huang Z, Luo Q. Genistein-inhibited cancer stem cell-like properties and reduced chemoresistance of gastric cancer. *Int J Mol Sci* 2014; 15: 3432-43.

[123] Yu D, Shin HS, Lee YS, Lee D, Kim S, Lee YC. Genistein attenuates cancer stem cell characteristics in gastric cancer through the downregulation of Gli1. *Oncol Rep* 2014; 31: 673-8.

[124] Wang P, Zhang K, Zhang Q, Mei J, Chen CJ, Feng ZZ, Yu DH. Effects of quercetin on the apoptosis of the human gastric carcinoma cells. *Toxicol In Vitro* 2012; 26: 221-8.

[125] Ekambaram G, Rajendran P, Devaraja R, Muthuvel R, Sakthisekaran D. Impact of naringenin on glycoprotein levels in N-methyl-N`-nitro-N-nitrosoguanidine-induced gastric carcinogenesis in rats. *Anticancer Drugs* 2008; 19: 885-90.

[126] Li YS, Wu LP, Li KH, Liu YP, Xiang R, Zhang SB, Zhu LY, Zhang LY. Involvement of nuclear factor κB (NF-κB) in the downregulation of cyclooxygenase-2 (COX-2) by genistein in gastric cancer cells. *J Int Med Res* 2011; 39: 2141-50.

[127] Azuine MA, Bhide SV. Adjuvant chemoprevention of experimental cancer: catechin and dietary turmeric in forestomach and oral cancer models. *J Ethnopharmacol* 1994; 44: 211-7.

[128] Tatsuta M, Iishi H, Baba M, Yano H, Uehara H, Nakaizumi A. Attenuation by genistein of sodium-chloride-enhanced gastric carcinogenesis induced by N-methyl-N`-nitro-N-nitrosoguanidine in Wistar rats. *Int J Cancer* 1999; 80: 396-9.

[129] Horie N, Hirabayashi N, Takahashi Y, Miyauchi Y, Taguchi H, Takeishi K. Synergistic effect of green tea catechins on cell growth and apoptosis induction in gastric carcinoma cells. *Biol Pharm Bull* 2005; 28: 574-9.

[130] Ma Y, Shi Y, Li W, Sun A, Zang P, Zhang P. Epigallocatechin-3-gallate regulates the expression of Kruppel-like factor 4 through myocyte enhancer factor 2A. *Cell Stress Chaperones* 2014; 19: 217-26.

[131] Tanaka T, Ishii T, Mizuno D, Mori T, Yamaji R, Nakamura Y, Kumazawa S, Nakayama T, Akagawa M. (-)-Epigallocatechin-3-gallate suppresses growth of AZ521 human gastric cancer cells by targeting the DEAD-box RNA helicase p68. *Free Radic Biol Med* 2011; 50: 1324-35.

[132] Horng SB, Kuo HH, Lin MY, Lin WW, Wang TC. Human gastric cells resistant to (-)-epigallocatechin gallate show cross-resistance to several environmental pollutants. *Food Chem Toxicol* 2007; 45: 2171-8.

[133] Okabe S, Ochiai Y, Aida M, Park K, Kim SJ, Nomura T, Suganuma M, Fujiki H. Mechanistic aspects of green tea as a cancer preventive: effect of components on human stomach cancer cell lines. *Jpn J Cancer Res* 1999; 90: 733-9.

[134] Hibasami H, Komiya T, Achiwa Y, Ohnishi K, Kojima T, Nakanishi K, Akashi K, Hara Y. Induction of apoptosis in human stomach cancer cells by green tea catechins. *Oncol Rep* 1998; 5: 527-9.

[135] Onoda C, Kuribayashi K, Nirasawa S, Tsuji N, Tanaka M, Kobayashi D, Watanabe N. (-)-Epigallocatechin-3-gallate induces apoptosis in gastric cancer cell lines by down-regulating survivin expression. *Int J Oncol* 2011; 38: 1403-8.

[136] Ran ZH, Xu Q, Tong JL, Xiao SD. Apoptotic effect of epigallocatechin-3-gallate on the human gastric cancer cell line MKN45 via activation of the mitochondrial pathway. *World J Gastroenterol* 2007; 13: 4255-9.

[137] Isemura M, Saeki K, Kimura T, Hayakawa S, Minami T, Sazuka M. Tea catechins and related polyphenols as anti-cancer agents. *Biofactors* 2000; 13: 81-5.

[138] Ma JP, Qiao X, Pan S, Shen H, Zhu GF, Hou AJ. New isoprenylated flavonoids and cytotoxic constituents from Artocarpus tonkinensis. *J Asian Nat Prod Res* 2010; 12: 586-92.

[139] Kanno S, Tomizawa A, Hiura T, Osanai Y, Shouji A, Ujibe M, Ohtake T, Kimura K, Ishikawa M. Inhibitory effects of naringenin on tumor growth in human cancer cell lines and sarcoma S-180-implanted mice. *Biol Pharm Bull* 2005; 28: 527-30.

[140] Duarte N, Lage H, Abrantes M, Ferreira MJ. Phenolic compounds as selective antineoplastic agents against multidrug-resistant human cancer cells. *Planta Med* 2010; 76: 975-80.

[141] Kang GD, Kim DH. Ponciretin attenuates ethanol-induced gastric damage in mice by inhibiting inflammatory responses. *Int Immunopharmacol* 2017; 43: 179-86.

[142] Nagao T, Abe F, Kinjo J, Okabe H. Antiproliferative constituents in plants 10. Flavones from the leaves of Lantana montevidensis Briq. and consideration of structure-activity relationship. *Biol Pharm Bull* 2002; 25: 875-9.

[143] Matsukawa Y, Marui N, Sakai T, Satomi Y, Yoshida M, Matsumoto K, Nishino H, Aoike A. Genistein arrests cell cycle progression at G2-M. *Cancer Res* 1993; 53: 1328-31.

[144] Wu K, Yuan LH, Xia W. Inhibitory effects of apigenin on the growth of gastric carcinoma SGC-7901 cells. *World J Gastroenterol* 2005; 11: 4461-4.

[145] Yang Y, Hu Y, Gu HY, Lu N, Liu W, Qi Q, Zhao L, Wang XT, You QD, Guo QL. Oroxylin A induces G2/M phase cell-cycle arrest via inhibiting Cdk7-mediated expression of Cdc2/p34 in human gastric carcinoma BGC-823 cells. *J Pharm Pharmacol* 2008; 60: 1459-63.

[146] Chen F, Qin X, Xu G, Gou S, Jin X. Reversal of cisplatin resistance in human gastric cancer cells by wogonin-conjugated Pt(IV) prodrug via attenuating Casein Kinase 2-mediated Nuclear Factor-κB pathways. *Biochem Pharmacol* 2017; 135: 50-68.

[147] Xiao W, Wu K, Yin M, Han S, Ding Y, Qiao A, Lu G, Deng B, Bo P, Gong W. Wogonin inhibits tumor-derived regulatory molecules by suppressing STAT3 signaling to promote tumor immunity. *J Immunother* 2015; 38: 167-84.

[148] Yoshimizu N, Otani Y, Saikawa Y, Kubota T, Yoshida M, Furukawa T, Kumai K, Kameyama K, Fujii M, Yano M, Sato T, Ito A, Kitajima M. Anti-tumour effects of nobiletin, a citrus flavonoid, on gastric cancer include: antiproliferative effects, induction of apoptosis and cell cycle deregulation. *Aliment Pharmacol Ther* 2004; 20 Suppl 1: 95-101.

[149] Moon JY, Cho SK. Nobiletin induces protective autophagy accompanied by ER-stress mediated apoptosis in human gastric cancer SNU-16 cells. *Molecules* 2016; 21: E914.

[150] Minagawa A, Otani Y, Kubota T, Wada N, Furukawa T, Kumai K, Kameyama K, Okada Y, Fujii M, Yano M, Sato T, Ito A, Kitajima M. The citrus flavonoid, nobiletin, inhibits peritoneal dissemination of human gastric carcinoma in SCID mice. *Jpn J Cancer Res* 2001; 92: 1322-8.

[151] Dong Y, Cao A, Shi J, Yin P, Wang L, Ji G, Xie J, Wu D. Tangeretin, a citrus polymethoxyflavonoid, induces apoptosis of human gastric cancer AGS cells through extrinsic and intrinsic signaling pathways. *Oncol Rep* 2014; 31: 1788-94.

[152] Choi EJ, Oh HM, Wee H, Choi CS, Choi SC, Kim KH, Han WC, Oh TY, Kim SH, Jun CD. Eupatilin exhibits a novel anti-tumor activity through the induction of cell cycle arrest and differentiation of gastric carcinoma AGS cells. *Differentiation* 2009; 77: 412-23.

[153] Lee S, Lee M, Kim SH. Eupatilin inhibits H(2)O(2)-induced apoptotic cell death through inhibition of mitogen-activated protein kinases and nuclear factor-kappaB. *Food Chem Toxicol* 2008; 46: 2865-70.

[154] Kim MJ, Lee HH, Seo MJ, Kang BW, Park JU, Kim KS, Kim GY, Joo WH, Choi YH, Cho YS, Jeong YK. Identification of 5-hydroxy-3,6,7,8,3`,4`-hexamethoxyflavone from Hizikia fusiforme involved in the induction of the apoptosis mediators in human AGS carcinoma cells. *J Microbiol Biotechnol* 2012; 22: 1665-72.

[155] Yoshida M, Sakai T, Hosokawa N, Marui N, Matsumoto K, Fujioka A, Nishino H, Aoike A. The effect of quercetin on cell cycle progression and growth of human gastric cancer cells. *FEBS Lett* 1990; 260: 10-3.

[156] Borska S, Chmielewska M, Wysocka T, Drag-Zalesinska M, Zabel M, Dziegiel P. In vitro effect of quercetin on human gastric carcinoma: targeting cancer cells death and MDR. *Food Chem Toxicol* 2012; 50: 3375-83.

[157] Liao W, Chen L, Ma X, Jiao R, Li X, Wang Y. Protective effects of kaempferol against reactive oxygen species-induced hemolysis and its antiproliferative activity on human cancer cells. *Eur J Med Chem* 2016; 114: 24-32.

[158] Uddin SJ, Grice D, Tiralongo E. Evaluation of cytotoxic activity of patriscabratine, tetracosane and various flavonoids isolated from the Bangladeshi medicinal plant Acrostichum aureum. *Pharm Biol* 2012; 50: 1276-80.

[159] Kim DA, Jeon YK, Nam MJ. Galangin induces apoptosis in gastric cancer cells via regulation of ubiquitin carboxy-terminal hydrolase isozyme L1 and glutathione S-transferase P. *Food Chem Toxicol* 2012; 50: 684-8.

[160] Awale S, Linn TZ, Li F, Tezuka Y, Myint A, Tomida A, Yamori T, Esumi H, Kadota S. Identification of chrysoplenetin from Vitex negundo as a potential cytotoxic agent against PANC-1 and a panel of 39 human cancer cell lines (JFCR-39). *Phytother Res* 2011; 25: 1770-5.

[161] Zhou Y, Tian L, Long L, Quan M, Liu F, Cao J. Casticin potentiates TRAIL-induced apoptosis of gastric cancer cells through endoplasmic reticulum stress. *PLoS One* 2013; 8: e58855.

[162] Hsu HY, Tsang SF, Lin KW, Yang SC, Lin CN. Cell death induced by flavonoid glycosides with and without copper. *Food Chem Toxicol* 2008; 46: 2394-401.

[163] Liu YL, Zhang GQ, Yang Y, Zhang CY, Fu RX, Yang YM. Genistein induces G2/M arrest in gastric cancer cells by increasing the tumor suppressor PTEN expression. *Nutr Cancer* 2013; 65: 1034-41.

[164] Cui HB, Na XL, Song DF, Liu Y. Blocking effects of genistein on cell proliferation and possible mechanism in human gastric carcinoma. *World J Gastroenterol* 2005; 11: 69-72.

[165] Yan GR, Zou FY, Dang BL, Zhang Y, Yu G, Liu X, He QY. Genistein-induced mitotic arrest of gastric cancer cells by downregulating KIF20A, a proteomics study. *Proteomics* 2012; 12: 2391-9.

[166] Yanagihara K, Ito A, Toge T, Numoto M. Antiproliferative effects of isoflavones on human cancer cell lines established from the gastrointestinal tract. *Cancer Res* 1993: 53: 5815-21.

[167] Liu Y, Li X. Genistein induced apoptosis in human gastric carcinoma cells. *Adv Mat Res* 2011; 183-185: 1283-6.

[168] Zhou HB, Chen JJ, Wang WX, Cai JT, Du Q. Apoptosis of human primary gastric carcinoma cells induced by genistein. *World J Gastroenterol* 2004; 10: 1822-5.

[169] Zhang Y, Vareed SK, Nair MG. Human tumor cell growth inhibition by nontoxic anthocyanidins, the pigments in fruits and vegetables. *Life Sci* 2005; 76: 1465-72.

[170] Vareed SK, Reddy MK, Schutzki RE, Nair MG. Anthocyanins in Cornus alternifolia, Cornus controversa, Cornus kousa and Cornus florida fruits with health benefits. *Life Sci* 2006; 78: 777-84.

[171] Ma J, Fu NY, Pang DB, Wu WY, Xu AL. Apoptosis induced by isoliquiritigenin in human gastric cancer MGC-803 cells. *Planta Med* 2001; 67: 754-7.

[172] Lou C, Yang G, Cai H, Zou M, Xu Z, Li Y, Zhao F, Li W, Tong L, Wang M, Cai B. 2`,4`-Dihydroxychalcone-induced apoptosis of human gastric cancer MGC-803 cells via down-regulation of survivin mRNA. *Toxicol In Vitro* 2010; 24: 1333-7.

[173] Satomi Y. Inhibitory effects of 3`-methyl-3-hydroxy-chalcone on proliferation of human malignant tumor cells and on skin carcinogenesis. *Int J Cancer* 1993; 55: 506-14.

[174] Federico A, Dallio M, DI Fabio G, Zarrelli A, Zappavigna S, Stiuso P, Tuccillo C, Caraglia M, Loguercio C. Silybin-phosphatidylcholine complex protects human gastric and liver cells from oxidative stress. *In Vivo* 2015; 29: 569-75.

[175] Minakawa T, Toume K, Arai MA, Koyano T, Kowithayakorn T, Ishibashi M. Prenylflavonoids isolated from Artocarpus champeden with TRAIL-resistance overcoming activity. *Phytochemistry* 2013; 96: 299-304.

[176] Yamane T, Nakatani H, Kikuoka N, Matsumoto H, Iwata Y, Kitao Y, Oya K, Takahashi T. Inhibitory effects and toxicity of green tea polyphenols for gastrointestinal carcinogenesis. *Cancer* 1996; 77: 1662-7.

[177] Hamdan DI, Mahmoud MF, Wink M, El-Shazly AM. Effect of hesperidin and neohesperidin from bittersweet orange (Citrus aurantium var. bigaradia) peel on indomethacin-induced peptic ulcers in rats. *Environ Toxicol Pharmacol* 2014; 37: 907-15.

[178] Lu N, Gao Y, Ling Y, Chen Y, Yang Y, Gu HY, Qi Q, Liu W, Wang XT, You QD, Guo QL. Wogonin suppresses tumor growth in vivo and VEGF-induced angiogenesis through inhibiting tyrosine phosphorylation of VEGFR2. *Life Sci* 2008; 82: 956-63.

[179] Yan XM, Joo MJ, Lim JC, Whang WK, Sim SS, Im C, Kim HR, Lee SY, Kim IK, Sohn UD. The effect of quercetin-3-O-ß-D-glucuronopyranoside on indomethacin-induced gastric damage in rats via induction of mucus secretion and down-regulation of ICAM-1 expression. *Arch Pharm Res* 2011; 34: 1527-34.

[180] Zhou HB, Chen JM, Cai JT, Du Q, Wu CN. Anticancer activity of genistein on implanted tumor of human SG7901 cells in nude mice. *World J Gastroenterol* 2008; 14: 627-31.

[181] Takekawa S, Matsui T, Arakawa Y. The protective effect of the soybean polyphenol genistein against stress-induced gastric mucosal lesions in rats, and its hormonal mechanisms. *J Nutr Sci Vitaminol* (Tokyo) 2006; 52: 274-80.

[182] Motwani M, Rizzo C, Sirotnak F, She Y, Schwartz GK. Flavopiridol enhances the effect of docetaxel in vitro and in vivo in human gastric cancer cells. *Mol Cancer Ther* 2003; 2: 549-55.

[183] Motwani M, Delohery TM, Schwartz GK. Sequential dependent enhancement of caspase activation and apoptosis by flavopiridol on paclitaxel-treated human gastric and breast cancer cells. *Clin Cancer Res* 1999; 5: 1876-83.

[184] Jung CP, Motwani MV, Schwartz GK. Flavopiridol increases sensitization to gemcitabine in human gastrointestinal cancer cell lines and correlates with down-regulation of ribonucleotide reductase M2 subunit. *Clin Cancer Res* 2001; 7: 2527-36.

[185] Schwartz GK, Farsi K, Maslak P, Kelsen DP, Spriggs D. Potentiation of apoptosis by flavopiridol in mitomycin-C-treated gastric and breast cancer cells. *Clin Cancer Res* 1997; 3: 1467-72.

[186] Zheng X, Meng WD, Xu YY, Cao JG, Qing FL. Synthesis and anticancer effect of chrysin derivatives. *Bioorg Med Chem Lett* 2003; 13: 881-4.

[187] Ai XH, Zheng X, Tang XQ, Sun L, Zhang YQ, Qin Y, Liu HQ, Xia H, Cao JG. Induction of apoptosis of human gastric carcinoma SGC-7901 cell line by 5,7-dihydroxy-8-nitrochrysin in vitro. *World J Gastroenterol* 2007; 13: 3824-8.

[188] Bao XR, Liao H, Qu J, Sun Y, Guo X, Wang EX, Zhen YH. Synthesis, characterization and cytotoxicity of alkylated quercetin derivatives. *Iran J Pharm Res* 2016; 15: 329-35.

[189] Si HY, Li DP, Wang TM, Zhang HL, Ren FY, Xu ZG, Zhao YY. Improving the anti-tumor effect of genistein with a biocompatible superparamagnetic drug delivery system. *J Nanosci Nanotechnol* 2010; 10: 2325-31.

[190] Kikuchi H, Ohtsuki T, Koyano T, Kowithayakorn T, Sakai T, Ishibashi M. Death receptor 5 targeting activity-guided isolation of isoflavones from Millettia brandisiana and Ardisia colorata and evaluation of ability to induce TRAIL-mediated apoptosis. *Bioorg Med Chem* 2009; 17: 1181-6.

[191] Moghaddam G, Ebrahimi SA, Rahbar-Roshandel N, Foroumadi A. Antiproliferative activity of flavonoids: influence of the sequential methoxylation state of the flavonoid structure. *Phytother Res* 2012; 26: 1023-8.

[192] Cherng JM, Shieh DE, Chiang W, Chang MY, Chiang LC. Chemopreventive effects of minor dietary constituents in common foods on human cancer cells. *Biosci Biotechnol Biochem* 2007; 71: 1500-4.

[193] Wang WS, Ye YH, Zhou YW. New prenylchalcones from the hops of Humulus lupulus. *J Asian Nat Prod Res* 2008; 10: 485-90.

[194] Hosny M, Rosazza JP. New isoflavone and triterpene glycosides from soybeans. *J Nat Prod* 2002; 65: 805-13.

[195] Abe F, Nagao T, Okabe H. Antiproliferative constituents in plants 9. Aerial parts of Lippia dulcis and Lippia canescens. *Biol Pharm Bull* 2002: 25: 920-2.

[196] Hayes CJ, Whittaker BP, Watson SA, Grabowska AM. Synthesis and preliminary anticancer activity studies of C4 and C8-modified derivatives of catechin gallate (CG) and epicatechin gallate (ECG). *J Org Chem* 2006; 71: 9701-12.

Chapter 7

Epidemiological Evidence on Chemopreventive Role of Flavonoids in Upper Gastrointestinal Tract

Diets rich in fruits and vegetables have been suggested to be related to beneficial effects against carcinogenesis and these activities are attributed, at least in part, to flavonoids [1-8]. Although numerous *in vitro* studies with cell lines and *in vivo* investigations with animal models have shown protective action of different flavonoids against development and progression of various common malignancies, extrapolation of experimental findings to the human body includes several risks due to poor absorption, low bioavailability and extensive metabolism of these plant polyphenols in human organism. Therefore, in reality, only findings from human intervention trials and epidemiological studies can provide relevant knowledge about the chemopreventive action of dietary flavonoids in different populations [9]. Nevertheless, current epidemiological findings about associations between high dietary intake of flavonoids and cancer incidence are still limited and reveal rather inconclusive and controversial results [4, 7, 10-15]. There are several reasons for this discrepancy: different flavonoids vary in their chemical structures and biological activities, these plant secondary metabolites cannot be absorbed in their parent form and undergo an extensive bioconversion before entering the central circulation revealing thus only limited bioavailability, the concentrations used in experimental works are usually much higher compared to those achieved in human target tissues remaining questionable whether the protective effects observed in *in vitro* studies are feasible in humans [7, 11, 14]. However, due to the direct contacts and interactions of upper gastrointestinal tract with dietary components, including various plant-derived polyphenols, carcinogenesis in oral cavity, pharynx, esophagus, stomach and duodenum might be more closely related to the intake of flavonoids [7, 11]. Therefore, the current epidemiological findings about the role of flavonoid subclasses and individual compounds on the risk of cancer development in upper gastrointestinal tract are considered in this chapter.

7.1. Epidemiological Knowledge About the Role of Flavonoids in Oral, Esophageal and Gastric Carcinogenesis

The results of epidemiological studies about associations between high dietary consumption of flavonoids and the risk of development of upper gastrointestinal neoplasms are summarized in Tables 7.1 and 7.2. In addition to dietary intake, relationship between different biomarkers in body fluids, reflecting flavonoids intake and bioconversion, and cancer incidence rate is presented in Table 7.3. There is no doubt that identification of modifiable behavioral risk factors, such as diet, is highly attractive measure for cancer prevention opening new avenues in the fight against this highly fatal disease.

7.1.1. Oral and Pharyngeal Cancer

So far, there are only two case-control studies available about the role of flavonoids on the risk of oral and pharyngeal cancer (Table 7.1). In 1999, De Stefani et al. demonstrated a risk below unity for total flavonoids in Uruguayan population; however, this estimate was statistically insignificant [18]. Besides only a small number of oral cancer cases (33) encompassed in this work, the study was conducted in the time when knowledge about the content of flavonoids in food items was still very limited [16]. In the work performed among Italian population in 2007, Rossi et al. reported a significant inverse association between high dietary intake of total flavonoids, flavanones or flavonols, but not flavones, flavanols, isoflavones or anthocyanidins, and oral and pharyngeal tumorigenesis [4]. Stratifying the data by gender, age or BMI did not substantially change the risk patterns, although the inverse relationship of flavonols was somewhat stronger for never and ex-smokers than current smokers [4] (Table 7.1). The lack of association with consumption of isoflavones could be caused by low intake of soy foods as the major dietary sources of isoflavones among Italian population [4]. These trends were partly confirmed by a prospective cohort study with American adults (the NIH-AARP Diet and Healthy Study) showing that the high intake of flavanones was associated with a 42% reduced risk of malignant neoplasms in the oropharynx [8]. Moreover, high dietary consumption of anthocyanidins was also found to be related to the lower risk of carcinogenesis in the oral cavity in this large prospective study [8] (Table 7.2). It is clear that further research is needed about the possible relation between intake of flavonoids and the risk of cancer development in oral cavity by focusing on different flavonoid subclasses or individual flavonoids, and different populations.

7.1.2. Esophageal Cancer

Diet may be an important factor in the etiology of esophageal premalignant conditions but also initiation and progression of neoplasms [16, 17]. Whereas esophageal squamous cell carcinoma (ESCC) is located mainly in the upper third and middle of the esophagus, most esophageal adenocarcinoma (EAC) cases develop in the lower third of the esophagus near the gastroesophageal junction possessing different risk factors, including smoking, obesity and Barrett`s esophagus [8, 16].

High dietary consumption of total flavonoids was shown to be associated with a 60% reduction in esophageal cancer risk (unspecified histology) in Uruguayan population [18]. However, in a case-control study performed with American men, no significant association between intake of total flavonoids and susceptibility to carcinogenesis was described either for EAC or ESCC, both for white and black subjects [19]. Similar results of no consistent association were reported also in another multicenter study with American women and men [6]. Total flavonoids revealed no protective effects against ESCC also in a case-control study conducted with Italian population [3] (Table 7.1).

No association was found for high intake of flavones and EAC in American case-control studies [6, 19]. However, whereas no meaningful relationship was described between flavone intake and the risk of ESCC among American men [19] or Italian women and men [3], a 45% decreased risk of ESCC was still reported in an American multicenter case-control study performed with participants of both genders [6]. Regardless of histological subtypes, no significant association was observed between dietary consumption of flavonols and the risk of esophageal cancer for American participants [6, 19]; although a borderline significant inverse relationship was detected in the Italian study with ESCC, particularly for current tobacco smokers and non- or moderate alcohol drinkers [3]. At that, kaempferol revealed a statistically insignificant 33% reduction in the risk of ESCC [3]. These findings show that flavonols could avoid damage to the esophageal mucosa, including injury of oxidative stress caused by tobacco smoking. No consistent association was found between high dietary intake of flavanols and esophageal cancer, either EAC or ESCC, in American [6, 19] or Italian population [3] (Table 7.1).

Table 7.1. Epidemiological case-control studies on dietary intake of flavonoids and the risk of cancers in upper gastrointestinal tract

Flavonoid subclass	Certain compound	Cancer subsite, subtype[1]	Population	Gender[2]	Controls[3]	Cases/ controls	Intake comparison (low vs high, mg/day)[4]	Multi-variate-adjusted OR[5]	p for trend[6]	Comments[7]	Ref.
ORAL CAVITY / PHARYNX											
Total flavonoids			Italian	F, M	HB	805/ 2081	83.5 vs 204.0 (upper cutpoint) (Q5)	0.56 (0.40-0.78)	0.001*	19-79 y	[4]
Total flavonoids			Uruguayan	F, M	HB	33/ 393	(T3)	0.8 (0.5-1.4)			[18]
Anthocyanidins			Italian	F, M	HB	805/ 2081	5.3 vs 33.2 (upper cutpoint) (Q5)	0.86 (0.60-1.22)	0.13	19-79 y	[4]
Flavanols			Italian	F, M	HB	805/ 2081	23.3 vs 99.6 (upper cutpoint) (Q5)	0.84 (0.60-1.18)	0.15	19-79 y	[4]
Flavanones			Italian	F, M	HB	805/ 2081	10.2 vs 67.0 (upper cutpoint) (Q5)	0.51 (0.37-0.71)	<0.001*	19-79 y; risk pattern consistent of covariates age (<60 or ≥60 y), BMI (<20 or ≥20), smoking (never/ex-smokers or current smokers)	[4]
				M	HB	659/ 1302	(Q5)	0.81 (0.72-0.92)	*	19-79 y	
				F	HB	146/ 779	(Q5)	0.65 (0.51-0.83)	*	19-79 y	
Flavones			Italian	F, M	HB	805/ 2081	0.30 vs 0.67 (upper cutpoint) (Q5)	0.75 (0.55-1.04)	0.073	19-79 y	[4]

Flavonoid subclass	Certain compound	Cancer subsite, subtype[1]	Population	Gender[2]	Controls[3]	Cases/ controls	Intake comparison (low vs high, mg/day)[4]	Multi-variate-adjusted OR[5]	p for trend[6]	Comments[7]	Ref.
Flavonols			Italian	F, M	HB	805/ 2081	13.9 vs 29.9 (upper cutpoint) (Q5)	0.62 (0.43-0.89)	0.002*	19-79 y; risk pattern consistent of covariates age (<60 or ≥60 y), BMI (<20 or ≥20), stronger inverse association for never or ex-smokers than current smokers	[4]
				M	HB	659/ 1302	(Q5)	0.85 (0.74-0.97)	*	19-79 y	
				F	HB	146/ 779	(Q5)	0.68 (0.53-0.87)	*	19-79 y	
Isoflavones			Italian	F, M	HB	805/ 2081	0.0147 vs 0.0325 (upper cutpoint) (Q5)	0.90 (0.64-1.26)	0.34	19-79 y	[4]
ESOPHAGUS											
Total flavonoids			Uruguayan	F, M	HB	66/ 393	(T3)	0.4 (0.3-0.6)	*		[18]
Total flavonoids		EAC	American	F, M	PB	274/ 662	<63.81 vs ≥217.36 (Q4)	0.92 (0.63-1.37)	0.61	30-79 y	[6]
		ESCC	American	F, M	PB	191/ 662	<63.81 vs ≥217.36 (Q4)	0.87 (0.53-1.41)	0.42	30-79 y	

Table 7.1. (Continued)

Flavonoid subclass	Certain compound	Cancer subsite, subtype[1]	Population	Gender[2]	Controls[3]	Cases/ controls	Intake comparison (low vs high, mg/day)[4]	Multi-variate-adjusted OR[5]	p for trend[6]	Comments[7]	Ref.
Total flavonoids		EAC	American	M	PB	161/ 678	<43.0 vs >107 /1000 kcal (Q4)	0.71 (0.36-1.42)	0.42	30-79 y; white men	[19]
		ESCC	American	M	PB	114/ 678	<43.0 vs >107 /1000 kcal (Q4)	1.19 (0.50-2.81)	0.50	30-79 y; white men; no significant association also for black men	
Total flavonoids		ESCC	Italian	F, M	HB	304/ 743	96.5 vs 217.4 (upper cutpoint) (Q5)	0.99 (0.55-1.79)	0.634	36-77 y	[3]
Anthocyanidins		EAC	American	F, M	PB	274/ 662	<7.21 vs ≥18.48 (Q4)	0.43 (0.29-0.66)	0.06	30-79 y	[6]
		ESCC	American	F, M	PB	191/ 662	<7.21 vs ≥18.48 (Q4)	0.43 (0.26-0.70)	0.11	30-79 y	
Anthocyanidins		EAC	American	M	PB	161/ 678	<1.45 vs >4.76 /1000 kcal (Q4)	0.47 (0.24-0.91)	0.04*	30-79 y; white men	[19]
		ESCC	American	M	PB	114/ 678	<1.45 vs >4.76 /1000 kcal (Q4)	0.73 (0.32-1.67)	0.48	30-79 y; white men; no significant association also for black men	
Anthocyanidins		ESCC	Italian	F, M	HB	304/ 743	8.1 vs 41.2 (upper cutpoint) (Q5)	0.84 (0.46-1.54)	0.502	36-77 y	[3]
Flavanols		EAC	American	F, M	PB	274/ 662	<10.29 vs ≥130.70 (Q4)	1.02 (0.69-1.51)	0.32	30-79 y	[6]
		ESCC	American	F, M	PB	191/ 662	<10.29 vs ≥130.70 (Q4)	0.98 (0.60-1.59)	0.68	30-79 y	

Flavonoid subclass	Certain compound	Cancer subsite, subtype[1]	Population	Gender[2]	Controls[3]	Cases/ controls	Intake comparison (low vs high, mg/day)[4]	Multi-variate-adjusted OR[5]	p for trend[6]	Comments[7]	Ref.
Flavanols		EAC	American	M	PB	161/ 678	<10.3 vs >60.6 /1000 kcal (Q4)	1.22 (0.60-2.49)	0.69	30-79 y; white men	[19]
		ESCC	American	M	PB	114/ 678	<10.3 vs >60.6 /1000 kcal (Q4)	0.95 (0.36-2.52)	0.82	30-79 y; white men; no significant association also for black men	
Flavanols		ESCC	Italian	F, M	HB	304/ 743	32.6 vs 109.1 (upper cutpoint) (Q5)	1.06 (0.58-1.94)	0.773	36-77 y; no significant effect modification by smoking or drinking status	[3]
Flavanones		EAC	American	F, M	PB	274/ 662	<11.57 vs ≥49.53 (Q4)	0.56 (0.37-0.85)	0.003*	30-79 y	[6]
		ESCC	American	F, M	PB	191/ 662	<11.57 vs ≥49.53 (Q4)	0.48 (0.29-0.78)	0.002*	30-79 y	
Flavanones		EAC	American	M	PB	161/ 678	<9.30 vs >26.2 /1000 kcal (Q4)	0.99 (0.56-1.75)	0.94	30-79 y; white men	[19]
		ESCC	American	M	PB	114/ 678	<9.30 vs >26.2 /1000 kcal (Q4)	0.94 (0.47-1.90)	0.98	30-79 y; white men; inverse association for black men [0.57 (0.30-1.08), p 0.08]	
Flavanones		ESCC	Italian	F, M	HB	304/ 743	6.4 vs 58.8 (upper cutpoint) (Q5)	0.38 (0.23-0.66)	0.004*	36-77 y; stronger inverse relation for heavy drinkers	[3]

Table 7.1. (Continued)

Flavonoid subclass	Certain compound	Cancer subsite, subtype[1]	Population	Gender[2]	Controls[3]	Cases/ controls	Intake comparison (low vs high, mg/day)[4]	Multi-variate-adjusted OR[5]	p for trend[6]	Comments[7]	Ref.
Flavones		EAC	American	F, M	PB	274/ 662	<1.29 vs ≥2.63 (Q4)	0.84 (0.56-1.25)	0.81	30-79 y	[6]
		ESCC	American	F, M	PB	191/ 662	<1.29 vs ≥2.63 (Q4)	0.55 (0.34-0.89)	0.02*	30-79 y	
Flavones		EAC	American	M	PB	161/ 678	<2.08 vs >4.41 /1000 kcal (Q4)	0.81 (0.43-1.51)	0.53	30-79 y; white men	[19]
		ESCC	American	M	PB	114/ 678	<2.08 vs >4.41 /1000 kcal (Q4)	0.79 (0.36-1.73)	0.50	30-79 y; white men; no significant association also for black men	
Flavones		ESCC	Italian	F, M	HB	304/ 743	0.3 vs 0.7 (upper cutpoint) (Q5)	0.97 (0.57-1.67)	0.877	36-77 y; no significant effect modification by smoking status	[3]
Flavonols		EAC	American	F, M	PB	274/ 662	<8.31 vs ≥17.82 (Q4)	0.80 (0.54-1.18)	0.71	30-79 y	[6]
		ESCC	American	F, M	PB	191/ 662	<8.31 vs ≥17.82 (Q4)	0.97 (0.62-1.53)	0.70	30-79 y	
Flavonols		EAC	American	M	PB	161/ 678	<6.89 vs >15.9 /1000 kcal (Q4)	0.98 (0.47-2.01)	0.79	30-79 y; white men	[19]
		ESCC	American	M	PB	114/ 678	<6.89 vs >15.9 /1000 kcal (Q4)	1.09 (0.41-2.87)	0.50	30-79 y; white men; no significant association also for black men	

Flavonoid subclass	Certain compound	Cancer subsite, subtype[1]	Population	Gender[2]	Controls[3]	Cases/ controls	Intake comparison (low vs high, mg/day)[4]	Multi-variate-adjusted OR[5]	p for trend[6]	Comments[7]	Ref.
Flavonols		ESCC	Italian	F, M	HB	304/ 743	15.9 vs 31.9 (upper cutpoint) (Q5)	0.68 (0.38-1.24)	0.060	36-77 y; insignificant inverse relation for current smokers and non-/moderate drinkers	[3]
Flavonols	Kaempferol	ESCC	Italian	F, M	HB	304/ 743	(Q5)	0.67	0.052	36-77 y	[3]
Isoflavones			Chinese	F, M	HB	359/ 380	<8.0 vs >26.0 (T3)	0.37 (0.25-0.55)	<0.001*	Inverse association for Han and Uyghur minority adults	[21]
Isoflavones		EAC	American	F, M	PB	274/ 662	<0.27 vs ≥0.60 (Q4)	1.65 (1.02-2.65)	0.07	30-79 y	[6]
		ESCC	American	F, M	PB	191/ 662	<0.27 vs ≥0.60 (Q4)	0.72 (0.40-1.29)	0.11	30-79 y	
Isoflavones		EAC	American	M	PB	161/ 678	<0.005 vs >0.019 /1000 kcal (Q4)	0.65 (0.36-1.18)	0.20	30-79 y; white men	[19]
		ESCC	American	M	PB	114/ 678	<0.005 vs >0.019 /1000 kcal (Q4)	0.43 (0.20-0.93)	0.01*	30-79 y; white men; no inverse relation for black men	
Isoflavones	Daidzein		Chinese	F, M	HB	359/ 380	<3.6 vs >11.7 (T3)	0.38 (0.26-0.57)	0.001*	Inverse association for Han and Uyghur minority adults	[21]
Isoflavones	Genistein		Chinese	F, M	HB	359/ 380	<4.0 vs >13.0 (T3)	0.36 (0.25-0.54)	<0.001*	Inverse association for Han and Uyghur minority adults	[21]

Table 7.1. (Continued)

Flavonoid subclass	Certain compound	Cancer subsite, subtype[1]	Population	Gender[2]	Controls[3]	Cases/ controls	Intake comparison (low vs high, mg/day)[4]	Multi-variate-adjusted OR[5]	p for trend[6]	Comments[7]	Ref.
Isoflavones	Glycitein		Chinese	F, M	HB	359/ 380	<0.4 vs >1.1 (T3)	0.43 (0.29-0.62)	<0.001*	Inverse association for Han and Uyghur minority adults	[21]
STOMACH											
Total flavonoids			Spanish	F, M	HB	354/ 354	(Q4)	0.44 (0.25-0.78)	0.003*	31-88 y	[1]
Total flavonoids			Korean	F, M	HB	334/ 334	52.2 vs 152.3 (T3)	0.49 (0.31-0.76)	0.007*	35-75 y	[12]
				M	HB	208/ 208	52.2 vs 152.3 (T3)	0.70 (0.39-1.24)	0.393	35-75 y; no clear differences by *H. pylori* infection status or by smoking status	
				F	HB	126/ 126	52.2 vs 152.3 (T3)	0.33 (0.15-0.73)	0.001*	35-75 y; no clear differences by *H. pylori* infection status	
Total flavonoids		GAC, cardia	American	F, M	PB	248/ 662	<63.81 vs ≥217.36 (Q4)	1.32 (0.87-2.00)	0.07	30-79 y	[6]
		GAC, other	American	F, M	PB	341/ 662	<63.81 vs ≥217.36 (Q4)	1.08 (0.73-1.58)	0.50	30-79 y	
Anthocyanidins			Italian	F, M	HB	230/ 547	6.2 vs 21.5 (upper cutpoint) (Q5)	0.91 (0.56-1.47)	0.27	22-80 y	[5]

Flavonoid subclass	Certain compound	Cancer subsite, subtype[1]	Population	Gender[2]	Controls[3]	Cases/ controls	Intake comparison (low vs high, mg/day)[4]	Multi-variate-adjusted OR[5]	p for trend[6]	Comments[7]	Ref.
Anthocyanidins			Korean	F, M	HB	334/ 334	9.4 vs 35.1 (T3)	0.73 (0.46-1.15)	0.065	35-75 y	[12]
				M	HB	208/ 208	9.4 vs 35.1 (T3)	0.92 (0.51-1.67)	0.972	35-75 y; no clear differences by *H. pylori* infection status	
				F	HB	126/ 126	9.4 vs 35.1 (T3)	0.58 (0.27-1.25)	0.005*	35-75 y; strong inverse relation only for women negative for *H. pylori*	[12]
Anthocyanidins		GAC	Greek	F, M	HB	110/ 100		1.14 (0.72-1.80)	0.587		[2]
Anthocyanidins		GAC, cardia	American	F, M	PB	248/ 662	<7.21 vs ≥18.48 (Q4)	0.71 (0.46-1.10)	0.91	30-79 y	[6]
		GAC, other	American	F, M	PB	341/ 662	<7.21 vs ≥18.48 (Q4)	0.70 (0.47-1.03)	0.10	30-79 y	
Flavanols			Italian	F, M	HB	230/ 547	21.6 vs 79.2 (upper cutpoint) (Q5)	0.75 (0.45-1.23)	0.35	22-80 y	[5]
Flavanols			Korean	F, M	HB	334/ 334	3.2 vs 39.7 (T3)	0.58 (0.38-0.88)	0.001*	35-75 y	[12]
				M	HB	208/ 208	3.2 vs 39.7 (T3)	0.70 (0.41-1.21)	0.062	35-75 y; no clear differences by *H. pylori* infection status	

Table 7.1. (Continued)

Flavonoid subclass	Certain compound	Cancer subsite, subtype[1]	Population	Gender[2]	Controls[3]	Cases/ controls	Intake comparison (low vs high, mg/day)[4]	Multi-variate-adjusted OR[5]	p for trend[6]	Comments[7]	Ref.
				F	HB	126/ 126	3.2 vs 39.7 (T3)	0.36 (0.17-0.77)	0.004*	35-75 y; no clear differences by *H. pylori* infection status	
Flavanols		GAC	Greek	F, M	HB	110/ 100		1.04 (0.68-1.58)	0.871		[2]
Flavanols		GAC, cardia	American	F, M	PB	248/ 662	<10.29 vs ≥130.70 (Q4)	1.17 (0.77-1.78)	0.07	30-79 y	[6]
		GAC, other	American	F, M	PB	341/ 662	<10.29 vs ≥130.70 (Q4)	1.30 (0.88-1.92)	0.36	30-79 y	
Flavanones			Italian	F, M	HB	230/ 547	12.9 vs 56.8 (upper cutpoint) (Q5)	0.91 (0.54-1.51)	0.90	22-80 y	[5]
Flavanones			Korean	F, M	HB	334/ 334	0.7 vs 11.0 (T3)	0.66 (0.43-1.02)	0.032*	35-75 y	[12]
				M	HB	208/ 208	0.7 vs 11.0 (T3)	0.90 (0.52-1.56)	0.476	35-75 y; no clear differences by *H. pylori* infection status	
				F	HB	126/ 126	0.7 vs 11.0 (T3)	0.39 (0.18-0.86)	0.001*	35-75 y; no clear differences by *H. pylori* infection status	

Flavonoid subclass	Certain compound	Cancer subsite, subtype[1]	Population	Gender[2]	Controls[3]	Cases/ controls	Intake comparison (low vs high, mg/day)[4]	Multi-variate-adjusted OR[5]	p for trend[6]	Comments[7]	Ref.
Flavanones		GAC	Greek	F, M	HB	110/ 100		0.55 (0.31-0.96)	0.036*		[2]
Flavanones		GAC, cardia	American	F, M	PB	248/ 662	<11.57 vs ≥49.53 (Q4)	1.23 (0.81-1.87)	0.82	30-79 y	[6]
		GAC, other	American	F, M	PB	341/ 662	<11.57 vs ≥49.53 (Q4)	0.88 (0.60-1.28)	0.36	30-79 y	
Flavones			Italian	F, M	HB	230/ 547	0.3 vs 0.7 (upper cutpoint) (Q5)	0.83 (0.50-1.39)	0.45	22-80 y	[5]
Flavones			Korean	F, M	HB	334/ 334	0.6 vs 2.1 (T3)	0.51 (0.31-0.82)	0.004*	35-75 y	[12]
				M	HB	208/ 208	0.6 vs 2.1 (T3)	0.70 (0.38-1.29)	0.690	35-75 y; no clear differences by *H. pylori* infection status	
				F	HB	126/ 126	0.6 vs 2.1 (T3)	0.15 (0.06-0.38)	<0.0001*	35-75 y; no clear differences by *H. pylori* infection status	
Flavones		GAC	Greek	F, M	HB	110/ 100		0.70 (0.43-1.14)	0.150		[2]
Flavones		GAC, cardia	American	F, M	PB	248/ 662	<1.29 vs ≥2.63 (Q4)	1.09 (0.71-1.67)	0.15	30-79 y	[6]
		GAC, other	American	F, M	PB	341/ 662	<1.29 vs ≥2.63 (Q4)	1.01 (0.69-1.50)	0.98	30-79 y	

Table 7.1. (Continued)

Flavonoid subclass	Certain compound	Cancer subsite, subtype[1]	Population	Gender[2]	Controls[3]	Cases/ controls	Intake comparison (low vs high, mg/day)[4]	Multi-variate-adjusted OR[5]	p for trend[6]	Comments[7]	Ref.
Flavonols			Italian	F, M	HB	230/ 547	13.2 vs 32.3 (upper cutpoint) (Q5)	0.62 (0.38-1.02)	0.021*	22-80 y	[5]
Flavonols			Korean	F, M	HB	334/ 334	10.9 vs 30.8 (T3)	0.51 (0.32-0.82)	0.014*	35-75 y	[12]
				M	HB	208/ 208	10.9 vs 30.8 (T3)	0.59 (0.32-1.10)	0.264	35-75 y; no clear differences by *H. pylori* infection status	
				F	HB	126/ 126	10.9 vs 30.8 (T3)	0.51 (0.24-1.10)	0.017*	35-75 y; no clear differences by *H. pylori* infection status	
Flavonols		GAC	Greek	F, M	HB	110/ 100		0.77 (0.42-1.40)	0.392		[2]
Flavonols		GAC, cardia	American	F, M	PB	248/ 662	<8.31 vs ≥17.82 (Q4)	1.42 (0.93-2.17)	0.10	30-79 y	[6]
		GAC, other	American	F, M	PB	341/ 662	<8.31 vs ≥17.82 (Q4)	0.98 (0.67-1.46)	0.92	30-79 y	
Flavonols	Kaempferol		Italian	F, M	HB	230/ 547	(Q5)	0.61 (0.36-1.01)		22-80 y	[5]
Flavonols	Kaempferol		Spanish	F, M	HB	354/ 354	(Q4)	0.48 (0.26-0.88)	0.04*	31-88 y	[1]

Flavonoid subclass	Certain compound	Cancer subsite, subtype[1]	Population	Gender[2]	Controls[3]	Cases/ controls	Intake comparison (low vs high, mg/day)[4]	Multi-variate-adjusted OR[5]	p for trend[6]	Comments[7]	Ref.
Flavonols	Myricetin		Italian	F, M	HB	230/ 547	(Q5)	0.86 (0.53-1.40)		22-80 y	[5]
Flavonols	Myricetin		Spanish	F, M	HB	354/ 354	(Q4)	1.12 (0.67-1.85)	0.45	31-88 y	[1]
Flavonols	Quercetin		Spanish	F, M	HB	354/ 354	(Q4)	0.62 (0.35-1.10)	0.02*	31-88 y	[1]
Flavonols	Quercetin	GAC, cardia	Swedish	F, M	PB	81/ 1116	0.16-3.88 vs ≥11.89 (Q5)	0.76 (0.40-1.44)	0.35	40-79 y	[24]
		GAC, non-cardia	Swedish	F, M	PB	420/ 1116	0.16-3.88 vs ≥11.89 (Q5)	0.57 (0.40-0.83)	<0.001*	40-79 y	
				M	PB	665/ 347	0.16-3.88 vs 11.89-50.29 (Q5)	0.66 (0.42-1.04)		40-79 y	
				F	PB	155/ 359	0.16-3.88 vs 11.89-50.29 (Q5)	0.49 (0.25-0.94)		40-79 y; strong inverse relation for smokers [0.19 (0.60-0.56)*], not non-smokers	
		GAC, intestinal non-cardia	Swedish	F, M	PB	230/ 1116	0.16-3.88 vs ≥11.89 (Q5)	0.51 (0.32-0.82)	<0.001*	40-79 y	
		GAC, diffuse non-cardia	Swedish	F, M	PB	153/ 1116	0.16-3.88 vs ≥11.89 (Q5)	0.54 (0.31-0.92)	0.008*	40-79 y	
Isoflavones			Italian	F, M	HB	230/ 547	0.015 vs 0.0343 (upper cutpoint) (Q5)	0.88 (0.53-1.43)	0.23	22-80 y	[5]

Table 7.1. (Continued)

Flavonoid subclass	Certain compound	Cancer subsite, subtype[1]	Population	Gender[2]	Controls[3]	Cases/ controls	Intake comparison (low vs high, mg/day)[4]	Multi-variate-adjusted OR[5]	p for trend[6]	Comments[7]	Ref.
Isoflavones			Korean	F, M	HB	334/ 334	11.4 vs 43.7 (T3)	0.72 (0.46-1.12)	0.115	35-75 y	[12]
				M	HB	208/ 208	11.4 vs 43.7 (T3)	0.90 (0.52-1.54)	0.494	35-75 y; no clear differences by *H. pylori* infection status	
				F	HB	126/ 126	11.4 vs 43.7 (T3)	0.51 (0.24-1.08)	0.059	35-75 y; no clear differences by *H. pylori* infection status	
Isoflavones		GAC	Greek	F, M	HB	110/ 100		1.16 (0.73-1.84)	0.538		[2]
Isoflavones		GAC, cardia	American	F, M	PB	248/ 662	<0.27 vs ≥0.60 (Q4)	1.56 (0.93-2.60)	0.17	30-79 y	[6]
		GAC, other	American	F, M	PB	341/ 662	<0.27 vs ≥0.60 (Q4)	1.50 (0.96-2.37)	0.37	30-79 y	

[1]EAC, esophageal adenocarcinoma; ESCC, esophageal squamous cell carcinoma; GAC, gastric adenocarcinoma.
[2]F, female; M, male.
[3]HB, hospital-based; PB, population-based.
[4]Q4, quartile; Q5, quintile; T3, tertile.
[5]OR, odds ratio.
[6]Statistically significant effects (p for trend <0.05) are marked by asterisk.
[7]BMI, body mass index; *H. pylori*, *Helicobacter pylori*.

Table 7.2. Epidemiological prospective cohort studies on dietary intake of flavonoids and the risk of cancers in upper gastrointestinal tract

Flavonoid subclass	Certain compound	Cancer subsite, subtype[1]	Study[2]	Population[3]	Median follow-up (years)	Gender[4]	Cases/ cohort	Intake comparison (low vs high, mg/day)[5]	Multivariate-adjusted RR/HR[6]	p for trend[7]	Comments[8]	Ref.
ORAL CAVITY												
Total flavonoids			NIH-AARP	American	12	F, M	176/ 469008	<84.1 vs 438.0-4211.2 (Q5)	0.79 (0.64-0.96)	0.3	50-71 y	[8]
Antho-cyanidins			NIH-AARP	American	12	F, M	184/ 469008	<3.8 vs 20.7-376.6 (Q5)	0.78 (0.63-0.96)	0.04*	50-71 y	[8]
Flavanols			NIH-AARP	American	12	F, M	178/ 469008	<25.6 vs 346.7-3828.7 (Q5)	0.86 (0.70-1.05)	0.6	50-71 y	[8]
Flavanones			NIH-AARP	American	12	F, M	200/ 469008	<8.8 vs 66.0-978.9 (Q5)	0.90 (0.74-1.10)	0.5	50-71 y	[8]
Flavones			NIH-AARP	American	12	F, M	178/ 469008	<0.44 vs 1.46-29.02 (Q5)	0.86 (0.69-1.07)	0.4	50-71 y	[8]
Flavonols			NIH-AARP	American	12	F, M	193/ 469008	<10.4 vs 30.5-188.3 (Q5)	0.81 (0.65-0.99)	0.5	50-71 y	[8]
Isoflavones			NIH-AARP	American	12	F, M	180/ 469008	<0.16 vs 0.46-5.36 (Q5)	0.88 (0.61-0.98)	0.2	50-71 y	[8]
OROPHARYNX												
Total flavonoids			NIH-AARP	American	12	F, M	60/ 469008	<84.1 vs 438.0-4211.2 (Q5)	0.69 (0.50-0.96)	0.2	50-71 y	[8]
Antho-cyanidins			NIH-AARP	American	12	F, M	75/ 469008	<3.8 vs 20.7-376.6 (Q5)	0.74 (0.54-1.02)	0.4	50-71 y	[8]
Flavanols			NIH-AARP	American	12	F, M	66/ 469008	<25.6 vs 346.7-3828.7 (Q5)	0.87 (0.63-1.20)	0.5	50-71 y	[8]
Flavanones			NIH-AARP	American	12	F, M	51/ 469008	<8.8 vs 66.0-978.9 (Q5)	0.58 (0.41-0.82)	0.03*	50-71 y	[8]
Flavones			NIH-AARP	American	12	F, M	71/ 469008	<0.44 vs 1.46-29.02 (Q5)	0.95 (0.68-1.33)	0.9	50-71 y	[8]

Table 7.2. (Continued)

Flavonoid subclass	Certain compound	Cancer subsite, subtype[1]	Study[2]	Population[3]	Median follow-up (years)	Gen-der[4]	Cases/ cohort	Intake comparison (low vs high, mg/day)[5]	Multivariate-adjusted RR/HR[6]	p for trend[7]	Comments[8]	Ref.
Flavonols			NIH-AARP	American	12	F, M	79/ 469008	<10.4 vs 30.5-188.3 (Q5)	0.78 (0.56-1.08)	0.5	50-71 y	[8]
Isoflavones			NIH-AARP	American	12	F, M	78/ 469008	<0.16 vs 0.46-5.36 (Q5)	0.61 (0.43-0.88)	0.2	50-71 y	[8]
ESOPHAGUS												
Total flavonoids			EPIC	European	11	F, M	341/ 477312	(Q4)	0.92 (0.64-1.32)	0.917	35-70 y; significant inverse relation for current smokers [0.84 (0.71-0.99)*], not never or former smokers	[16]
Total flavonoids		EAC	NIH-AARP	American	12	F, M	154/ 469008	<84.1 vs 438.0-4211.2 (Q5)	1.03 (0.82-1.29)	0.8	50-71 y	[8]
		ESCC	NIH-AARP	American	12	F, M	47/ 469008	<84.1 vs 438.0-4211.2 (Q5)	0.86 (0.58-1.27)	0.8	50-71 y	
Antho-cyanidins			EPIC	European	11	F, M	341/ 477312	(Q4)	0.87 (0.59-1.28)	0.306	35-70 y; no significant effect modification by smoking status	[16]
Antho-cyanidins		EAC	NIH-AARP	American	12	F, M	153/ 469008	<3.8 vs 20.7-376.6 (Q5)	0.89 (0.70-1.11)	0.8	50-71 y	[8]
		ESCC	NIH-AARP	American	12	F, M	51/ 469008	<3.8 vs 20.7-376.6 (Q5)	0.68 (0.46-1.00)	0.3	50-71 y	

Flavonoid subclass	Certain compound	Cancer subsite, subtype[1]	Study[2]	Population[3]	Median follow-up (years)	Gen-der[4]	Cases/ cohort	Intake comparison (low vs high, mg/day)[5]	Multivariate-adjusted RR/HR[6]	p for trend[7]	Comments[8]	Ref.
Flavanols			EPIC	European	11	F, M	341/ 477312	(Q4)	0.84 (0.57-1.25)	0.988	35-70 y; significant inverse relation for current smokers [0.89 (0.80-0.99)*], not never or former smokers	[16]
Flavanols		EAC	NIH-AARP	American	12	F, M	162/ 469008	<25.6 vs 346.7-3828.7 (Q5)	1.10 (0.88-1.37)	0.6	50-71 y	[8]
		ESCC	NIH-AARP	American	12	F, M	51/ 469008	<25.6 vs 346.7-3828.7 (Q5)	0.95 (0.65-1.39)	0.9	50-71 y	
Flavanones			EPIC	European	11	F, M	341/ 477312	(Q4)	0.95 (0.68-1.32)	0.630	35-70 y; significant inverse relation for former smokers [0.89 (0.81-0.99)*], not current or never smokers	[16]
Flavanones		EAC	NIH-AARP	American	12	F, M	162/ 469008	<8.8 vs 66.0-978.9 (Q5)	0.99 (0.79-1.24)	0.8	50-71 y	[8]
		ESCC	NIH-AARP	American	12	F, M	34/ 469008	<8.8 vs 66.0-978.9 (Q5)	0.67 (0.43-1.05)	0.09	50-71 y	
Flavones			EPIC	European	11	F, M	341/ 477312	(Q4)	0.70 (0.47-1.06)	0.103	35-70 y; no significant effect modification by smoking status	[16]

Table 7.2. (Continued)

Flavonoid subclass	Certain compound	Cancer subsite, subtype[1]	Study[2]	Population[3]	Median follow-up (years)	Gen-der[4]	Cases/ cohort	Intake comparison (low vs high, mg/day)[5]	Multivariate-adjusted RR/HR[6]	p for trend[7]	Comments[8]	Ref.
Flavones		EAC	NIH-AARP	American	12	F, M	141/ 469008	<0.44 vs 1.46-29.02 (Q5)	0.93 (0.73-1.19)	0.7	50-71 y	[8]
		ESCC	NIH-AARP	American	12	F, M	48/ 469008	<0.44 vs 1.46-29.02 (Q5)	0.80 (0.53-1.23)	0.5	50-71 y	
Flavonols			EPIC	European	11	F, M	341/ 477312	(Q4)	0.88 (0.61-1.29)	0.445	35-70 y; significant inverse relation for current smokers [0.72 (0.56-0.92)*], not never or former smokers	[16]
Flavonols		EAC	NIH-AARP	American	12	F, M	160/ 469008	<10.4 vs 30.5-188.3 (Q5)	1.10 (0.86-1.40)	0.9	50-71 y	[8]
		ESCC	NIH-AARP	American	12	F, M	57/ 469008	<10.4 vs 30.5-188.3 (Q5)	1.17 (0.76-1.80)	0.7	50-71 y	
Isoflavones			EPIC	European	11	F, M	341/ 477312	(Q4)	0.71 (0.44-1.16)	0.213	35-70 y; no significant effect modification by smoking status	[16]
Isoflavones		EAC	NIH-AARP	American	12	F, M	184/ 469008	<0.16 vs 0.46-5.36 (Q5)	0.94 (0.72-1.21)	0.9	50-71 y	[8]
		ESCC	NIH-AARP	American	12	F, M	57/ 469008	<0.16 vs 0.46-5.36 (Q5)	1.19 (0.74-1.90)	0.6	50-71 y	[8]
STOMACH												
Total flavonoids			NLCS	Dutch	4.3	F, M	183/ 3123	12.7 vs 43.5 (Q5)	0.86 (0.47-1.57)	0.535	55-69 y	[27]

Flavonoid subclass	Certain compound	Cancer subsite, subtype[1]	Study[2]	Population[3]	Median follow-up (years)	Gen-der[4]	Cases/ cohort	Intake comparison (low vs high, mg/day)[5]	Multivariate-adjusted RR/HR[6]	p for trend[7]	Comments[8]	Ref.
Total flavonoids			FMC	Finnish	24	F, M	64/ 9959	<2.1 vs >4.8 (M), <2.4 vs >5.5 (F) (Q4)	1.15 (0.48-2.78)		15-99 y	[25]
Total flavonoids			FMC	Finnish	30	F, M	74/ 9865	4.3 vs 26.9 (M), 8.5 vs 39.5 (F) (Q4)	0.87 (0.44-1.75)	0.73		[26]
Total flavonoids			EPIC	European	11	M	395/ 142250	<200.4 vs >595.5 (Q4)	0.97 (0.67-1.41)	0.995	35-70 y; no significant effect modification by smoking status or by excluding cases diagnosed less than 2 y of follow-up	[10]
						F	288/ 335062	<200.7 vs >580.6 (Q4)	0.49 (0.30-0.80)	0.003*	35-70 y; significant inverse relation for ever-smokers [0.76 (0.63-0.93)*], not never-smokers; no effect modification by exclusion cases diagnosed less than 2 y of follow-up	

Table 7.2. (Continued)

Flavonoid subclass	Certain compound	Cancer subsite, subtype[1]	Study[2]	Population[3]	Median follow-up (years)	Gender[4]	Cases/ cohort	Intake comparison (low vs high, mg/day)[5]	Multivariate-adjusted RR/HR[6]	p for trend[7]	Comments[8]	Ref.
Total flavonoids		Cardia	NIH-AARP	American	12	F, M	105/ 469008	<84.1 vs 438.0-4211.2 (Q5)	1.02 (0.78-1.34)	0.8	50-71 y	[8]
		Non-cardia	NIH-AARP	American	12	F, M	124/ 469008	<84.1 vs 438.0-4211.2 (Q5)	1.11 (0.86-1.44)	0.5	50-71 y	
Total flavonoids		Cardia	EPIC	European	11	M	145/ 142250	(Q4)	0.93 (0.78-1.12)		35-70 y	[10]
						F	56/ 335062	(Q4)	0.84 (0.64-1.11)		35-70 y	
Total flavonoids		Non-cardia	EPIC	European	11	M	159/ 142250	(Q4)	0.98 (0.81-1.18)		35-70 y	[10]
						F	164/ 335062	(Q4)	0.85 (0.70-1.03)		35-70 y	
Total flavonoids		Intestinal	EPIC	European	11	M	130/ 142250	(Q4)	0.92 (0.74-1.14)		35-70 y	[10]
						F	73/ 335062	(Q4)	0.70 (0.52-0.94)	*	35-70 y	
Total flavonoids		Diffuse	EPIC	European	11	M	100/ 142250	(Q4)	1.13 (0.88-1.45)		35-70 y	[10]
						F	117/ 335062	(Q4)	0.94 (0.74-1.19)		35-70 y	
Anthocyanidins			EPIC	European	11	M	395/ 142250	<11.7 vs >32.8 (Q4)	0.98 (0.68-1.41)	0.680	35-70 y; no significant effect modification by smoking status	[10]

Flavonoid subclass	Certain compound	Cancer subsite, subtype[1]	Study[2]	Population[3]	Median follow-up (years)	Gen-der[4]	Cases/ cohort	Intake comparison (low vs high, mg/day)[5]	Multivariate-adjusted RR/HR[6]	p for trend[7]	Comments[8]	Ref.
						F	288/ 335062	<13.9 vs >38.7 (Q4)	0.71 (0.44-1.16)	0.184	35-70 y; no significant effect modification by smoking status	
Anthocyanidins		Cardia	NIH-AARP	American	12	F, M	111/ 469008	<3.8 vs 20.7-376.6 (Q5)	1.05 (0.80-1.39)	0.9	50-71 y	[8]
		Non-cardia	NIH-AARP	American	12	F, M	115/ 469008	<3.8 vs 20.7-376.6 (Q5)	0.94 (0.72-1.23)	0.9	50-71 y	
Anthocyanidins		Cardia	EPIC	European	11	M	145/ 142250	(Q4)	1.04 (0.87-1.23)		35-70 y	[10]
						F	56/ 335062	(Q4)	0.89 (0.69-1.15)		35-70 y	
Anthocyanidins		Non-cardia	EPIC	European	11	M	159/ 142250	(Q4)	1.00 (0.87-1.16)		35-70 y	[10]
						F	164/ 335062	(Q4)	0.90 (0.79-1.04)		35-70 y	
Anthocyanidins		Intestinal	EPIC	European	11	M	130/ 142250	(Q4)	0.96 (0.81-1.13)		35-70 y	[10]
						F	73/ 335062	(Q4)	0.92 (0.73-1.16)		35-70 y	
Anthocyanidins		Diffuse	EPIC	European	11	M	100/ 142250	(Q4)	1.29 (1.04-1.60)		35-70 y	[10]
						F	117/ 335062	(Q4)	0.86 (0.75-0.99)		35-70 y	

Table 7.2. (Continued)

Flavonoid subclass	Certain compound	Cancer subsite, subtype[1]	Study[2]	Population[3]	Median follow-up (years)	Gen-der[4]	Cases/ cohort	Intake comparison (low vs high, mg/day)[5]	Multivariate-adjusted RR/HR[6]	p for trend[7]	Comments[8]	Ref.
Flavanols			EPIC	European	11	M	395/ 142250	<26.1 vs >199.9 (Q4)	0.98 (0.68-1.40)	0.721	35-70 y; no significant effect modification by smoking status	[10]
						F	288/ 335062	<21.1 vs >241.6 (Q4)	0.55 (0.34-0.88)	0.035*	35-70 y; significant inverse relation for ever-smokers [0.82 (0.70-0.97)*], not never-smokers	
Flavanols		Cardia	NIH-AARP	American	12	F, M	108/ 469008	<25.6 vs 346.7-3828.7 (Q5)	1.04 (0.80-1.36)	0.8	50-71 y	[8]
		Non-cardia	NIH-AARP	American	12	F, M	129/ 469008	<25.6 vs 346.7-3828.7 (Q5)	1.19 (0.92-1.54)	0.4	50-71 y	
Flavanols		Cardia	EPIC	European	11	M	145/ 142250	(Q4)	0.93 (0.84-1.04)		35-70 y	[10]
						F	56/ 335062	(Q4)	0.91 (0.79-1.05)		35-70 y	
Flavanols		Non-cardia	EPIC	European	11	M	159/ 142250	(Q4)	1.04 (0.94-1.16)		35-70 y	[10]
						F	164/ 335062	(Q4)	0.93 (0.84-1.02)		35-70 y	
Flavanols		Intestinal	EPIC	European	11	M	130/ 142250	(Q4)	1.01 (0.89-1.14)		35-70 y	[10]
						F	73/ 335062	(Q4)	0.83 (0.72-0.95)	*	35-70 y	

Flavonoid subclass	Certain compound	Cancer subsite, subtype[1]	Study[2]	Population[3]	Median follow-up (years)	Gen-der[4]	Cases/ cohort	Intake comparison (low vs high, mg/day)[5]	Multivariate-adjusted RR/HR[6]	p for trend[7]	Comments[8]	Ref.
Flavanols		Diffuse	EPIC	European	11	M	100/ 142250	(Q4)	1.04 (0.90-1.19)		35-70 y	[10]
						F	117/ 335062	(Q4)	1.01 (0.89-1.14)		35-70 y	
Flavanones			EPIC	European	11	M	395/ 142250	<6.1 vs >29.2 (Q4)	0.91 (0.64-1.28)	0.296	35-70 y; significant inverse relation for ever-smokers [0.91 (0.86-0.97)*], not never-smokers	[10]
						F	288/ 335062	<7.5 vs >29.3 (Q4)	1.01 (0.68-1.50)	0.802	35-70 y; no significant effect modification by smoking status	
Flavanones		Cardia	NIH-AARP	American	12	F, M	117/ 469008	<8.8 vs 66.0-978.9 (Q5)	0.87 (0.68-1.13)	0.8	50-71 y	[8]
		Non-cardia	NIH-AARP	American	12	F, M	119/ 469008	<8.8 vs 66.0-978.9 (Q5)	0.99 (0.76-1.30)	0.9	50-71 y	
Flavanones		Cardia	EPIC	European	11	M	145/ 142250	(Q4)	0.90 (0.83-0.97)	*	35-70 y	[10]
						F	56/ 335062	(Q4)	0.92 (0.81-1.05)		35-70 y	
Flavanones		Non-cardia	EPIC	European	11	M	159/ 142250	(Q4)	0.97 (0.89-1.06)		35-70 y	[10]
						F	164/ 335062	(Q4)	0.99 (0.90-1.09)		35-70 y	
Flavanones		Intestinal	EPIC	European	11	M	130/ 142250	(Q4)	0.93 (0.85-1.02)		35-70 y	[10]

Table 7.2. (Continued)

Flavonoid subclass	Certain compound	Cancer subsite, subtype[1]	Study[2]	Population[3]	Median follow-up (years)	Gen-der[4]	Cases/ cohort	Intake comparison (low vs high, mg/day)[5]	Multivariate-adjusted RR/HR[6]	p for trend[7]	Comments[8]	Ref.
						F	73/ 335062	(Q4)	1.10 (0.92-1.32)		35-70 y	[10]
Flavanones		Diffuse	EPIC	European	11	M	100/ 142250	(Q4)	0.97 (0.85-1.09)		35-70 y	[10]
						F	117/ 335062	(Q4)	0.96 (0.87-1.07)		35-70 y	
Flavanones	Hesperetin		FMC	Finnish	30	F, M	74/ 9865	0 vs 15.4 (M), 3.2 vs 26.8 (F) (Q4)	0.88 (0.43-1.80)	0.67		[26]
Flavanones	Naringenin		FMC	Finnish	30	F, M	74/ 9865	0 vs 4.7 (M), 0.9 vs 7.7 (F) (Q4)	0.94 (0.47-1.88)	0.78		[26]
Flavones			EPIC	European	11	M	395/ 142250	<1.2 vs >4.6 (Q4)	0.86 (0.60-1.23)	0.473	35-70 y; significant inverse relation for ever-smokers [0.90 (0.82-0.98)*], not never-smokers	[10]
						F	288/ 335062	<1.3 vs >4.3 (Q4)	0.59 (0.38-0.93)	0.118	35-70 y; significant inverse relation for ever-smokers [0.82 (0.71-0.94)*], not never-smokers	

Flavonoid subclass	Certain compound	Cancer subsite, subtype[1]	Study[2]	Population[3]	Median follow-up (years)	Gen-der[4]	Cases/ cohort	Intake comparison (low vs high, mg/day)[5]	Multivariate-adjusted RR/HR[6]	p for trend[7]	Comments[8]	Ref.
Flavones		Cardia	NIH-AARP	American	12	F, M	88/ 469008	<0.44 vs 1.46-29.02 (Q5)	0.99 (0.73-1.34)	0.9	50-71 y	[8]
		Non-cardia	NIH-AARP	American	12	F, M	113/ 469008	<0.44 vs 1.46-29.02 (Q5)	1.06 (0.80-1.40)	0.8	50-71 y	[8]
Flavones		Cardia	EPIC	European	11	M	145/ 142250	(Q4)	0.90 (0.79-1.03)		35-70 y	[10]
						F	56/ 335062	(Q4)	0.83 (0.65-1.07)		35-70 y	
Flavones		Non-cardia	EPIC	European	11	M	159/ 142250	(Q4)	0.94 (0.84-1.06)		35-70 y	[10]
						F	164/ 335062	(Q4)	0.94 (0.82-1.07)		35-70 y	
Flavones		Intestinal	EPIC	European	11	M	130/ 142250	(Q4)	0.96 (0.83-1.10)		35-70 y	[10]
						F	73/ 335062	(Q4)	0.97 (0.76-1.24)		35-70 y	
Flavones		Diffuse	EPIC	European	11	M	100/ 142250	(Q4)	0.99 (0.85-1.16)		35-70 y	[10]
						F	117/ 335062	(Q4)	0.92 (0.78-1.09)		35-70 y	
Flavonols			EPIC	European	11	M	395/ 142250	<15.0 vs >33.7 (Q4)	0.93 (0.63-1.37)	0.577	35-70 y; no significant effect modification by smoking status	[10]

Table 7.2. (Continued)

Flavonoid subclass	Certain compound	Cancer subsite, subtype[1]	Study[2]	Population[3]	Median follow-up (years)	Gender[4]	Cases/ cohort	Intake comparison (low vs high, mg/day)[5]	Multivariate-adjusted RR/HR[6]	p for trend[7]	Comments[8]	Ref.
						F	288/ 335062	<13.9 vs >35.2 (Q4)	0.45 (0.27-0.75)	0.005*	35-70 y; significant inverse relation for ever-smokers [0.75 (0.58-0.97)*], not never-smokers	
Flavonols		Cardia	NIH-AARP	American	12	F, M	104/ 469008	<10.4 vs 30.5-188.3 (Q5)	1.08 (0.80-1.45)	0.8	50-71 y	[8]
		Non-cardia	NIH-AARP	American	12	F, M	126/ 469008	<10.4 vs 30.5-188.3 (Q5)	1.25 (0.94-1.65)	0.4	50-71 y	
Flavonols		Cardia	EPIC	European	11	M	145/ 142250	(Q4)	0.91 (0.71-1.17)		35-70 y	[10]
						F	56/ 335062	(Q4)	0.85 (0.60-1.20)		35-70 y	
Flavonols		Non-cardia	EPIC	European	11	M	159/ 142250	(Q4)	0.98 (0.76-1.26)		35-70 y	[10]
						F	164/ 335062	(Q4)	0.90 (0.71-1.13)		35-70 y	
Flavonols		Intestinal	EPIC	European	11	M	130/ 142250	(Q4)	0.95 (0.71-1.28)		35-70 y	[10]
						F	73/ 335062	(Q4)	0.72 (0.49-1.06)		35-70 y	

Flavonoid subclass	Certain compound	Cancer subsite, subtype[1]	Study[2]	Population[3]	Median follow-up (years)	Gen-der[4]	Cases/ cohort	Intake comparison (low vs high, mg/day)[5]	Multivariate-adjusted RR/HR[6]	p for trend[7]	Comments[8]	Ref.
Flavonols		Diffuse	EPIC	European	11	M	100/ 142250	(Q4)	0.97 (0.69-1.37)		35-70 y	[10]
						F	117/ 335062	(Q4)	1.04 (0.78-1.40)		35-70 y	
Flavonols	Kaempferol		NLCS	Dutch	4.3	F, M	183/ 3123	2.4 vs 12.4 (Q5)	0.84 (0.49-1.45)	0.337	55-69 y	[27]
Flavonols	Kaempferol		FMC	Finnish	30	F, M	74/ 9865	0.1 vs 0.8 (M), 0.2 vs 0.9 (F) (Q4)	1.14 (0.59-2.22)	0.98		[26]
Flavonols	Myricetin		FMC	Finnish	30	F, M	74/ 9865	0 vs 0.11 (M), 0.03 vs 0.20 (F) (Q4)	1.16 (0.59-2.26)	0.29		[26]
Flavonols	Quercetin		NLCS	Dutch	4.3	F, M	183/ 3123	8.4 vs 30.3 (Q5)	1.08 (0.56-2.05)	0.972	55-69 y	[27]
Flavonols	Quercetin		FMC	Finnish	30	F, M	74/ 9865	1.5 vs 3.9 (M), 1.8 vs 4.7 (F) (Q4)	1.03 (0.52-2.07)	0.82		[26]
Isoflavones			EPIC	European	11	M	395/ 142250	<0.3 vs >1.1 (Q4)	0.77 (0.50-1.18)	0.388	35-70 y; no significant effect modification by smoking status	[10]
						F	288/ 335062	<0.3 vs >1.1 (Q4)	1.05 (0.61-1.82)	0.697	35-70 y; no significant effect modification by smoking status	

Table 7.2. (Continued)

Flavonoid subclass	Certain compound	Cancer subsite, subtype[1]	Study[2]	Population[3]	Median follow-up (years)	Gender[4]	Cases/ cohort	Intake comparison (low vs high, mg/day)[5]	Multivariate-adjusted RR/HR[6]	p for trend[7]	Comments[8]	Ref.
Isoflavones			TS	Japanese	15.5	M	441/ 14219	17.6 vs 75.5 (Q4)	0.81 (0.60-1.09)	0.098	≥35 y	[28]
						F	237/ 16573	20.1 vs 72.6 (Q4)	0.60 (0.37-0.98)	0.001*	≥35 y	
Isoflavones		Cardia	NIH-AARP	American	12	F, M	128/ 469008	<0.16 vs 0.46-5.36 (Q5)	0.99 (0.73-1.34)	0.7	50-71 y	[8]
		Non-cardia	NIH-AARP	American	12	F, M	114/ 469008	<0.16 vs 0.46-5.36 (Q5)	0.73 (0.54-0.98)	0.4	50-71 y	
Isoflavones		Cardia	EPIC	European	11	M	145/ 142250	(Q4)	0.94 (0.78-1.14)		35-70 y	[10]
						F	56/ 335062	(Q4)	1.13 (0.85-1.50)		35-70 y	
Isoflavones		Non-cardia	EPIC	European	11	M	159/ 142250	(Q4)	0.91 (0.76-1.09)		35-70 y	[10]
						F	164/ 335062	(Q4)	1.00 (0.83-1.19)		35-70 y	
Isoflavones		Intestinal	EPIC	European	11	M	130/ 142250	(Q4)	0.86 (0.70-1.07)		35-70 y	[10]
						F	73/ 335062	(Q4)	1.08 (0.80-1.47)		35-70 y	
Isoflavones		Diffuse	EPIC	European	11	M	100/ 142250	(Q4)	0.95 (0.75-1.20)		35-70 y	[10]
						F	117/ 335062	(Q4)	0.90 (0.71-1.13)		35-70 y	

Flavonoid subclass	Certain compound	Cancer subsite, subtype[1]	Study[2]	Population[3]	Median follow-up (years)	Gen-der[4]	Cases/ cohort	Intake comparison (low vs high, mg/day)[5]	Multivariate-adjusted RR/HR[6]	p for trend[7]	Comments[8]	Ref.
Isoflavones	Genistein		JPHC	Japanese	5	M	899/ 39569	9.2 vs 42.3 (Q4)	1.00 (0.81-1.24)	0.96	40-69 y	[29]
						F	350/ 45312	9.4 vs 41.8 (Q4)	1.07 (0.77-1.50)	0.6	40-69 y; significant positive relation for ever users of EFH [2.80 (0.93-8.39), p-trend 0.03*], not never users	
Isoflavones	Genistein	Upper third, incl. cardia	JPHC	Japanese	5	M	104/ 39569	9.2 vs 42.3 (Q4)	2.00 (0.97-4.12)	0.2	40-69 y	[29]
						F	33/ 45312	9.4 vs 41.8 (Q4)	1.43 (0.52-3.95)	0.4	40-69 y	
Isoflavones	Genistein	Distal	JPHC	Japanese	5	M	609/ 39569	9.2 vs 42.3 (Q4)	0.97 (0.74-1.26)	0.9	40-69 y	[29]
						F	237/ 45312	9.4 vs 41.8 (Q4)	1.00 (0.66-1.53)	0.9	40-69 y	

[1]EAC, esophageal adenocarcinoma; ESCC, esophageal squamous cell carcinoma.

[2]EPIC, The European Prospective Investigation into Cancer and Nutrition Cohort; FMC, The Finnish Mobile Clinic Health Examination Survey; JPHC, The Japan Public Health Center-based prospective study; NIH-AARP, the prospective cohort study consisting of AARP (the American Association of Retired Persons) members; NLCS, The Netherlands Cohort Study; TS, The Takayama Study.

[3]European, participants from ten European countries.

[4]F, female; M, male.

[5]Q4, quartile; Q5, quintile.

[6]HR, hazard ratio; RR, relative risk.

[7]Statistically significant effects (p for trend <0.05) are marked by asterisk.

[8]EFH, exogenous female hormones; age at baseline of the study.

Table 7.3. Epidemiological studies on biomarkers of flavonoids and the risk of cancers in upper gastrointestinal tract

Flavonoid subclass	Certain compound	Cancer subsite, subtype	Bio-marker	Study[1]	Popula-tion	Gen-der[2]	Cases/ controls	Multivariate-adjusted OR[3]	p for trend[4]	Comments[5]	Ref.
ESOPHAGUS											
Flavanols	Epicatechin (EC)		Urinary	SCS	Chinese	M	42/209	0.99 (0.32-3.04)	0.91	45-64 y	[23]
Flavanols	Epigallocatechin (EGC)		Urinary	SCS	Chinese	M	42/209	0.91 (0.36-2.31)	0.86	45-64 y; significant inverse relation after exclusion cases diagnosed less than 4 y of follow-up [0.58 (0.19-1.79)*]	[23]
STOMACH											
Flavanols	Epicatechin (EC)		Plasma	JPHC	Japanese	M	331/331	1.35 (0.80-2.30)	0.18	40-69 y	[30]
						F	163/163	0.78 (0.37-1.62)	0.47	40-69 y	
Flavanols	Epicatechin (EC)		Urinary	SCS	Chinese	M	190/563	1.78 (0.92-3.44)	0.17	45-64 y; OR close to unity after exclusion cases diagnosed less than 4 y of follow-up	[23]
Flavanols	Epicatechin gallate (ECG)		Plasma	JPHC	Japanese	M	331/331	1.22 (0.69-2.15)	0.55	40-69 y	[30]
						F	163/163	0.25 (0.08-0.73)	0.02*	40-69 y	

Flavonoid subclass	Certain compound	Cancer subsite, subtype	Bio-marker	Study[1]	Population	Gender[2]	Cases/ controls	Multivariate-adjusted OR[3]	p for trend[4]	Comments[5]	Ref.
Flavanols	Epigallocatechin (EGC)		Plasma	JPHC	Japanese	M	331/331	2.06 (1.23-3.45)	0.003*	40-69 y	[30]
						F	163/163	0.98 (0.45-2.15)	0.89	40-69 y	
Flavanols	Epigallocatechin (EGC)		Urinary	SCS	Chinese	M	190/563	0.82 (0.49-1.38)	0.77	45-64 y; significant inverse relation after exclusion cases diagnosed less than 4 y of follow-up [0.52 (0.28-0.97)*]	[23]
Flavanols	Epigallocatechin (EGC)	Cardia	Urinary	SCS	Chinese	M		0.22 (0.05-0.90)	*	45-64 y; more than 4 y between urine collection and diagnosis	[23]
		Non-cardia	Urinary	SCS	Chinese	M		0.73 (0.38-1.41)		45-64 y; more than 4 y between urine collection and diagnosis	
Flavanols	Epigallocatechin gallate (EGCG)		Plasma	JPHC	Japanese	M	331/331	1.48 (0.89-2.44)	0.08	40-69 y	[30]
						F	163/163	0.54 (0.25-1.16)	0.19	40-69 y	
Isoflavones	Daidzein		Plasma	KMCC	Korean	F, M	131/393	0.21 (0.08-0.58)	0.0009*	>30 y; significant inverse relation for *H. pylori* positive subjects [0.53 (0.31-0.92), p trend 0.024*]	[32]

Table 7.3. (Continued)

Flavonoid subclass	Certain compound	Cancer subsite, subtype	Bio-marker	Study[1]	Population	Gender[2]	Cases/ controls	Multivariate-adjusted OR[3]	p for trend[4]	Comments[5]	Ref.
Isoflavones	Daidzein		Plasma	JPHC	Japanese	F, M	483/483	1.11 (0.74-1.66)	0.60	40-69 y	[31]
						M	324/324	1.04 (0.64-1.70)	0.97	40-69 y	
						F	159/159	1.00 (0.46-2.15)	0.9	40-69 y	
Isoflavones	Daidzein	Differentiated	Plasma	JPHC	Japanese	F, M	284/284	1.41 (0.82-2.44)	0.3	40-69 y	[31]
		Undifferentiated	Plasma	JPHC	Japanese	F, M	147/147	0.68 (0.30-1.57)	0.4	40-69 y	
		Distal	Plasma	JPHC	Japanese	F, M		1.38 (0.85-2.24)	0.20	40-69 y	
Isoflavones	Equol		Plasma	KMCC	Korean	F, M	131/393	0.50 (0.27-0.90)	0.0228*	>30 y	[32]
Isoflavones	Genistein		Plasma	KMCC	Korean	F, M	131/393	0.54 (0.31-0.93)	0.0346*	>30 y; significant inverse relation for *H. pylori* positive subjects [0.54 (0.31-0.95), p trend 0.040*]	[32]
Isoflavones	Genistein		Plasma	JPHC	Japanese	F, M	483/483	0.96 (0.64-1.44)	0.90	40-69 y	[31]
						M	324/324	0.88 (0.54-1.44)	0.70	40-69 y	
						F	159/159	1.09 (0.48-2.47)	0.80	40-69 y	
Isoflavones	Genistein	Differentiated	Plasma	JPHC	Japanese	F, M	284/284	1.10 (0.66-1.84)	0.6	40-69 y	[31]
		Undifferentiated	Plasma	JPHC	Japanese	F, M	147/147	0.52 (0.19-1.40)	0.10	40-69 y	
		Distal	Plasma	JPHC	Japanese	F, M		1.21 (0.74-1.97)	0.40	40-69 y	

[1]JPHC, The Japan Public Health Center-Based Prospective Study; KMCC; The Korean Multicenter Cancer Cohort Study; SCS, The Shanghai Cohort Study.
[2]F, female; M, male.
[3]OR, odds ratio.
[4]Statistically significant effects (p for trend <0.05) are marked by asterisk.
[5]*H. pylori*, *Helicobacter pylori*.

Although Bobe et al. did not detect any association between intake of flavanones and susceptibility to EAC in a population-based case-control study performed among white American men [19], Petrick et al. described a significant 44% decrease in the risk of this esophageal cancer subtype in an American multicenter work with both male and female subjects [6]. An inverse association between greater flavanone intake and ESCC incidence rate was observed in black American men [19], American women and men (52% lowering in the risk) [6], and Italian subjects (62% risk reduction) [3]. The later inverse association was stronger for heavy drinkers consuming more than six drinks per day [3] (Table 7.1). Thus, flavanones can account, at least in part, for the protective effects of diets rich in citrus fruits against esophageal tumorigenesis; indeed, about 90% of flavanone intake comes from citrus fruits [20].

Dietary intake of isoflavones was associated with a suggestive reduction in the risk of ESCC among American subjects in a multicenter case-control study [6]. However, Bobe et al. observed that this decrease in ESCC incidence was restricted to white men (a significant 57% lowering in the risk), but not black men, and was attenuated by further adjusting for dietary fiber, raw vegetables and fruits [19]. Nevertheless, a significant inverse relation was shown between intake of soy foods and the susceptibility to esophageal tumorigenesis in north-west China with a remarkably higher habitual consumption of soy products than in Western countries [21]. At that, total as well as specific isoflavones (genistein, daidzein and glycitein) were significantly inversely associated with the risk of esophageal cancer (about 60% decrease), independent on ethnic groups (Han or Uyghur minority adults) [21]. Although data about the histological subtypes of tumor was not presented in this study, relation between high dietary intake of isoflavones and reduced esophageal cancer incidence rate suggests that foods rich in soy products might be chemopreventive against esophageal malignant transformation, an effect probably depending on the amounts of isoflavones consumed in different populations (Table 7.1).

Some inconsistencies reveal in results about the intake of anthocyanidins on esophageal tumorigenesis. Whereas Petrick et al. demonstrated a 57% reduction in the risk of both EAC and ESCC among American women and men indicating an important role of increased intake of foods rich in anthocyanidins on lowering the incidence rate of this fatal cancer [6], Bobe et al. observed a significant 53% protection only against EAC, but not ESCC, in white American men and this effect was attenuated by adjusting the results for dietary fiber, raw vegetables and fruits [19]. Moreover, no meaningful relationship with ESCC emerged for anthocyanidins also in Italian population [3] (Table 7.1). Inverse association between increased intake of anthocyanidins and susceptibility to EAC appeared also in studies with precursor lesion of adenocarcinoma, i.e., Barrett`s esophagus. Petrick et al. demonstrated about 50% decrease in the risk of Barrett`s esophagus with greater dietary consumption of food items containing plenty of anthocyanidins [22]. These results suggest that anthocyanidins may be important in different stages of EAC development influencing both formation of precursor lesions as well as initiation and progression of malignant transformation. Therefore, adequate dietary intake of anthocyanidins can potentially lead to lower incidence (and also mortality) of this fatal malignancy.

The findings presented above clearly indicate that chemopreventive properties of different flavonoid subclasses vary in their activities on different histological subtypes of esophageal cancer, i.e., EAC and ESCC. Furthermore, despite these promising results obtained from the case-control works, two large prospective cohort studies performed with participants from ten European countries (the European Prospective Investigation into Cancer and Nutrition) or American adults (the NIH-AARP Diet and Healthy Study) revealed no significant relationships between dietary consumption of any flavonoid subclasses and the risk of esophageal cancer in general, as well as EAC or ESCC considered separately [8, 16]. However, within the European cohort, increased intake of total flavonoids, flavonols or flavanols tended to be inversely associated with a reduced incidence of esophageal neoplasms among current tobacco smokers, but not in former or never smokers, and greater consumption of flavanones was found to be related to a lower esophageal cancer risk among former smokers [16] (Table 7.2). Oxidative stress resulting from tobacco smoking is suggested to be a risk factor of esophageal carcinogenesis and chemopreventive action of dietary flavonoids might include their antioxidant activities.

To date, only one epidemiological work about the association between esophageal cancer risk and biomarker (urinary level) of flavonoids has been carried out. No overall relation was found for urinary concentration of tea polyphenols in a nested case-control study performed with men in Shanghai [23]. However, exclusion of cancer cases diagnosed less than four years after enrollment revealed a significant protective effect for epigallocatechin (EGC) against esophageal tumorigenesis with a 42% risk reduction, whereas epicatechin (EC) remained unrelated to esophageal tumorigenesis [23] (Table 7.3). As EGC is considered to be a more direct biomarker for reflecting tea consumption, EC is rather unspecific with several other dietary sources besides tea, including chocolate, apples and wine. At that, the protective role of EGC against esophageal neoplasms was observed only in subjects with low levels of serum carotenes (OR 0.44, 95% confidence interval (CI) 0.14-1.35) [23].

As the epidemiological findings published to date about relationship between dietary intake of flavonoids and the risk of esophageal cancer are limited and still rather inconclusive and also somewhat conflicting, further (and preferably prospective) studies are certainly needed to clarify the potential chemoprotective role of different flavonoids in esophageal carcinogenesis. These studies might open new strategies for prevention of esophageal malignancies by a modifiable behavioral factor, i.e., making appropriate changes in the diet, being especially important in regions with a relatively high esophageal cancer incidence.

7.1.3. Gastric Cancer

Similarly to esophagus, also gastric mucosa is in direct contact and interaction with food components, including phytochemicals from plant-derived food items. Therefore, it is tempting to expect that anticarcinogenic effects of flavonoids observed in *in vitro* and *in vivo* experimental studies could emerge also in the human`s body, as assessed in epidemiological works.

Increased dietary intake of total flavonoids was shown to be inversely associated with gastric cancer risk in a case-control study performed with Spanish population revealing a 56% decrease in stomach tumor incidence [1]. Similarly, Woo et al. found up to 70% reduction in the susceptibility to gastric tumorigenesis among Korean subjects [12]. However, this relationship was attenuated by additional adjustment for consumption of fruits and vegetables indicating that beneficial effects may be closely related to intake of these plant-based food products. Moreover, significant protective effects observed with greater intake of total flavonoids appeared only in women, but not in men, and did not depend on the infection status of *Helicobacter pylori* (*H. pylori*) [12]. No consistent relation was found between intake of total flavonoids and either gastric adenocarcinoma (GAC) in cardia or other GAC among American men and women [6] (Table 7.1).

Although no meaningful protective effect was described for greater dietary consumption of flavones against tumorigenesis in stomach among Italian [5], Greek [2], or American population [6], a statistically significant inverse association was still found in Korean subjects, especially in women, with more than 80% lowering of gastric cancer incidence, an estimate being similar in *H. pylori* positive and negative participants [12]. Similar trend of the protective role was emerged for intake of flavonols in Korean women (about 50% decrease in the gastric cancer risk); however, this beneficial effect was attenuated by further adjustment for fruits and vegetables consumption [12]. A 38% reduction in gastric cancer incidence with increased intake of flavonols was reported also in an Italian case-control study and regarding specific flavonols, a stronger inverse association was found for kaempferol than myricetin intake [5]. Dietary consumption of kaempferol and also quercetin, but not myricetin, was significantly inversely related to gastric cancer risk also in Spanish population [1]. Furthermore, increased intake of quercetin was strongly inversely associated with non-cardia GAC, but not GAC in cardia, among Swedish subjects with a 43% lowering in the risk of tumorigenesis, whereas the estimates were similar for both histological types of non-cardia GAC, i.e., diffuse and intestinal neoplasms [24]. This protective action was particularly strong for females exposed to oxidative stress resulting from tobacco smoking [24]. Infection with *H. pylori* is suggested to be a necessary factor for tumorigenesis in gastric non-cardia by inducing oxidative stress and inflammatory condition. Therefore, antioxidant properties of dietary flavonoids may be highly important to decrease the gastric cancer burden. Moreover, it can be only speculated that about 2-3 times higher gastric cancer incidence in men than in women might be somewhat related to the stronger beneficial effects of abundant food flavonol quercetin under oxidative stress conditions in female than in male subjects. On the other hand, no meaningful relationship with total flavonols was observed with GAC in Greek population [2] or with GAC in cardia and non-cardia among American men and women [6] (Table 7.1).

Some inconsistencies can be found also in the epidemiological findings from case-control studies about the role of dietary intake of flavanones on gastric cancer risk. Whereas no significant association was observed in Italian [5] or American population [6], Lagiou et al. described a 45% decrease in GAC incidence among Greek subjects [2] and Woo et al. reported about 60% reduction in gastric tumorigenesis in Korean women, but not in men, regardless of their infection status with *H. pylori* [12]. It is possible that flavanones can account, at least in part, for the plausible protective effect of citrus fruits against gastric tumorigenesis. Similar statistically significant inverse association with gastric cancer incidence was observed also for increased intake of flavanols in Korean women (about 65% decrease) with no considerable difference among *H. pylori* positive or negative subjects [12]. On the contrary, no meaningful relationship was found between higher intake of flavanols and gastric tumorigenesis in Greek [2], Italian [5] or American population [6] (Table 7.1).

No statistically significant association between dietary intake of isoflavones and gastric cancer risk was reported in case-control studies performed among Italian [5], Greek [2], Korean [12] or American subjects [6], regardless remarkable differences in the amount of isoflavones consumed in these populations. Also, epidemiological findings published to date about dietary intake of anthocyanidins do not allow exhibit any clear trends in the risk of gastric neoplasms pointing possibly to no or only little role of these phytochemicals in etiology of this cancer type [2, 5, 6]. Although some risk reduction was reported by Woo et al. in Korean women, this association was attenuated by additional adjustment for fruits and vegetables consumption [12] (Table 7.1). Altogether, these data show some contrast to the possible protective role of anthocyanidins in oral and esophageal tumorigenesis discussed in the previous sections.

Results from the prospective cohort studies performed with Finnish or Dutch subjects revealed no meaningful relationships between dietary consumption of total flavonoids or some specific compounds (flavonols kaempferol, myricetin, quercetin; flavanones hesperetin, naringenin) and tumorigenesis in the stomach [25-27]. However, these cohort studies (the Finnish Mobile Clinic Health Examination Survey and the

Netherlands Cohort Study) were published before 2002 when the food composition data were very limited compared to those available today. A latter study with prospective cohort design (the European Prospective Investigation into Cancer and Nutrition Cohort) found a significant protective effect against gastric carcinogenesis for increased dietary intake of total flavonoids and some flavonoid subclasses (flavones, flavonols, flavanols) in women, but not in men, among participants from ten European countries [10]. This inverse association was not related to the anatomical site of the tumor, i.e., cardia or non-cardia, but appeared particularly with intestinal neoplasms for higher consumption of total flavonoids and flavanols [10]. Regarding smoking status, intake of total flavonoids, flavones, flavonols or flavanols was significantly inversely related to the risk of gastric cancer in female smokers, but not non-smokers [10], probably resulting from the antioxidant action of these phytochemicals under oxidative stress caused by tobacco smoking. However, no association was observed for isoflavones intake in this prospective study with participants from European countries [10]. On the contrary, the results from a Japanese prospective cohort study (the Takayama Study) revealed a statistically significant 40% reduction in gastric cancer risk for increased intake of isoflavones among female, but not male subjects; an effect probably related to the remarkably higher doses of soy foods consumed in Asian countries than in European populations [28]. However, intake of genistein (and also daidzein), the two most abundant isoflavones in soy foods, revealed no associations with gastric neoplasms in either Japanese women or men, regardless histological type of tumor. On the contrary, in women, a positive relationship with gastric cancer risk (a 2.8-fold increase) was found for higher genistein intake among subjects ever consumed exogenous female hormones, but not among never users [29]. These inconsistent results clearly show that further research is needed to clarify the role of isoflavones on gastric tumorigenesis. Furthermore, a recent prospective study with American retired persons (the NIH-AARP Diet and Health Study) found no associations between the intake of total flavonoids or any flavonoid subclasses and the risk of cancer development in the cardia or non-cardia area of the stomach, showing thus that the impact of flavonoids on gastric carcinogenesis depends largely on the underlying population. As this study did not differentiate the risk data between male and female participants, the protective effects of total flavonoids, flavones, flavonols and flavanols reported for the European women remain still unproven among the American counterparts [8] (Table 7.2).

Biomarker studies with different body fluids could provide some more insight in relationship between flavonoids and the risk of carcinogenesis considering also the issues of bioavailability and metabolic conjugation of these polyphenols. Studies of flavonoids in plasma demonstrated a statistically significant inverse association for green tea flavonoid epicatechin gallate (ECG) only among Japanese women with a 75% reduction in gastric cancer risk; also other studied flavanols (EC and epigallocatechin gallate (EGCG)) were found to be inversely, although insignificantly, related to gastric neoplasms among female subjects [30]. On the contrary, in Japanese men, a high plasma concentration of EGC was related to a twofold increased risk of gastric tumorigenesis and other tea polyphenols (EC, ECG, EGCG) similarly suggested an elevated risk [30]. These relationships were affected by smoking status showing risk estimates below unity for past or never smokers, but not for current smokers [30]. Although urinary levels of EC revealed a suggestive increase in gastric cancer risk among Chinese men, this trend was attenuated by excluding the cases diagnosed less than four years after enrollment [23]. Moreover, lengthening the duration of follow-up revealed a statistically significant inverse relation with gastric tumorigenesis for urinary EGC with a 48% decrease in the cancer risk [23]. This protective effect was stronger for neoplasms in cardia than non-cardia, and was more prominent among non-smoking and -drinking men and those with low levels of serum carotenes [23] (Table 7.3).

Levels of two isoflavones, genistein and daidzein, in plasma displayed no associations with gastric tumorigenesis in Japanese subjects and these findings were not significantly influenced by gender, infection status with *H. pylori*, or anatomical location and histological subtype of neoplasm [31]. In contrast, higher plasma concentrations of these isoflavones as well as equol were associated with a significant decrease in the risk of gastric cancer among Korean subjects and this reduction was observed in *H. pylori* positive participants [32]. The authors explained this protective effect by antioxidant and antiinflammatory properties of isoflavones [32] (Table 7.3). These inconsistencies in understanding the role of isoflavones on gastric cancer risk clearly point to the need for further research.

7.2. Some Reasons for Inconsistencies of Epidemiological Findings

Epidemiological studies about the chemopreventive role of flavonoids were long hindered by the lack of food composition data to estimate consumption of these phytochemicals [1-3]. In 2003, the United States Department of Agriculture (USDA) made available a database for dietary flavonoids promoting the observational investigations with flavonoids; later on, these tables have been repetitively updated [11]. However, there are several inherent limitations in composition data of flavonoids, including variability in their content related to conditions of plant cultivation (climatic, soil, growing), harvesting (degree of ripeness) and storage, or cultural differences in processing and preparation of foods [3, 4, 6, 8, 14, 19, 22]. These problems are probably most prominent for flavones which are consumed predominantly through aromatic herbs, spices and garnishes [3, 19]. Moreover, food intake assessments do not usually consider the consumption of flavonoids in the form of dietary supplements [6, 16, 22]. All these constraints call in question the adaptability of flavonoid databases created based on foods utilized in US to the studies performed in European or Asian countries [4-6, 22]. Furthermore, various populations consume different amounts of flavonoids and it is possible that protective effects of these plant secondary metabolites against cancer emerge only in doses above certain threshold, a phenomenon especially feasible for soy foods rich in isoflavones [21, 32]. Indeed, the dietary intake levels of isoflavones are considerably higher in Asian than in Western subjects and inverse association with tumorigenesis in upper gastrointestinal tract were observed only in studies with Asian participants [10, 29, 32]. Discrepancies in epidemiological findings may result also from timing of exposure to flavonoids as the beneficial effects of isoflavones might be more evident when the intervention is started at early ages [13].

After ingestion, flavonoids undergo an extensive metabolism in small intestine and liver as a result of which different conjugates with potentially changed biological activities enter the systemic circulation and reach target tissues [33]. Therefore, estimation of flavonoids in the body fluids (plasma, serum, urine) may provide a more relevant measure considering also individual differences in bioavailability and biotransformation, including intestinal microflora or previous consumption of antibiotics [6, 7, 13, 16, 22, 29, 31, 32]. However, these biomarkers may reflect only short-term and not long-term intake of food items rich in flavonoids with plasma peak levels reached within some hours following to ingestion and most excreted from the body after a few days, revealing their limited usefulness [6, 8, 22, 23, 30-32]. Therefore, assessment of nutritional intake remains a crucial tool in epidemiological works, especially in studies of cancer development in upper gastrointestinal tract due to the direct contact and interaction of ingested phytochemicals with epithelial surfaces of these anatomical sites.

As in the case of all observational works, potential biases must be taken into account by interpreting the findings. Studies with case-control design are vulnerable to recall bias due to the problems in remembering the past dietary intake [11, 13]. This is especially relevant for tumors in esophagus or stomach as patients might have been suffered from intestinal discomfort and dysphagia already years prior to diagnosis, leading to modifications in their usual dietary habits [2, 3, 6, 10, 11, 16, 21, 22]. Another type of bias inherent in case-control studies includes the selection of participants. It is possible that hospital controls might have different dietary habits than those of the general population [1, 3, 4]. Information bias is probably unlikely in the association studies between flavonoids intake and the risk of upper gastrointestinal tract tumors as the potential advantageous effects of flavonoid rich food products are not widely known in general public [21]. Although prospective cohort studies can probably avoid these biases, it still remains possible that prospectively designed investigations with too short follow-up times (less than 10 years) might not be able to detect the true effects [11, 13, 29].

Finally, even by controlling various established risk factors, unmeasured or unknown residual confounding could not be ruled out [10, 16, 21, 24, 28]. While isoflavones can be protective against gastric carcinogenesis, soy foods as the major dietary sources of these flavonoids might contain plenty of salt or N-nitroso compounds, if consumed in the form of fermented products [28, 29]. Both of these non-isoflavone aspects of soy foods are well-known risk factors for gastric neoplasms [28, 29]. Tobacco smoking and alcohol consumption are suggested to be risk determinants for tumors in upper gastrointestinal tract. Not only antioxidant action of flavonoids but also their modulation of enzymatic pathways can reduce the level of oxidative stress and mucous damage resulted from smoking or drinking [7, 10, 12, 18, 30]. Results of association studies might vary also for tumors in different anatomical subsites and histological subtypes [6, 10, 21]. Moreover, infection status with *H. pylori* may substantially affect the role of dietary flavonoids on gastric tumorigenesis [12, 14, 30]. It is also possible that increased intake of flavonoids can just be a marker of some other potentially anticarcinogenic factors related to a healthy lifestyle and high consumption of plant-based foods [1, 6, 22, 24].

7.3. Current Status and Further Perspectives

The epidemiological studies performed to date about the role of flavonoids on carcinogenesis in upper gastrointestinal tract are still limited today and reveal rather inconsistent results. However, as tumors in oral cavity, pharynx, esophagus and stomach have different risk factors, certain flavonoids and their subclasses can affect the risk of these malignancies also in different ways.

High dietary exposure to citrus flavonoids, i.e., flavanones, seems to provide some protection against tumorigenesis in the oropharynx.

Studies with case-control design revealed significant protective role of anthocyanidins against esophageal adenocarcinoma (and its precursor lesion, Barrett`s esophagus) and chemopreventive effect of high dietary intake of isoflavones (characteristic to Asian countries) against esophageal tumorigenesis. Although these results were not confirmed in prospective cohort studies, it is possible that doses of soy products consumed by European and American subjects were too low for emergence of the protective effect of isoflavones on the health of esophagus. Therefore, further studies in different populations with varying habitual intake of flavonoids are definitely needed.

Despite several inconsistencies in the epidemiological findings published to date about association between dietary intake of flavonoids and the risk of gastric cancer, some interesting trends become still evident. First, protective effects of flavonoids against tumorigenesis in stomach seem to be stronger among women than men; a feature that has been shown for different subclasses (flavones, flavonols, flavanols, isoflavones) or individual compounds (quercetin, green tea flavanols) in either case-control, prospective cohort or biomarker studies. This characteristic can provide some explanation to the higher prevalence of gastric neoplasms in male than female subjects (about two- to threefold difference), as women may be more protected against malignant transformation through daily consumption of plant-derived food items rich in various flavonoids. Biological reason for such gender difference is still unclear but could comprise intervention of flavonoids in action of female sex hormones. Moreover, it has been shown that bioavailability of some polyphenols is higher in women than in men [10]. Second, concerning dietary consumption of isoflavones both null, negative as well as positive relationships with the risk of gastric cancer have been observed. Besides varying amounts of isoflavones daily consumed by different populations (subjects in Western countries have considerably lower intake of isoflavones as compared to their counterparts in Asian regions), also the food sources of these phytochemicals might be important. Indeed, some dietary sources of isoflavones can contain plenty of salt or N-nitroso compounds (like fermented soy products) which in themselves are known as risk factors for gastric tumorigenesis [28, 32]. Third, in spite the causative role of infection with *H. pylori* in the development of cancer in gastric non-cardia, only limited studies have stratified the epidemiological results of flavonoids intake by infection status and those few works have not displayed any significant differences. Last but not least, to date, no association studies can be found for intake of flavonoids with the risk of development of duodenal tumors.

With no doubt, further large-scale epidemiological studies in diverse populations focusing on tumors in different anatomical locations of upper gastrointestinal tract and also various histological subtypes are needed to clarify the role of flavonoids on carcinogenesis in upper digestive tract. Based on the current promising findings it can be only expected that these studies might open new avenues in the cancer prevention strategies through recommendation of certain changes in the daily diet, being a modifiable factor in human`s behavioral pattern. Such changes can be especially important for subjects at-risk due to higher prevalence of oral, esophageal or gastric tumors in family history or being exposed to the risk factors of these malignancies.

References

[1] Garcia-Closas R, Gonzalez CA, Agudo A, Riboli E. Intake of specific carotenoids and flavonoids and the risk of gastric cancer in Spain. *Cancer Causes Control* 1999; 10: 71-5.

[2] Lagiou P, Samoli E, Lagiou A, Peterson J, Tzonou A, Dwyer J, Trichopoulos D. Flavonoids, vitamin C and adenocarcinoma of the stomach. *Cancer Causes Control* 2004; 15: 67-72.

[3] Rossi M, Garavello W, Talamini R, La Vecchia C, Franceschi S, Lagiou P, Zambon P, Dal Maso L, Bosetti C, Negri E. Flavonoids and risk of squamous cell esophageal cancer. *Int J Cancer* 2007; 120: 1560-4.

[4] Rossi M, Garavello W, Talamini R, Negri E. Bosetti C, Dal Maso L, Lagiou P, Tavani A, Polesel J, Barzan L, Ramazzotti V, Franceschi S, La Vecchia C. Flavonoids and the risk of oral and pharyngeal cancer: a case-control study from Italy. *Cancer Epidemiol Biomarkers Prev* 2007; 16: 1621-5.

[5] Rossi M, Rosato V, Bosetti C, Lagiou P, Parpinel M, Bertuccio P, Negri E, La Vecchia C. Flavonoids, proanthocyanidins, and the risk of stomach cancer. *Cancer Causes Control* 2010; 21: 1597-604.

[6] Petrick JL, Steck SE, Bradshaw PT, Trivers KF, Abrahamson PE, Engel LS, He K, Chow WH, Mayne ST, Risch HA, Yaughan TL, Gammon MD. Dietary intake of flavonoids and oesophageal and gastric cancer: incidence and survival in the United States of America (USA). *Br J Cancer* 2015; 112: 1291-300.

[7] Xie Y, Huang S, Su Y. Dietary flavonols intake and risk of esophageal and gastric cancer: A meta-analysis of epidemiological studies. *Nutrients* 2016; 8: 91.

[8] Sun L, Subar AF, Bosire C, Dawsey SM, Kahle LL, Zimmerman TP, Abnet CC, Heller R, Graubard BI, Cook MB, Petrick JL. Dietary flavonoid intake reduces the risk of head and neck but not esophageal or gastric cancer in US men and women. *J Nutr* 2017; 147: 1729-38.

[9] Ding Y, Yao H, Yao Y, Fai LY, Zhang Z. Protection of dietary polyphenols against oral cancer. *Nutrients* 2013; 5: 2173-91.

[10] Zamora-Ros R, Agudo A, Lujan-Barroso L, Romieu I, Ferrari P, Knaze V, Bueno-de-Mesquita HB, Leenders M, Travis RC, Navarro C, Sanchez-Cantalejo E, Slimani N, Scalbert A, Fedirko V, Hjartaker A, Engeset D, Skeie G, Boeing H, Förster J, Li K, Teucher B, Agnoli C, Tumino R, Mattiello A, Saieva C, Johansson I, Stenling R, Redondo ML, Wallström P, Ericson U, Khaw KT, Mulligan AA, Trichopoulou A, Dilis V, Katsoulis M, Peeters PH, Igali L, Tjonneland A, Halkjaer J, Touillaud M, Perquier F, Fagherazzi G, Amiano P, Ardanaz E. Bredsdorff L, Overvad K, Ricceri F, Riboli E, Gonzalez CA. Dietary flavonoid and lignan intake

and gastric adenocarcinoma risk in the European Prospective Investigation into Cancer and Nutrition (EPIC) study. *Am J Clin Nutr* 2012; 96: 1398-408.

[11] Woo HD, Kim J. Dietary flavonoid intake and risk of stomach and colorectal cancer. *World J Gastroenterol* 2013; 19: 1011-9.

[12] Woo HD, Lee J, Choi IJ, Kim CG, Lee JY, Kwon O, Kim J. Dietary flavonoids and gastric cancer risk in a Korean population. *Nutrients* 2014; 6: 4961-73.

[13] Golpour S, Rafie N, Safavi SM, Miraghajani M. Dietary isoflavones and gastric cancer: A brief review of current studies. *J Res Med Sci* 2015; 20: 893-900.

[14] Bo Y, Sun J, Wang M, Ding J, Lu Q, Yuan L. Dietary flavonoid intake and the risk of digestive tract cancers: a systematic review and meta-analysis. *Sci Rep* 2016; 6: 24836.

[15] Cui L, Liu X, Tian Y, Xie C, Li Q, Cui H, Sun C. Flavonoids, flavonoid subclasses, and esophageal cancer risk: a meta-analysis of epidemiologic studies. *Nutrients* 2016; 8: E350.

[16] Vermeulen E, Zamora-Ros R, Duell EJ, Lujan-Barroso L, Boeing H, Aleksandrova K, Bueno-de-Mesquita HB, Scalbert A, Romieu I, Fedirko V, Touillaud M, Fagherazzi G, Perquier F, Molina-Montes E, Chirlaque MD, Vicente Arguelles M, Amiano P, Barricarte A, Pala V, Mattiello A, Saieva C, Tumino R, Ricceri F, Trichopoulou A, Vasilopoulou E, Ziara G, Crowe FL, Khaw KT, Wareham NJ, Lukanova A, Grote VA, Tjonneland A, Halkjaer J, Bredsdorff L, Overvad K, Siersema PD, Peeters PH, May AM, Weiderpass E, Skeie G, Hjartaker A, Landberg R, Johansson I, Sonestedt E, Ericson U, Riboli E, Gonzalez CA. Dietary flavonoid intake and esophageal cancer risk in the European prospective investigation into cancer and nutrition cohort. *Am J Epidemiol* 2013; 178: 570-81.

[17] Jiao L, Kramer JR, Rugge M, Parente P, Verstovsek G, Alsarraj A, El-Serag HB. Dietary intake of vegetables, folate, and antioxidants and the risk of Barrett`s esophagus. *Cancer Causes Control* 2013; 24: 1005-14.

[18] De Stefani E, Ronco A, Mendilaharsu M, Deneo-Pellegrini H. Diet and risk of cancer of the upper aerodigestive tract – II. Nutrients. *Oral Oncol* 1999; 35: 22-6.

[19] Bobe G, Peterson JJ, Gridley G, Hyer M, Dwyer JT, Brown LM. Flavonoid consumption and esophageal cancer among black and white men in the United States. *Int J Cancer* 2009; 125: 1147-54.

[20] Rossi M, Bosetti C, Negri E, Lagiou P, La Vecchia C. Flavonoids, proanthocyanidins, and cancer risk: a network of case-control studies from Italy. *Nutr Cancer* 2010; 62: 871-7.

[21] Tang L, Lee AH, Xu F, Zhang T, Lei J, Binns CW. Soya and isoflavone intakes associated with reduced risk of oesophageal cancer in north-west China. *Public Health Nutr* 2015; 18: 130-4.

[22] Petrick JL, Steck SE, Bradshaw PT, Chow WH, Engel LS, He K, Risch HA, Vaughan TL, Gammon MD. Dietary flavonoid intake and Barrett`s esophagus in western Washington State. *Ann Epidemiol* 2015; 25: 730-5.

[23] Sun CL, Yuan JM, Lee MJ, Yang CS, Gao YT, Ross RK, Yu MC. Urinary tea polyphenols in relation to gastric and esophageal cancers: a prospective study of men in Shanghai, China. *Carcinogenesis* 2002; 23: 1497-503.

[24] Ekström AM, Serafini M, Nyren O, Wolk A, Bosetti C, Bellocco R. Dietary quercetin intake and risk of gastric cancer: results from a population-based study in Sweden. *Ann Oncol* 2011; 22: 438-43.

[25] Knekt P, Järvinen R, Seppänen R, Hellövaara M, Teppo L, Pukkala E, Aromaa A. Dietary flavonoids and the risk of lung cancer and other malignant neoplasms. *Am J Epidemiol* 1997; 146: 223-30.

[26] Knekt P, Kumpulainen J, Järvinen R, Rissanen H, Heliövaara M, Reunanen A, Hakulinen T, Aromaa A. Flavonoid intake and risk of chronic diseases. *Am J Clin Nutr* 2002; 76: 560-8.

[27] Goldbohm RA, Hertog MG, Brants HA, van Poppel G, van den Brandt PA. Intake of flavonoids and cancer risk: a prospective cohort study. In: Armado R, Andersson H, Bardocz S, Serra F, eds. *Polyphenols in food.* Office for Official Publications of the European Communities 1998, pp. 159-66.

[28] Wada K, Tsuji M, Tamura T, Konishi K, Kawachi T, Hori A, Tanabashi S, Matsushita S, Tokimitsu N, Nagata C. Soy isoflavone intake and stomach cancer risk in Japan: From the Takayama study. *Int J Cancer* 2015; 137: 885-92.

[29] Hara A, Sasazuki S, Inoue M, Iwasaki M, Shimazu T, Sawada N, Yamaji T, Tsugane S, Japan Public Health Center-Based Prospective Study Group. Isoflavone intake and risk of gastric cancer: a population-based prospective cohort study in Japan. *Am J Clin Nutr* 2012; 95: 147-54.

[30] Sasazuki S, Inoue M, Miura T, Iwasaki M, Tsugane S, Japan Public Health Center-based Prospective Study Group. Plasma tea polyphenols and gastric cancer risk: a case-control study nested in a large population-based prospective study in Japan. *Cancer Epidemiol Biomarkers Prev* 2008; 17: 343-51.

[31] Hara A, Sasazuki S, Inoue M, Miura T, Iwasaki M, Sawada N, Shimazu T, Yamaji T, Tsugane S, Japan Public Health Center-Based Prospective Study Group. Plasma isoflavone concentrations are not associated with gastric cancer risk among Japanese men and women. *J Nutr* 2013; 143: 1293-8.

[32] Ko KP, Park SK, Park B, Yang JJ, Cho LY, Kang C, Kim CS, Gwack J, Shin A, Kim Y, Kim J, Yang HK, Kang D, Chang SH, Shin HR, Yoo KY. Isoflavones from phytoestrogens and gastric cancer risk: a nested case-control study within the Korean Multicenter Cancer Cohort. *Cancer Epidemiol Biomarkers Prev* 2010; 19: 1292-300.

[33] Sak K. Intake of individual flavonoids and risk of carcinogenesis: Overview of epidemiological evidence. *Nutr Cancer* 2017; 69: 1119-50.

Chapter 8

CONCLUSIONS AND FURTHER PERSPECTIVES

Prevention is probably the most promising strategy to reduce the global burden of cancer and save the lives. Malignant neoplasms in the sites of upper gastrointestinal tract (oral cavity, esophagus, stomach) represent serious medical problems being associated with a high mortality rate and often leading to a decreased quality of life, besides psychosocial challenges for patients and high financial costs for the whole society. As mucosal epithelia of upper digestive tract are in the direct contact with various food constituents, including phytochemicals from plant-derived products, evidence-based dietary modifications may provide an important approach in the fight against gastrointestinal cancers.

Numerous studies have demonstrated that plant secondary metabolites, flavonoids, can suppress different cancer hallmarks in experimental models of oral, esophageal and gastric cancer, i.e., the respective cell lines or xenografts-bearing laboratory animals. Flavonoids inhibit the growth and proliferation of malignant cells, induce the cell cycle arrest, trigger apoptosis, block angiogenesis, and restrain invasive and migratory potential. At that, the green tea flavanol epigallocatechin gallate (EGCG), flavone luteolin and flavonol quercetin seem to be the most widely investigated flavonoids in the oral, esophageal and gastric cancer models. However, despite the immense amount of studies, biological activities of only a limited selection of structurally diverse flavonoids have been examined to date, especially regarding to the high variety of these polyphenols in the nature. Therefore, studies of anticancer properties of expanded set of flavonoid molecules with varied structural moieties must be continued. Moreover, most conclusions about the anticancer effects of flavonoids are still done based on *in vitro* cell line or *in vivo* animal model studies, whereas extrapolation of these results to the human body faces several obstacles, such as issues of absorption, metabolism and bioavailability. To confirm the chemopreventive potency and prove the efficacy of experimental leading compounds as potential novel drug candidates, it is necessary to move on with human intervention studies and clinical trials. At that, it is of critical importance to thoroughly understand the molecular mechanisms of action of different flavonoids in gastrointestinal carcinogenesis. Within these studies, cellular signaling pathways might provide novel molecular targets for further interfering with cancer development and progression.

Through the regular diet, human beings are daily exposed to hundreds of bioactive phytochemicals indicating that additive and/or synergistic action of different combinations of flavonoids may be responsible for the health benefits attributed to the intake of plant-based food items. Therefore, studies of flavonoid mixtures on cancer cells derived from tumors of upper gastrointestinal tract deserve more attention to better implement the chemopreventive potential of these polyphenols in the future. Furthermore, some experimental studies have shown that combined administration of flavonoids and clinically approved chemotherapeutic agents can substantially affect the efficacy of drugs by both potentiating but also antagonizing their anticancer responses. However, the current knowledge about such chemomodulating effects of different flavonoids in the upper gastrointestinal tract cancer models is still rather limited and certainly needs further unraveling. Likewise, effects of different flavonoids on the therapeutic efficacy of irradiation of oral, esophageal or gastric tumors have to be clarified in the respective preclinical models.

While numerous experimental investigations have been devoted to analyze the action of different flavonoids in oral, esophageal and gastric cancer models, only couple of studies has dealt with potential effects of these plant polyphenols in duodenal cancer models. This situation can probably be caused by the rare duodenal carcinogenesis in humans. In 1989, Fujita et al. reported in their paper published in the Japanese Journal of Cancer Research that EGCG could suppress N-ethyl-N`-nitro-N-nitrosoguanidine (ENNG)-induced duodenal tumorigenesis in male mice by reducing both the tumor incidence as well as the average number of tumors per rodent. Quite contrarily, in 2002, Matsukawa et al. described in their article in the Environmental Health and Preventive Medicine that quercetin even enhanced the tumorigenicity of ENNG in mice duodenum and promoted the development of neoplasms. In the light of these latter results, the safety issue of flavonoids should still remain in the focus of any further research.

Last but not least, a fundamental advantage of natural flavonoids over the conventional pharmaceutical chemotherapeutics (with typically a single molecular target) is the multitargeted action of phytochemicals. Flavonoids can simultaneously affect multiple molecular targets and exhibit pleiotropic cellular effects against development and progression of malignant neoplasms representing thus attractive candidates for prevention and treatment of cancer as a dynamic and heterogeneous system of aberrant cells. Therefore, it can be expected that the multitargeted action of flavonoids or purposely selected combinations of different flavonoids on the continuously evolving malignant cell complex might be the key for future cancer therapy provided us by the nature. On account of the direct contacts of ingested flavonoids with mucosal epithelia of organ sites of the upper digestive tract, the potential protection obtained from plant-derived flavonoids against upper gastrointestinal cancers is especially feasible.

About the Author

Katrin Sak, MSc, PhD, Dip. Nut. Med.
Head of the NGO Praeventio,
Tartu, Estonia
Email: katrin.sak.001@mail.ee

The author of this book, Dr. Katrin Sak, was born in Viljandi, Estonia, in 1975. After studies in the University of Tartu, Estonia she received a bachelor degree (BSc) in chemistry in 1996 and a master degree (MSc) in bioorganic chemistry in 1997. In 2001, Katrin Sak was graduated with doctor degree (PhD) in bioorganic chemistry. During 2001-2004, Dr. Sak did postdoctoral research at the Free University of Brussels, Belgium and in 2012, she acquired diploma in cancer nutrition in the Health Schools Australia.

Dr. Sak is the author of more than 70 international articles published in peer-reviewed journals; most of them dealing anticancer activities of various flavonoids in the cellular models of different cancer types. She has published three chapters of books, including Nutraceuticals Nanotechnology in the Agri-Food Industry, Vol. 4 (Elsevier, 2016) and Food Bioconversion Handbook of Food Bioengineering, Vol. 2 (Elsevier, 2017). Also, Dr. Sak is the author of the book "Food and Cancer" (in Estonian, 2013) and numerous popular-scientific articles published in the Estonian health journals.

The major research topics of Katrin Sak involve different anticancer actions of natural and semisynthetic flavonoids in diverse malignant systems, structure-activity relationships, engaged cellular signaling pathways and molecular mechanisms; besides the metabolic bioconversion of these food polyphenols in the human body and bioactivities of various metabolites.

Katrin Sak is the head of the NGO Praeventio.

INDEX

#

A

B

C

D

E

M

N

O

P

Q

R

S

T

V

W

X